ALLEZEIT WACH 1842

J. Bereiter-Hahn
O. R. Anderson W.-E. Reif (Eds.)

Cytomechanics

The Mechanical Basis of Cell Form and Structure

With 115 Figures

Springer-Verlag
Berlin Heidelberg New York
London Paris Tokyo

Professor Dr. JÜRGEN BEREITER-HAHN
Arbeitsgruppe Kinematische Zellforschung
FB Biologie der Universität
Senckenberganlage 27
D-6000 Frankfurt 1

Professor Dr. O. ROGER ANDERSON
Columbia University
Biological Oceanography
Lamont-Doherty Geological Observatory
Palisades, NY 10964, U.S.A.

Professor Dr. WOLF-ERNST REIF
Universität Tübingen
Institut und Museum für Geologie und Paläontologie
Sigwartstr. 10
D-7400 Tübingen

For explanation of the cover motif see legend to Fig. 2, page 238

ISBN-13: 978-3-642-72865-5 e-ISBN-13: 978-3-642-72863-1
DOI: 10.1007/978-3-642-72863-1

Library of Congress Cataloging-in-Publication Data. Cytomechanics : the mechanical basis of cell form and structure. Includes bibliographies. 1. Cells--Morphology. 2. Cytoskeleton. 3. Biomechanics. I. Bereiter-Hahn, J. (Jürgen) II. Anderson, O. Roger, 1937– . III. Reif, Wolf-Ernst. QH611.C97 1987 574.87'2 87-23340 ISBN 0-387-18123-7 (U.S.)

Softcover reprint of the hardcover 1st edition 1987

Typesetting, printing, and bookbinding: Konrad Triltsch, Graphischer Betrieb, 8700 Würzburg.
2131/3130-543210

Preface

One goal of fundamental biological research is to understand living organisms in terms of a machine theory. This includes the subtle interplay and control of various parts as well as their self-reproduction. Whether this research is reductionistic and how close it approaches reality depends on our understanding of what a "machine" is, and not on this basic paradigm itself. At the end of the 19th century many biologists assumed they could understand cell shape and function by mimicking cellular behavior with droplets of immiscible fluids, e.g., oil in various salt solutions, or uptake of lacquer threads by chloroform droplets similar to the phagocytotic process. Surface tension was regarded as the main factor determining shape, locomotion, spreading, phagocytosis, and cell division, a view which still dominated d'Arcy Thompson's famous book on *Growth and Form* (1942), (see also W.-E. Reif, Biol. Zentralbl. 1986, 105: 593–598). Meanwhile electron microscopy and immuno-fluorescence techniques have revealed a detailed picture of cell structures prohibiting the description of cell behavior by means of simple physical parameters. On the other hand the development of appropriate simple model systems may be a valuable tool to promote our understanding of dynamic processes at the supramolecular level (cf. Chap. I.3).

Despite the steadily growing insight into the molecular and fine structural basis of cellular organization, we are far from comprehending the dynamical events and the mechanical forces involved in the generation of shape and motility of cells. Genetic information determines the composition of molecules comprising cytoskeletal elements, membranes, and receptors. The supramolecular arrangement of these components represents a self-assembly process controlled by physicochemical and mechanical interactions. This general hypothesis demarcates the aim of studying cellular mechanics. The description and evaluation of mechanical properties of cells and their organelles as well as of the forces exerted by them, thus being responsible for generation and maintenance of cell and tissue architecture, is the scope of this book on *Cytomechanics*. This perspective is necessarily holistic, because a description of forces acting in a system must regard the synergism of all forces interacting with each other, i.e., the force closure or "mechanical continuum"; otherwise the mechanical properties are not adequately described.

Measuring mechanical parameters at the level of cells or of subcellular components is often very tedious, and the methods developed by engineers in most cases cannot be used due to the small dimensions. Therefore qualitative descriptions of trajectories or of force relations based on microscope observations, such as, for example, polarization microscopy, prevail. The situation is different with respect to mechanical properties of tissues and blood vessels. The mechanics of muscle and bone also are well established [Fung YC (1981), Biomechanics – Mechanical properties of living tissues. Springer 433 pp]. Therefore these fields have not been included here.

The interest of one of the editors (JBH) in the field of cytomechanics developed during a series of interdisciplinary discussions with a group of architects, technicians, and biologists. One of the first outcomes of these discussions was the appreciation of the principle "pneu" underlying the production of shape in living cells [Bereiter-Hahn J (1977), Die Zelle ein Pneu? in Otto F. (ed) Pneus in Natur und Technik, IL 9: 152–156]. In May 1985 a symposium on cell mechanics was organized by this group in consultation with the editors of this volume. This symposium was sponsored by the Sonderforschungsbereich 230: Natürliche Konstruktionen, Leichtbau in Architektur und Natur of the Deutsche Forschungsgemeinschaft. The contributions to this symposium form the backbone of the present publication, which is presented as an introduction to the field, and by no means covers all the widespread information obtained with various methods. The objective of this publication is to stimulate the further development of a scientific field of growing importance. Thus it is not possible to review the whole literature starting with the work of early cell biologists who probed the mechanical properties of cells by micromanipulation (e.g., R. Chambers, J.N. Kopac) and culminating with a compilation of metabolic effects of mechanical stimulation of cells. Emphasis is placed on the role of mechanical properties in generation of shape and cytoplasmic motion, and on the basic principles and components determining mechanical properties.

Thanks are due to Prof. M. Schliwa (Berkeley) and Dr. K. Beck (Basel) for valuable suggestions during the planning phase of this book and to Prof. Otto (Stuttgart) for providing a forum for this interdisciplinary dialog.

August 1987

JÜRGEN BEREITER-HAHN
O. ROGER ANDERSON
WOLF-ERNST REIF

Robert Day Allen (1927 – 1986)

We have lost one of our great pioneers in the fields of cell motility and cytoplasmic structure and function. When Bob Allen was informed early in 1985 that he had pancreatic cancer and would have only a few more weeks to live, he refused to give up and fought the disease valiantly (Higgins 1985). By sheer defiance, he lived vigorously for another year which he spent active in research, visiting his friends the world over, and traveling to scientific meetings and to Woods Hole, where he spent his last summer as productive as he had done every summer since the early 1950s. Despite great efforts to overcome his fatal illness, he died at his home in Hanover, New Hampshire on March 23, 1986.

Bob Allen received his Ph.D. from the University of Pennsylvania, Philadelphia, where he worked in the laboratory of L. V. Heilbrunn studying the events during fertilization and artificial activation of clam eggs. At this early state in his career he had already developed an interest in cell motility and structure, and the biophysical parameters of cytoplasm. All of these factors could be even better studied in amoebae which became the topic of his work during the next decade. He combined optical techniques designed to study the living cell with in vitro methods used to gain access to the free cytoplasm. The biophysical approaches he used included centrifugation, micromanipulation – one of his special skills – and especially measuring techniques in light microscopy. Working at marine stations in France he collaborated with Georges Nomarski on polarization and interference contrast microscopy. Rather than just observing and recording the behavior of the cells and their cytoplasm he always strived toward quantitation and, therefore, studied contractive forces, thixotropy, dichroism, and streaming velocities. The work of this era led to the formulation of the frontal contraction theory of amoeboid movement, one of the two concepts dominating this field.

His studies of motility in a variety of cells such as slime molds, plant cells, marine eggs, foraminifers, and all kinds of amoebae led to the organization, together with Noburo Kamiya, of a conference on "Primitive Motile Systems in Cell Biology." This meeting set the pace for the development of the whole field of cell motility and has influenced our work up to the present day.

The analysis of the bidirectional organelle movement in axons, first reported from his laboratory in 1965, and in the pseudopods of foraminifers became the main challenge of his scientific life. In these systems as well as in keratocytes he and his collaborators proved that organelles move along individual microtubules, a conclusion which could only be drawn form combined light and electron microscope studies including a most powerful technique he had developed. Bob Allen was not only involved in the design of differential interference contrast microscopes in the 1960s, but also developed a phase modulation microscope to measure the different polarization properties of biological specimens individually, and with colleagues he built a remarkable laser scanning phase modulation microscope. His most important contribution to microscopy originated from observations made fortuitously when he taught his annual microscopy course at Woods Hole.

Merging video and computer technology with state-of-the-art light microscopy he could greatly enhance the contrast of microscope images. This now permits the visualization of living, unstained cellular components of a size well beyond the limits of resolution of conventional optical microscopy. The fact that microtubules and synaptic vesicles are among the biological objects which became now visible made those for the first time amenable to direct observation. And the results were good for several surprises: single microtubules transport organelles in both directions; small vesicles are not moving in a saltatory, but in a continuous fashion; and finally, microtubules are capable of active movements of an unexpected variety. The video-enhanced contrast microscopy according to Allen (AVEC) will certainly yield several more startling discoveries when applied to various other cellular systems. The long, but fascinating way to these achievements can be followed in the article which he wrote during the last weeks of his life and which reads like an autobiography (Allen 1987).

Bob Allen started his teaching career at the University of Michigan in 1956 and moved to Princeton in 1960 where he began work on most of the many questions which he pursued for the next 2 decades. In 1966 he assumed chairmanship of the Biology Department of the State University of New York at Albany and used the chance to shape it according to his ideas. Since 1974 he had the Eastman professorship for biology at Dartmouth College.

He founded in 1981 the journal *Cell Motility and the Cytoskeleton* which grew rapidly to become the leading journal in the field. Its success is in part attributable to his foresighted idea to introduce videodisk supplements which enabled us for the first time to adequately communicate in a scientific journal findings of the extraordinary complexity and multidimensionality that characterized cellular and intracellular movements.

We will miss Bob Allen sorely, that tall person sitting in the front row of scientific meetings all over the world, giving thought-provoking comments, pleading for the purity of scientific arguments, and defending fairness and honesty within the scientific community. We will miss the concerts he gave each summer at Woods Hole as a gifted cellist. And we will definitely miss all the developments of optical equipment which he had proposed, but could no longer realize.

The symposium on Cytomechanics held at Stuttgart in 1985 was one in a series of meetings he attended during his last year and contrary to the predicitions of his doctors. We, who were there and all those who worked with him will remember him, while for all those who did not know him personally, a close friend put it in the shortest and most concise way when he wrote: "He was creative, inventive, productive and artistic. He could also be demanding, difficult and controversial" (Rebhun 1986).

References

Allen RD (1987) Microtubules as intracellular engines. Sci Am 256:42 – 49

Higgins R (1985) A new window into living cells. The Boston Globe, Aug. 12, pp 37 – 38

Rebhun LJ (1986) Robert Day Allen (1927 – 1986): An Appreciation. Cell Motil Cytoskel 6:248 – 255 and 674 – 678

DIETER G. WEISS
Institut für Zoologie
Technische Universität München
D-8046 Garching, FRG

Contents

I. General Principles

I.1 Mechanical Principles of Architecture of Eukaryotic Cells
J. BEREITER-HAHN (With 7 Figures) 3

1.1 Introduction 3
1.2 Basic Mechanical Parameters of Cells 3
1.3 Cellular Viscosity 6
1.4 Elasticity, Contractile Forces, and Surface Tension . . 7
1.5 The Structural Basis of Cell Mechanics 11
1.5.1 Actin and Actin-Based Structures 12
1.5.2 Membrane-Associated Actin Fibrils 14
1.5.3 Microtubules and Related Structures 15
1.5.4 Intermediate Filaments and Related Structures . . . 18
1.6 Aspects of Cytoplasmic Architecture 19
1.6.1 Localization of Organelles 20
1.6.2 Interaction of Cytoskeletal Elements in Generating Cell Shape 21
1.6.3 Cytoplasmic Streaming 23
1.7 Physiological Effects of Mechanical Stresses 24
1.7.1 Mechanical Aspects of Morphogenesis During Embryo Development 24
1.7.2 Influences of Mechanical Stresses on Cellular Metabolism 24
References 25

I.2 Evaluation of Cytomechanical Properties
Y. HIRAMOTO (With 10 Figures) 31

2.1 Introduction 31
2.2 Physical Structure of the Cell 31
2.3 Mechanical Properties of the Cell Surface 32
2.3.1 Relationship Between the Surface Force and the Internal Pressure of the Cell 32
2.3.2 Direct Measurement of the Internal Pressure 33
2.3.3 Indirect Measurements of the Surface Force and the Internal Pressure 33

2.3.3.1 Compression Method 34
2.3.3.2 Suction Method 34
2.3.3.3 Stretching Method 36
2.3.3.4 Sessile Drop Method 36
2.3.4 Elasticity and Viscoelasticity of the Cell Surface . . . 37
2.4 Mechanical Properties of the Endoplasm 38
2.4.1 Measurements of Mechanical Properties of the Endoplasm 38
2.4.1.1 Centrifuge Method 39
2.4.1.2 Magnetic Particle Method 39
2.4.1.3 Capillary Method 41
2.4.1.4 Brownian Movement Method 43
2.4.1.5 Diffusion Method 43
2.4.2 Relationship Between the Mechanical Properties and Submicroscopic Structure of the Endoplasm 44
References 45

I.3 Use of Finite Element Methods in Cytomechanics: Study of the Mechanical Stability of the Skeletal Basal Plate of *Callimitra* a Biomineralizing Protozoan
G. L. Steucek, R. R. Archer, and O. R. Anderson (With 3 Figures) 47

3.1 Introduction 47
3.2 *Callimitra* Architecture 48
3.3 Finite Element Approach 49
3.4 Further Applications of FEM and Their Implications . 52
References 53

I.4 Mechanics and Hydrodynamics of Rotating Filaments
R. Jarosch (With 23 Figures) 54

4.1 The Molecular Basis of Filament Rotation 54
4.2 Longitudinal (Screw-Mechanical) Effects 56
4.2.1 Waving and Screwing 56
4.2.2 The Oscillation 59
4.2.3 Control of Polymerization and Depolymerization . . 59
4.2.4 The Translocation of Particles 62
4.2.5 Crossbridges 65
4.3 Lateral (Hydrodynamic) Effects 68
4.3.1 Pattern of Flows 68
4.3.2 Flows Adjacent to a Wall 69
4.3.3 Flows and Molding of an Adjacent Liquid Surface . . 70
4.3.4 Rolling Motions and Self-Arrangements 70
References 72

II. The Supramolecular Level

II.1 **Mechanical Concepts of Membrane Dynamics: Diffusion and Phase Separation in Two Dimensions**
K. Beck (With 6 Figures) 79

1.1 Introduction 79
1.2 Translational Diffusion in Fluid Phase Membranes . . 80
1.2.1 Net Transport by Diffusion: The Einstein-Smoluchowski Equation 80
1.2.2 Diffusion Modeled as a Stochastic Random Walk: The Free Volume Model 81
1.2.3 Diffusion Modeled by Continuum Hydromechanics: The Saffman-Delbrück Model 85
1.2.4 Diffusion in Biological Membranes 88
1.3 Fluid-Solid Phase Separation in Two Dimensions . . 90
1.3.1 Effective Medium and Percolation Theory 90
1.3.2 Phase Separation in Lipid Monolayers 92
1.3.3 Phase Separation in Biological Membranes 95
1.4 Concluding Comments 96
References 96

II.2 **Implications of Microtubules in Cytomechanics: Static and Motile Aspects**
D. G. Weiss, G. M. Langford, and R. D. Allen (With 4 Figures) 100

2.1 Microtubule Structure: Statics and Elasticity 100
2.1.1 Substructure of Microtubules 100
2.1.2 Rigidity of Microtubules 102
2.1.3 Integration of Microtubules into the Cytoskeleton . . 102
2.2 Microtubule-Associated Dynamics: Motion and Tension 103
2.2.1 Elongation of Microtubules 103
2.2.2 Shortening of Microtubules 104
2.2.3 Treadmilling of Microtubules 104
2.2.4 Organelle Movement Along Microtubules 104
2.2.5 Gliding of Microtubules 106
2.2.6 Sliding of Microtubules 107
2.2.7 Movement of Axostyle Microtubules 108
2.2.8 Complex Interactions of Microtubules 108
2.2.9 Contraction of Microtubule Arrays 108
2.3 Conclusions 109
References 110

II.3 The Nature and Significance of ATP-Induced Contraction of Microtubule Gels
R. C. WEISENBERG (With 2 Figures) 114

3.1 Introduction 114
3.2 Microtubule Gelation-Contraction 114
3.2.1 In Vitro Experiments 114
3.2.2 Significance of Microtubule Gelation-Contraction in Living Cells 116
3.2.2.1 Mitotic Spindle 116
3.2.2.2 Axonal Transport 117
References 119

II.4 Generation of Propulsive Forces by Cilia and Flagella
M. E. J. HOLWILL and P. SATIR (With 1 Figure) . . . 120

4.1 Introduction 120
4.2 Hydrodynamic Interactions 122
4.3 Passive Elastic Properties 125
4.4 Active Mechanical Properties 126
4.5 Conclusions 128
References 129

II.5 The Cortical Cytoplasmic Actin Gel
T. P. STOSSEL, P. A. JANMEY, and K. S. ZANER (With 5 Figures) 131

5.1 Historical Background 131
5.2 The Assembly of Actin and Actin-Binding Proteins Regulating Actin Assembly 134
5.3 The Rheology of Actin and Its Modulation by Actin-Binding Proteins and Other Factors 136
5.4 Actin Gelation in the Cell 144
5.5 Regulation of the Actin Sol/Gel Transformation in the Cell 148
References 149

II.6 Dynamic Organization and Force Production in Cytoplasmic Strands
K.-E. WOHLFARTH-BOTTERMANN (With 5 Figures) . . 154

6.1 Nature and Locomotory Phenomena of Physarum Plasmodia 154
6.2 The Generation of Hydrostatic Pressure Flow 154
6.3 Contractile Activities as Measured by Tensiometry . . 157
6.4 Analysis of Morphological Alterations Induced by Stretch Experiments 158
6.5 Nature and Implications of the Contraction Cycle . . 161

6.6 The Widely Unknown Regulation 162
6.7 Cytomechanical Implications 162
References 165

III. Mechanical Factors Determining Morphogenesis of Protists

III.1 Determination of Body Shape in Protists by Cortical Structures
K. HAUSMANN and M. MULISCH (With 13 Figures) . . 169

1.1 Introduction 169
1.2 Intracellular Cortex Structures 169
1.3 Extracellular Cortex Structures 174
1.4 Concluding Remarks 179
References 181

III.2 Morphogenetic Forces in Diatom Cell Wall Formation
A.-M. M. SCHMID (With 11 Figures) 183

2.1 Introduction 183
2.2 Possible Functions of the Diatom Cell Wall 186
2.3 Preconditions of Valve Formation 188
2.3.1 Mitosis and Cleavage 188
2.3.2 The Molding Surface: The Plasmalemma 190
2.3.3 The Mold for the Valve Outline 191
2.4 Valve Formation 192
2.4.1 The Silica Deposition Vesicle (SDV) 192
2.4.2 The Role of the Nucleus, Microtubule Center, and Microtubules 192
2.4.3 The Molding System for the Valve Pattern 194
2.4.4 Mechanisms for Mechanical Stabilization of the Valve 195
2.4.5 The Organic Coat and Valve Release 196
2.5 Conclusions 197
References 198

III.3 The Cytoskeletal and Biomineralized Supportive Structures in Radiolaria
O. R. ANDERSON, M. L. MOSS, and R. SKALAK (With 4 Figures) 200

3.1 Introduction 200
3.2 Cytoskeletal Organization of the Axopodia 201
3.3 Biomineralization and Skeletal Morphogenesis . . . 202
3.3.1 Analysis of Growth Phases 202
3.3.2 Finite Element Analysis 203

3.3.2.1 FEM Descriptors 204
3.3.2.2 FEM Results 206
3.3.2.3 Limitations and Implications of FEM Analysis with Radiolaria 210
References 211

IV. Mechanical Factors Determining Plant Cell Morphogenesis

IV.1 Mechanical and Hydraulic Aspects of Plant Cell Growth
D. J. COSGROVE (With 3 Figures) 215

1.1 Introduction 215
1.2 Directionality of Cell Growth 215
1.2.1 Patterns of Expansion 216
1.2.2 Wall Architecture 216
1.2.3 Multinet Growth 217
1.2.4 The Wall Matrix 219
1.3 Wall Loosening and Expansion 219
1.3.1 Physics of Wall Expansion 220
1.3.2 Stress Relaxation 221
1.3.3 Molecular Models of Wall Loosening 222
1.4 Water Uptake and Turgor Maintenance 223
1.4.1 Physics of Water Uptake 223
1.4.2 Restriction of Growth by Water Transport 224
1.4.3 Solute Uptake 225
1.5 Summary 226
References 226

IV.2 Plant Cytomechanics and Its Relationship to the Development of Form
P. M. LINTILHAC (With 4 Figures) 230

2.1 Introduction 230
2.2 The Logic of Development 230
2.2.1 The Role of the Genome in the Development of Form 230
2.2.2 The Role of the Environment in the Development of Form 231
2.3 The Architecture of Plant Form 232
2.3.1 Division and Growth. The Basic Events 232
2.3.2 Growth as a Source of Mechanical Stress 232
2.3.3 Factors Affecting Stress Distribution in Embryonic Plant Organs 233
2.3.4 The Role of Stress in the Generation of Form 234
2.4 The Ultrastructural Basis of Cell Behavior 234

2.4.1 The Role of Cytomechanics in the Development of Form . . . 234
2.5 Other Responses to Mechanical Stimuli. Reaction Wood . . . 235
2.5.1 Tropic Responses . . . 235
2.6 Meiosis as a Mechanically-Induced Process . . . 236
2.6.1 The Sporangium as a Stress-Focusing Device . . . 236
2.6.2 Isotropic Stress as a Developmental Effector . . . 237
References . . . 241

IV.3 Mechanical Properties of the Cyclamen Stalk and Their Structural Basis
K. DIERKS and L. HAFNER (With 5 Figures) . . . 242

3.1 Anatomy of the Cyclamen Persicum Flower Stalk . . . 242
3.2 Internal Hydrostatic Pressure . . . 243
3.3 Behavior Under Ultimate Load . . . 244
3.4 Summary . . . 246
References . . . 246

V. Mechanical Forces Determining the Shape of Metazoan Cells

V.1 Forces Shaping an Erythrocyte
R. E. WAUGH and R. M. HOCHMUTH (With 2 Figures) 249

1.1 Introduction . . . 249
1.2 Membrane Elasticity . . . 250
1.2.1 Shear Elasticity . . . 250
1.2.1.1 Molecular Basis of Shear Elasticity . . . 252
1.2.1.2 Metabolic, pH, and Ionic Effects . . . 253
1.2.2 Area Elasticity . . . 253
1.2.2.1 Molecular Basis of Area Elasticity . . . 254
1.2.3 Bending Elasticity . . . 255
1.2.3.1 Molecular Basis of Bending Rigidity . . . 255
1.3 Membrane Viscosity . . . 256
1.3.1 Molecular Basis of Membrane Viscosity . . . 257
1.4 Erythrocyte Shape . . . 257
References . . . 258

V.2 Hydrostatic Pressure in Metazoan Cells in Culture: Its Involvement in Locomotion and Shape Generation
J. BEREITER-HAHN and R. STROHMEIER (With 5 Figures) . . . 261

2.1 Introduction . . . 261
2.2 Osmotic Equations Applied to Cells . . . 261

2.3 Physical State of Cell Water 262
2.4 Solute Leakage 262
2.5 Osmotic Behavior of Cytogel 263
2.6 Generation of Intracellular Hydrostatic Pressure . . . 263
2.6.1 Osmotic Behavior of Cells in Culture 264
2.6.2 Determination of Hydrostatic Pressure in Culture Cells 264
2.6.3 "Visualization" of Tension in the Cortical Fibrillar-Meshwork-Plasma Membrane Complex 266
2.7 Functional Significance of Hydrostatic Pressure in Wall-Free Cells 268
2.7.1 Cell Shape 268
2.7.2 Cell Locomotion 269
2.7.3 Integration of Cells into Tissues 270
2.7.4 Hydraulic Interaction of Organelles 270
References 270

V.3 The Transmission of Forces Between Cells and Their Environment
M. OPAS (With 2 Figures) 273

3.1 Introduction 273
3.2 Focal Contact: Subcellular Level 273
3.3 Traction: Cellular Level 276
3.4 Adhesion: Supracellular Level 280
3.5 Conclusions 281
References 282

Subject Index 287

List of Contributors

You will find the addresses at the beginning of the respective contribution

Allen, R. D. 100
Anderson, O. R. 47, 200
Archer, R. R. 47
Beck, K. 79
Bereiter-Hahn, J. 3, 261
Cosgrove, D. J. 215
Dierks, K. 242
Hafner, L. 242
Hausmann, K. 169
Hiramoto, Y. 31
Hochmuth, R. M. 249
Holwill, M. E. J. 120
Janmey, P. A. 131
Jarosch, R. 54
Langford, G. M. 100
Lintilhac, P. M. 230
Moss, M. L. 200
Mulisch, M. 169
Opas, M. 273
Satir, P. 120
Schmid, A.-M. M. 183
Skalak, R. 200
Steucek, G. L. 47
Stossel, T. P. 131
Strohmeier, R. 261
Waugh, R. E. 249
Weisenberg, R. C. 114
Weiss, D. G. 100
Wohlfarth-Bottermann, K.-E. 154
Zaner, K. S. 131

List of Contributors

I. General Principles

I.1 Mechanical Principles of Architecture of Eukaryotic Cells

JÜRGEN BEREITER-HAHN[1]

1.1 Introduction

Living cells are soft bodies of a characteristic form, but endowed with a capacity for a steady turnover of their structures. This dynamic character is the main difference between cells and most objects engineers are concerned with. Despite these dynamics, an overall shape can be maintained and structural continuity assures continuity of force transmission throughout ontogeny. Dynamics is the basis for the high reactivity of cells to internal and external signals. For instance, migrating cells colliding with others may either immediately change their direction of locomotion or they may adhere to each other and start to form an epithelium. Both events require structural reorganization. This is determined by three factors: (1) the commitment of the cell to a certain pattern of behavior controlled by the genetic material, (2) the mechanical constraints inherent in the cell structure itself, and (3) mechanical interaction of the cell with its environment. Therefore, continuity of tension transmission can be regarded as a form of information processing (Albrecht-Bühler 1985 and Chap. IV.2). A historical overview of the development of ideas on the "skeletal structures" in cells is given by Stossel et al. (Chap. II.5). The most consequent approach to such an understanding of cell architecture, at the present, is that of Ingber and Jamieson (1985) describing cells as tensegrity structures, i.e., structures which generate their own tensional forces and exhibit an architectural integrity independent of gravity. A discussion of whether their model holds true in detail for a description of cell architecture would excess the scope of these introductory remarks.

1.2 Basic Mechanical Parameters of Cells

Mechanical properties of cells can be described as a molecular continuum. However, at the level of supramolecular structures, we approach the limits of the validity of continuum mechanics. This is the case in studying mobility of molecules in membranes (Chap. II.1) or for actin and myosin interaction at the molecular level at high time resolution. In this book, emphasis is placed on the gross behavior of cells and their structural components such as stress fibers or

---

[1] Cinematic Cell Research Group, J. W. Goethe-University, Senckenberganlage 27, D-6000 Frankfurt a. M., FRG

on extracted actomyosin gel which can well be described efficiently in terms of continuum mechanics.

At least one mechanical parameter, i.e., viscoelasticity, is common to all living cells and soft tissues. To understand the meaning of this term, some definitions may be helpful: deformation of a body by an externally applied force depends on material characteristics and body size. If the length of deformation is constant, then the strain is related to the force (F) acting per unit of area (A) in the section perpendicular to the vector axis of the force. F/A is called stress (σ). The basic unit for the force is the newton (N), for the stress, newton per square meter ($N\ m^{-2}$) or pascal (Pa). Instead of the newton, the dyne (dyn) is frequently used: 1 dyn = 10^{-5} N; 1 dyn cm^{-2} = 0.1 Pa.

The deformation of a body resulting from the action of stress is termed strain. The type of deformation may be compression or expansion of volume (σ_V), increase or decrease of length (σ_E), or distortion and shear (σ_S). In these three cases the stress/strain relation may be formulated as follows:

$$\sigma_V = M \cdot \frac{V - V_0}{V_0} \tag{1}$$

$$\sigma_E = E \cdot \frac{L - L_0}{L_0} \tag{2}$$

$$\sigma_s = G \cdot \tan\alpha\,. \tag{3}$$

M is the modulus of compression; E is the elasticity modulus (Young's modulus); G is the shear modulus or modulus of rigidity; V and L are the volume and the length of the body, respectively, after application of stress. The torsion or a shear is measured by the angle α. These relations are valid only for small deformations, i.e., for a certain range of stress. In this case the incremental stresses are related to the strains by the material constants M, E, and G. If these yield stresses are exceeded, the stress/strain relation is no longer linear.

Equation (2) represents Hooke's law. Stretching ($L - L_0 = \Delta l$) of a body is always accompanied by a "contraction" (Δu) perpendicular to the direction of stretch. The ratio

$$\Delta u / \Delta l = \nu \tag{4}$$

is called the Poisson ratio and may be much larger than 1 for fibrous materials (Oster et al. 1983).

The Poisson ratio relates M, E, and G to each other:

$$G = \frac{E}{2(1+\nu)} = \frac{3M(1-2\nu)}{2(1+\nu)}\,. \tag{5}$$

The unit of M, E, and G is $N\ m^{-2}$. Isotropic materials exhibit a constant E and G regardless of the direction of the applied stresses. In anisotropic material the elasticity properties vary with the direction of deformation to the axis of anisotropy.

Let us consider again the concept of "viscoelasticity". In an elastic body at the limits of yield stresses, strain follows instantaneously the increment or decrement of stress. In a viscoelastic body (cf. Fig. 6 in Chap. I.2), deformation continues after the stress has reached a constant value (creep). However, the forces induced by a constant extermal load may disappear with time (relaxation). Under cyclic loading and unloading the stress/strain relationship differs in the loading and the unloading phase (hysteresis).

Creep, relaxation, and hysteresis of a viscoelastic body are phenomena inherent to the viscous element. In its simplest form viscosity may be described as a shear flow with a uniform velocity gradient (dv/dx) perpendicular to the flow axis (x). When the shear stress is τ, then Newton's viscosity equation is

$$\tau = \eta \cdot \frac{dv}{dx}, \tag{6}$$

where η is the viscosity coefficient. The unit of η is Ns m^{-2} (Pa · s), or in the CGS system the poise (P) (1 P = 10 Ns m^{-2}).

In a viscous fluid stress is related to the rate of strain. If this relation is linear, the fluid is called Newtonian. The streaming characteristic of such a fluid is described by Hagen-Poiseuille's law (see Chap. V.2), which is valid only for condition of laminar flow. This condition is fulfilled at low Reynold's numbers ($R \ll 1$)

$$R = \frac{v \cdot l \cdot \varrho}{\eta}, \tag{7}$$

where v = flow velocity, l = a characteristic length, ϱ = density of the fluid and η = viscosity.

For cytoplasm, R is in the range of 10^{-4} to 10^{-5}, while that of the vacuolar sap in a plant cell about 0.05 (Nothnagel and Webb 1982). The velocity of a particle driven through a viscous fluid is described by Stokes' law [Eq. (9) in Chap. I.2.4.1.1]).

Kinematic viscosity (ν) is achieved by the ratio of the viscosity coefficient and density. The unit of ν is the Stokes (St) (1 St = 1 cm^2 s^{-1}).

The viscoelastic properties of a body can be represented by models with elastic and viscous elements in a row or in series. The present description is based on that given by Fung (1981, 1984), where more detailed information can be obtained. In Fig. 1 the elastic elements are represented by springs, the viscous elements by dashpots. A technical example for a dashpot is the vibration absorber of a mechanical balance. Three basic arrangements are most often used for the description of the viscoelastic behavior of biological material (e.g. Fung 1984; Oster et al. 1985, Chap. I.2.4.2.3), the Maxwell body, the Voigt body, and the Kelvin body (Fig. 1). The Maxwell body comprises an elastic element in series with a viscous element, while in the Voigt model a spring and a dashpot are parallel, and in the Kelvin model a Maxwell body is in parallel with a spring. All these models exhibit the viscoelastic features of creep, relaxation, and hysteresis. The differences become apparent when subjected to loading cycles. With increasing cycle frequency the dashpot in the Maxwell body will move less and less at the same load, therefore, hysteresis decreases. In the Voigt body the dashpot takes up more and more load with increasing frequency, resulting in an increase in hysteresis. In the Kelvin body hysteresis

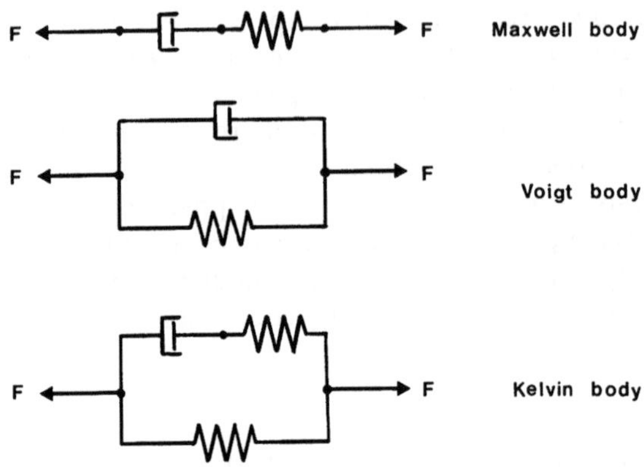

Fig. 1. The three basic models of viscoelastic bodies

reaches a maximum at a characteristic frequency. Further extensions of these models may be necessary to represent the behavior of a cell or tissue (e.g. Fig. 8 in Chap. I.2). For instance, complications arise due to nonlinearity of elasticity behavior, non- Newtonian viscosity, and active contractions. However, the introduction of an actively contractile element as proposed by Oster et al. (1985) represents only a qualitative description and cannot be measured separately from Hookean elasticity (cf. next section and Chap. V.2.6).

1.3 Cellular Viscosity

Extensive studies of the organization of the cytoskeleton revealed the apparently structureless ground cytoplasm to be in fact a highly structured matrix composed of interconnected fibrils and filaments, with interstitial spaces (for overview see e.g. Porter 1984). From this it follows that the cytomatrix cannot be regarded simply as a viscous fluid, but rather as a mixed phase body composed of a fibrillar network penetrated by a solution (Dembo et al. 1986). As a consequence viscosity is different in the various phases. Pollard (1979) gave a vivid description of this situation: "If one takes concentrations of 20–30% cells (*E. coli* suspension), breaks them open (yielding a curious looking mass), and upends the tube containing the mass, the broken cells will effectively never flow at all." From flow behavior, a viscosity of 1000 Ns m^{-2} was calculated. Estimating viscosity according to the Einstein-v. Smoluchovsky relation (see Chap. II.1.2.1) by measuring diffusion rates of sucrose, dextran, and β-galactosidase, the apparent viscosity was only about three to four times that of water (see also Dembo and Harlow 1986). Cytoplasm is inhomogeneous at all levels of organization and furthermore it may be anisotropic. Therefore, the viscosity value obtained, depends significantly on the method of evaluation. The smallest viscosity range may be investigated by fluorescence polarization, which reflects "microviscosity" of the solvent in the immediate vicinity of the fluorescent probe (see, e.g., Chen et al. 1977; Mastro and Keith 1984). Until now, this method was primarily used for studies of membrane viscosity, however, it can easily be applied to studies of viscosity in cytoskeletal fibrils. The network shear viscosity, which is determined by various methods of strain induction (e.g., the magnetic ball method or deformation by suction with a micropipette, etc., see Chap. I.2), is in the range of $10^3 - 10^6$ Pa · s (see Table 1). In this Chapter, the term viscosity is applied to shear viscosity. This is necessary since only theoretical considerations, but no experimental data, exist for "dilatation viscosity", i.e., the frictional resistance to the speed of network contraction, or expansion (Dembo and Harlow 1986; Dembo et al. 1986). Macrorheology may be applied not only to cytoplasm, but can also be used to describe the behavior of whole tissues (Phillips 1984) where, depending on their mutual adhesiveness, mobile subunits can slide past one another. In such a "fluid," shear stresses are soon relaxed as the subunits slide, and initial cell stretching (the elastic response) disappears. A widely discussed question is whether cytoplasm behaves like a Newtonian fluid. Certainly it does not, because apparent viscosity increases with decreasing shear rate (Valberg and Albertini 1985, Chap. I.2.4) and elastic properties add to its behavior (Sato et al. 1983; see next section). A non-Newtonian character of cytoplasm is obvious from a variety of evidence; e.g., from studies of the mechanical behavior of chick embryo cells when probed with a vibrating needle (Kucera and de Ribaupierre 1982), from investigations of actomyosin gel with the falling ball method (McLean-Fletcher and Pollard 1980), or from the velocity profile of streaming endoplasm in plasmodial strands of slime molds (Kamiya and Kuroda 1958). On the other hand, Newtonian behavior is shown by the linear velocity profile of vacuolar *sap* in characean algae (e.g. Nothnagel and Webb 1982). Regarding the higher complexity of non-Newtonian models, Dembo and Harlow (1986) emphasized the use of the simpler Newtonian model permitting a reasonable approach to the function of contractile networks.

1.4 Elasticity, Contractile Forces, and Surface Tension

If cytoplasm resembles a three-dimensional fibrillar meshwork, its viscoelastic properties depend on the mean free length of the fibrils, their elasticity; and with respect to crosslinks their binding strength, length, and elasticity. The type of binding may range from van der Waal's forces, and ionic attractions to covalent bindings. In the latter case, the elastic properties, which in the case of cyclic loading are independent of frequency, will dominate the behavior. In the first case, the viscous properties are prevalent. However, at higher frequencies, exceeding the relaxation time of the viscous fluid, the behavior approaches that of an elastic body. As explicated by Stossel et al. (1982) a solution of pure actin filaments represents a sol, because stress of nearly any magnitude does not evoke a constant degree of strain. This creeping shows that the F-actin solution does not have a modulus of rigidity, characteristic of Hookean gels (linear stress/strain relation); but rather it has a pseudoequilibrium compliance as is characteristic of entangled polymers. In accord with crosslinking, osmotic swelling of a gel may add to its elastic properties (Oster 1984; Oster and Odell 1983,

Table 1. Apparent viscosity of the cytoplasm and cellular components

Specimen	Method	Particle diameter μm	Apparent viscosity Pa · s	Reference
Fibroblasts				
Chick	Actin diffusion	2.3×10^{-3}	17.9	Kreis et al. 1982 [b]
	BSA diffusion	3.6×10^{-3}	10.1	
	Vimentin diffusion	3.3×10^{-3}	21.7	Geiger et al. 1985 [b]
	α-Actinin diffusion	7.8×10^{-3}	9.3	
	Magnetic dragging	2–5	~100	Crick and Hughes 1950 [a]
Human	BSA diffusion	3.6×10^{-3}	6.1	
	IgG diffusion	3.5×10^{-3}	5.0	
	Photobleaching recovery	7.2×10^{-3}	8.0	Wojcieszyn et al. 1981 [a]
Mouse	PCAOL-ESR	0.3×10^{-3}	0.2	
3T3 and SV 40 3T3	PCAOL diffusion	0.3×10^{-3}	0.2–0.3	Mastro and Keith 1984
3T3 Bleb endoplasm	Brownian motion	0.1	1–20	Albrecht-Bühler 1982
Newt cells in culture	Brownian motion	0.5–1.0	2.4	Taylor 1963
BHK cells	PCAOL-ESR	0.3×10^{-3}		Mastro and Keith 1984
300 mOsmol			0.3	
550 mOsmol			1.0	
Human neutrophils	Micropipette aspiration	Inner diameter: 3	~1000	Sung et al. 1982 [a]
Mouse "macrophage like" cells	Ovalbumin diffusion	2.4×10^{-3}	2.7	Wang et al. 1982 [a]
Pulmonary macrophages	Magnetometry of twisted particles	0.3–0.7	$12-27 \times 10^{4}$	Valberg and Albertini 1985
Squid axoplasm	Magnetic dragging	83–213	$10^{6}-10^{7}$	Sato et al. 1984 [a]
	Methylene blue diffusion	0.37×10^{-3}	0.45	Hodgkin and Keynes 1956 [b]
Sea urchin eggs	Magnetic dragging	5–9	~1000	Hiramoto 1969 [a]
	Centrifugation of nucleus	~10	~300	Hiramoto 1967 (cited from Chap. I. 2.4.2)
Chara				
endoplasm	Brownian motion		5–2.5	Nothnagel and Webb 1982
Cell sap	Brownian motion		0.1–0.25	Cited from Nothnagel and Webb 1982
Nitella				
endoplasm	Brownian motion	0.3–0.4	0.6	Kamiya and Kuroda 1965
Physarum				
endoplasm	Magnetic dragging	4–16	8–10	Sato et al. 1983
	Capillary method	1–5	18–50	
			0.3	Kamiya and Kuroda 1958a (cited from Chap. 12.4.1.3)

Table 1. (Continued)

Specimen	Method	Particle diameter μm	Apparent viscosity Pa.s	Reference
Amoeba	BSA diffusion	3.6×10^{-3}	0.2	Mastro and Keith 1984
	Photobleaching recovery	7.2×10^{-3}	0.1 – 0.2	Wang et al. 1982[a]
	Magnetic dragging	5 – 9	1 – 30	Yagi 1961[a]
E. coli cytoplasm	Efflux of isolated cytoplasm	$> 10^3$	> 1000	Pollard 1979
Acanthamoeba	Falling ball method	640		
Actin (1.5 mg ml^{-1})				
G-actin			0.1 – 0.2	McLean Fletcher and
F-actin			> 1200	Pollard 1980
Airway mucus	Oscillating particle analysis	50 – 150	750	King and Macklem 1977[a]

[a] Cited from Valberg and Albertini (1985).
[b] Cited from Mastro and Keith (1984).
Abbreviations: ESR, electron-spin resonance; PCAOL 2,2,5,5-tetramethyl-3-methanolpyrroline-N-oxyl.

1984a, b). Normal viscosimetry, e.g., using an Ostwald viscosimeter or the falling ball method (McLean-Fletcher and Pollard 1980), ignores elasticity contributions. These can be monitored by a sensitive goniorheometer, which then provides data for calculation of crosslinking density in a gel using models from rubber elasticity theory (Jen et al. 1982). In undisrupted cells the elastic component of cytoplasm is represented by a minimum stress (yield stress) needed to start motion of a particle through the cytoplasm (Sato et al. 1983). Qualitatively, the elastic component in a stress/strain relationship has been demonstrated in amoebae by the vanishing of strain birefringence due to elastic recoil after stretch release (Taylor et al. 1973, 1976).

One possible mechanism of crosslinking in actin gels is the actin-myosin interaction, either stable (in rigor state) or oscillating (during contraction). This implies that in a contracting system the contraction force which can be measured, e.g., by tensiometry (see Chap. II.6), always adds to elasticity (Oster et al. 1985). Determination of elasticity is possible only if one assumes that of the contraction force is constant while an external deforming force is applied. The correctness of this assumption cannot be proven, therefore, elasticity in a contractile gel cannot be separated from contractility. Even inhibition of active contraction does not allow separation of the two parameters because of changes in crosslinking.

The parameters of measurement are tension or pressure (Table 2). In many instances these do not allow calculation of elastic moduli or of stresses. For this purpuse, assumptions on the structural organization of the elements responsible for these parameters have to be made. An example may explain this difficulty: if a fibroblast exerts a certain tension on its substrate (see e.g. Harris et al. 1980, 1981) then this tension could be generated by contractions located either in the

Table 2. Stresses exerted by cells and cellular structures

Specimen	Stress (Nm^{-2})	Reference
Skeletal muscle		
rabbit	3.9×10^5	Gabella 1967
frog	2.9×10^5	Prosser and Brown 1961 [a]
Smooth muscle (cow)	0.7×10^5	Weber 1958
Mytilus muscle	5×10^5	Abbott and Lowy 1958 [a]
Cleavage furrow of sea urchin egg	$10^4 - 10^5$	Hiramoto 1967
Fibroblast traction	$10^4 - 10^5$	Harris et al. 1980
Fibroblast sheets (traction)	3.4×10^3	James and Taylor 1969
Sarcoma cell sheets (glycerinated)	2.2×10^2	Hoffmann-Berling 1956
Fish melanophores (aggregating stress)	$< 21 \times 10^5$	Schliwa and Bereiter-Hahn 1975
Gastrulation (Triturus)	3.8×10^4	James and Taylor 1969
Amoeba pseudopodium	2.9×10^3	Goldacre 1961
Particle transport on cell surface	10^{-1}	Dembo and Harris 1981
Particle transport at ecto/endoplasm interface (*Nitella*)	1	Dembo and Harris 1981
Motive force for cyclosis (*Nitella*)	$0.1 - 0.4$	Nothagel and Webb 1982
Tension of isometrically contracting *Physarum* cytoplasmic strand	$1 - 3 \times 10^3$	Wohlfarth-Bottermann Chap. II. 6 Matsumura et al. 1980
Motive force, slug of Dictyostelium	5.3×10^5	Inouyé 1984
Actomyosin thread from skeletal muscle	0.25×10^5	Portzehl 1951
from *Physarum* cytoplasm (Isometric contr.)	$10^2 - 10^3$	Matsumura et al. 1980
Yield strength		
F-actin	$0.6 - 2.9 \times 10^2$	McLean-Fletcher and
G-actin	< 2	Pollard 1980
Carchesium stalk (tearing force)	$2 - 20 \times 10^5$	Rahat et al. 1969
Blood clot contraction	1.5×10^2	Jen et al. 1982
	1.2×10^4	Rubenstein 1962 [a]

[a] Cited from James and Taylor (1969); see also Table 1, Chap. V.2

Table 3. Forces exerted by cells and cellular structures

Specimen	Force (N)	Reference
1 Dynein arm	max. 1×10^{-12}	Holwill and Satir Chap. II.4 Kamimura and Takahashi 1981
1 Actin/myosin crossbridge	$2 - 5 \times 10^{-12}$	Dembo and Harris 1981
Tension force exerted by a single fibroblast	1.65×10^{-7}	James and Taylor 1969
Force exerted by the mitotic spindle in newt cells in culture	2.5×10^{-8}	Taylor 1963
Cleavage furrow (sea urchin egg)	3×10^{-8}	Hiramoto 1978, cited from: Dembo and Harlow 1986
Force on single bead in *Nitella* at ecto/endoplasm interface	3×10^{-14}	Dembo and Harris 1981

cortical fibrillar meshwork or in the stress fibers oriented in parallel to the tension vector or by the whole cytomatrix. The decision as to which of these systems is responsible for production of force determines the cross-sectional area to which the force has to be related. Furthermore, the morphological data for evaluation of this area are subjected to a high degree of uncertainty.

Historically, the role ascribed to surface tension for control of cell shape and cell–cell interaction lost more and more of its significance as the intimate connection of the plasma membrane with the cytoskeleton became apparent. Regarding these recent discoveries, Harvey's (1931) definition of 'surface tension of cells' is still valid: "It seems unlikely that the tension is a true surface tension between non-miscible fluids, but the behaviour of deformed spherical cells shows very clearly that a surface force exists which can best be referred to as the 'tension at the surface' without implying either elastic tension or surface tension." Surface tension sensu strictu can be applied only for isolated vesicles. In cells, elastic and viscous properties of the membrane-bound cytoskeleton determine the "tension at the surface" (see Chaps. I.2.3, II.1, V.1, and V.2.7.3). Certain observations are consistent with the absence of true surface tension: i.e., the differential resistivity of certain areas of the cell surface against poking with a needle (Petersen et al. 1981) and the widespread appearance, of calcium-induced surface waves in eggs (Cheer et al. 1986).

Surface tension in the broad sense, as determined by the sessile drop method (Chap. I.2.3.3.4), seems to be directly involved in sorting of embryonal tissues (Steinberg 1978). Furthermore, the structure of tissues bears a certain resemblance to that of bubbles and foams, which allows the determination of surface forces keeping the "bubbles" together (ΔP) according to Laplace's law, assuming a bilayered interface:

$$\Delta \mathrm{P} = \frac{2\gamma}{\mathrm{r}}, \tag{8}$$

where γ=surface tension, r=radius of the interface between the "bubbles." For embryonic chick tissues γ is in the range of $10^{-3} - 3\times10^{-2}\,\mathrm{Nm^{-1}}$ (cited from Stein and Gordon 1982).

1.5 The Structural Basis of Cell Mechanics

Cytoplasm is composed of an abundance of densely interwoven filaments including interstitial spaces containing the freely mobile small metabolites and solutes. These filamentous structures are collectively termed "cytoskeleton" or "cytoskeletal networks-" They retain their complex appearance and preserve the overall shape of cells after extraction of other cytoplasmic components with nonionic detergents (see e.g. Ben Zeev et al. 1979; Penman et al. 1982; Weber and Osborn 1982). Its structural pattern has been elucidated first by "classic" electron microscopy, then by immunofluorescence, which started with the investigations of Lazarides and Weber (1974), and more recently by high voltage electron microscopy of whole cells and the quick-freezing, deep-etching technique developed by Heuser and Kirschner (1980).

Biochemically the components of these filaments seem to be well explored, their interaction with each other and with a steadily increasing number of newly detected associated proteins is a field of intensive research activities. For a detailed description of the filamentous constituents of the cytoskeleton the reader is referred to the excellent review by Schliwa (1986). There are four main types of filaments: (1) the actin-containing microfilaments (associated, e.g., with α-actinin, myosin, tropomyosin, filamin, vinculin and others; see also Chap. II.5), (2) the heterogeneous group of intermediate filaments, (3) the microtubules consisting of tubulin and several microtubule-associated proteins (MAPS), and (4) spasminlike filaments in some protists, which will not be discussed in this chapter. In cytoplasm these filamentous structures develop some of their mechanical properties by interaction with the cell membrane, and by interacting with the extracellular matrix via the cell membrane (see Chap. V.3). In addition, at least the nucleus, plastids, and mitochondria exhibit mechanical resistivity conferred by their own membrane and matrix components. Our knowledge of the nuclear matrix is just developing (Penman et al. 1982; Pouchelet et al. 1986), while that of chloroplasts and mitochondria still requires very basic studies (Bereiter-Hahn and Vöth 1983).

1.5.1 Actin and Actin-Based Structures

The most remarkable property of actin (with associated proteins) is its capability for a consecutive series of actions:

1. Assembly from mono- or oligomeric subunits to two-polar filament strands wound around each other (see Chap. II.5.2). For factors controlling assembly of actin see Chap. II.5.4.
2. Formation of interconnections among the filaments either in parallel or end-to-side as well as possible connections to other cytoplasmic structures such as the cell membrane, phagosomes (Cain and Kraus 1981), or exocytotic vesicles (Drenckhahn and Mannherz 1983).
3. Conversion to "cytomusculature" by alignment and association with myosin, probably in its phosphorylated form.
4. Disassembly of the actin filament into its subunits, probably as a result of its contraction.

The factors controlling this actin behavior have been discussed by Condeelis (1983). A discussion of the complex events taking place during this cycle is beyond the scope of this introduction.

Assembly of actin subunits is accompanied by a considerable increase in shear viscosity of the resulting sol (see Table 1 and Fig. 3 in Chap. II.5.2). This can be increased further by gelation due to crosslinking by proteins (see e.g. Rockwell et al. 1984) such as α-actinin or by superprecipitation (contraction). Shortening of filament length by capping proteins or severing reduces viscosity (see Chap. II.5.2).

With respect to its length, bending resistivity of a single filament can supposedly be low. Actin crystals, however, as occurring in flatworm spines (Cohen et al. 1982) or bundles of filaments crosslinked laterally with each other and to the cell membrane, seem to exhibit considerable stiffness (De Rosier et al. 1982). This was shown first for the acrosomal processes of various sperm cells (Tilney 1977) and for microvilli of the intestinal brushborder (Mooseker 1974; Mooseker and Tilney 1975). Assembly of an actin gel on a plasmalemma matrix seems to be involved in thinning the leading edge of a protruding lamella of migrating cells without myosin being engaged (Strohmeier and Bereiter-Hahn 1984). Myosin-driven sliding of actin filaments past one another is a generally accepted mechanism of force generation in cells. A dense and continuous layer of actomyosin fibrils (Fig. 2), which is crosslinked to the cell membrane, can cause intracellular hydrostatic pressure by isometric contraction (see Chap. V.2). Anisotropy may be induced in such a fibrillar meshwork by local fixation (Isenberg and Wohlfahrt-Bottermann 1976), as is the case of formation of focal contacts (Izzard and Lochner 1980, see Chap. V.3). Focal contacts are the points where stress fiber formation starts (see Fig. 1, Chap. V.3). Figure 3 shows a model of how the development of a fibrillar bundle, a stress fiber, could take place. The net is fixed at one point and pulled at some distance. Pulling in the model replaces the contraction force generated by the actomyosin gel itself. The end distant from the fixed point has either to be anchored within the cell, or at another part of the plasma membrane (Couchman et al. 1983; Rees et al. 1977). If it is not so anchored it will splice and smoothly transmit tension

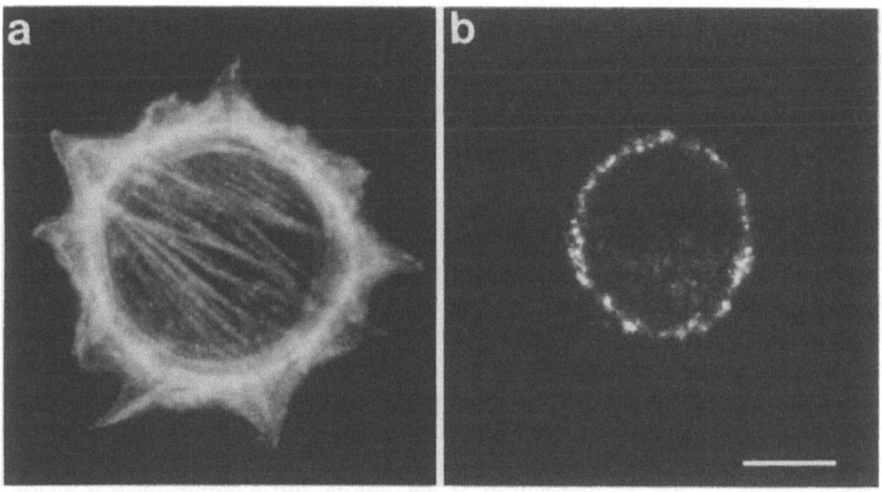

Fig. 2a, b. Example of the arrangement of actin fibrils. TRITC-phalloidin staining for F-actin in a single XTH-2 cell. Fluoromicrographs taken with a laser-scanning microscope at two focus levels. **a** Close to the substratum. Stress fibers traverse the basal part of the cell. At the periphery a ring of actin bundles marks the transition region between cell body and peripheral lamella. **b** 3.7 μm above **a**, only the dome-shaped central part of the cell is visible. Actin fibrils are restricted to the subplasmalemmal region. (Photographs by Dr. E. Stelzer, EMBL, Heidelberg) *Bar* = 10 μm

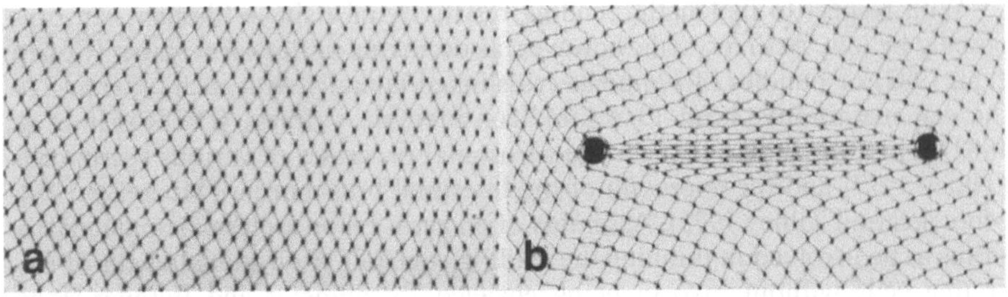

Fig. 3. Model experiment showing the transformation of an isodiametric network (**a**) into an anisotropic fibrillar structure (**b**) by generation of tension between two fixed points. Note that the lateral connection lies in register

to the more isotropic parts of the net (cf. Fig. 2, Chap. V.3). Long and thin fibrils can be considered to evolve by sliding of network filaments due to viscous relaxation. This behavior would be modeled more appropriately by pulling strands out of a viscous fluid such as honey.

In the example (shown in Fig. 3) the cross points of the net are stable connections at constant distances from each other. The fibrillar structure, resulting from local stretching of the net, exhibits a periodicity of cross-striations representing the connection sites (Fig. 3b). The model resembles the periodicity of myosin, tropomyosin, and α-actinin arrangement in stress fibers (Kreis and Birchmeier 1980). Whether the organization of stress fibers in fact evolves from regular and fixed crosslinks in the isotropic gel cannot be decided from the present experimental evidence. Primarily, stress fibers have been found in cell cultures. Recently, stress fibers, or at least actin bundles similar to stress fibers have been found under in situ conditions (Byers et al. 1980; Drenckhahn and Mannherz 1983; Kukulis et al. 1985).

1.5.2 Membrane-Associated Actin Fibrils

Membrane-associated fibrils allow transmission of tension to the extracellular matrix (see Chap. V.3) or from one cell to its neighbors throughout a continuous epithelium. Furthermore, this type of association provides a basis for several cytomechanical phenomena including (1) lateral receptor distribution in the membrane, (2) formation of a variety of surface protrusions like ruffles, microspikes, microvilli, or microridges, (3) the action of the cleaving furrow in cytokinesis, (4) cytotic processes (Cain and Kraus 1981; Drenckhahn and Mannherz 1983), and (5) generation of cell shape and locomotion (see below and Chap. V.2). The interconnection of the plasmalemma with the lamina enclosing the nucleus by the network of cytoskeletal fibrils may well be involved in the control of macromolecular synthesis (see Sect. 1.7.2).

An isotropic subplasmalemmal fibrillar meshwork has already been mentioned (Fig. 2); it constitutes a significant structural and dynamic component of many cells. The involvement of membrane-associated fibrils (see e.g. Byers and

Branton 1985) in shaping mammalian erythrocytes is discussed by Waugh et al. (Chap. V.1).

Adherens-type junctions [including zonae and zonulae adhaerentes of tadpole epidermis cells (Kunzenbacher et al. 1982), fasciae adhaerentes in cardiac myocytes, and focal contacts] are classified as a single family of contacts, characterized by the connection of actin filaments to the plasma membrane through a vinculin-rich plaque structure (Geiger et al. 1984a, b; Volberg et al. 1986). It is conceivable that these junctions in living cells are dynamic structures. This has been impressively documented for focal contacts of cultured fibroblasts (e.g. Couchman et al. 1983; Izzard and Lochner 1980; Izzard et al. 1986). This implies that various components of the junction can move and the entire structure can be induced to develop and grow or to wane and disappear (Kreis et al. 1984); Geiger et al. (1984b) provided a hypothesis describing developmental steps of local linkage of actin to membrane receptors. Straight bundles running in parallel to a row of junctions (e.g. Kunzenbacher et al. 1982; Tilney 1983) or emanating from a focal contact require for their orientation not only crosslinking but also tension. Linking to the plasma membrane provides a fixed point from which contraction can start organizing a straight fibril (Fig. 2a). Involvement of oriented actin bundles in the formation of cell boundaries has been shown for reaggregating epidermis cells (Bereiter-Hahn and Strohmeier 1987).

Actin filaments associated with membranes as in coated pits, stereocilia of hair cells, microvilli, and acrosomal processes exhibit the same polarity, i.e., the arrowheads of decorating HMM point away from the attachment sites of the membrane (Condeelis 1981). This further indicates that in all these cases filament assembly occurs at the membrane-linked end of the filaments (Tilney 1983). The resulting densely packed bundles have properties of a liquid crystal. The determinants of their formation have been discussed by De Rosier and Tilney (1982).

1.5.3 Microtubules and Related Structures

In many cells, such as animal cells in culture, cytoplasmic microtubules (for review see e.g. Dustin 1984) radiate from the centriole region in the perinuclear area to the cell periphery where they bend back or sideways (Fig. 4). They are subjected to a steady assembly and disassembly process involving a subtle equilibrium state with their heterodimeric subunit (see e.g. Salmon 1984). Assembly might well proceed against an externally applied pressure (Hill 1981). Regarding intratubular arrangement of the heterodimer and the site of preferential assembly, microtubules are polar structures (see e.g. Euteneur and McIntosh 1981). Microtubule organizing centers (MTOC) include the centrioles, some more or less undefined electron dense material and, during nuclear division, the kinetochore region of chromosomes. In addition, centrioles determine the number of protofilaments composing the microtubules (Evans et al. 1985). By elegant microinjection studies Schulze and Kirschner (1986) showed that microtubules elongate at their distal ends until they shift in-

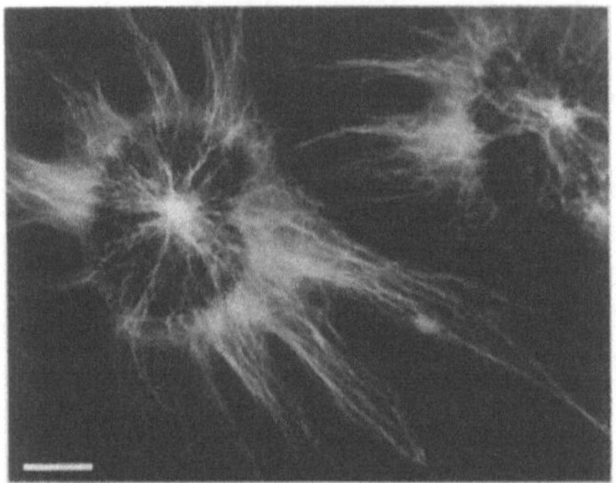

4

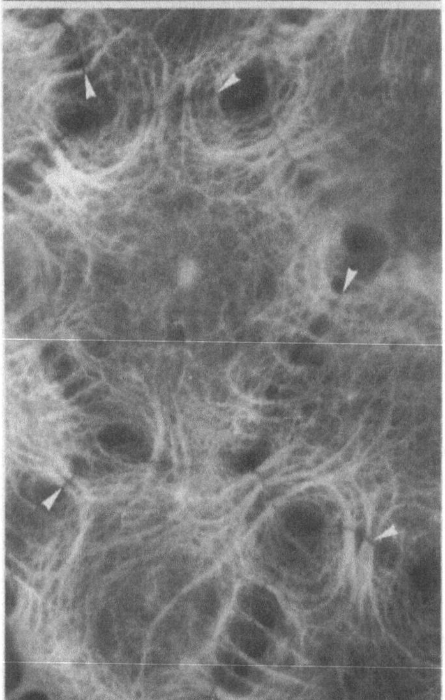

5

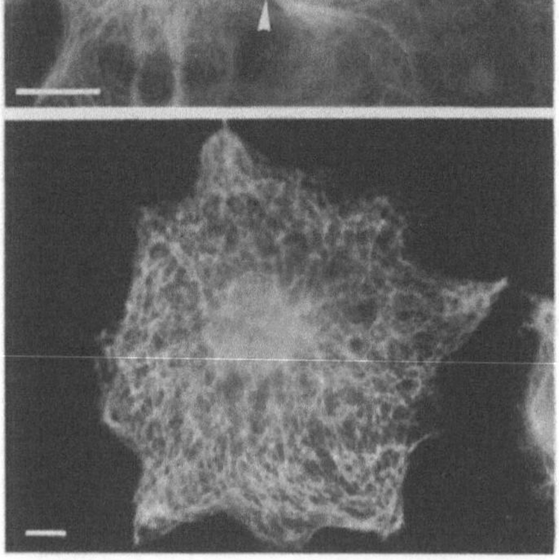

6

Figs. 4 – 6. Examples for the arrangement of microtubules, vimentin and cytokeratin fibrils.
Fig. 4. Microtubules in *Xenopus laevis* melanophores in culture shown by indirect immunofluorescence. The pigment is in an aggregated state (kidney-shaped *dark area*). The long microtubules emanate from the central, pericentriolar area toward the tips of the cellular processes
Fig. 5. Tonofilaments in epidermis cells of *Xenopus laevis* tadpoles in culture as shown by indirect immunofluorescence. The desmosomes (*arrowheads*) provide the structural basis of mechanical coupling of adjacent cells by tonofibrils. *Bar* = 10 μm
Fig. 6. Vimentin fibrils in a single 3T3 cell shown by indirect immunofluorescence. In this case the fibrils are perferentially arranged in a radial direction, in other cells loops around the nucleus may be prevailing. *Bar* = 10 μm

to the shrinking phase. At the same time new microtubules are nucleated by the centrosomes. The elongation rate of microtubules in green monkey kidney fibroblasts in culture has been determined to be about 3.6 μm min^{-1}. The turnover time for the entire microtubular array was estimated to be ca. 15 min. Schulze's and Kirschner's microinjection studies gave no indication for treadmilling. They showed the coexistence of a majority of microtubules in a slowly growing phase, and a minority in a rapidly depolymerizing phase (Kristofferson et al. 1986), a behavior which is called "dynamic instability" (Mitchison and Kirschner 1984). In fact, not all cytoplasmic microtubules in interphase cells participate in the rapid turnover, about 10% of the population seem to remain stable for at least 2 h. The functional significance of this behavior is a matter of speculation (Schulze and Kirschner 1986). In this context it is also unclear why there is a heterogeneous reaction of microtubules within the same cell against various monoclonal antibodies (Draber et al. 1986). Despite their rapid turnover, microtubules represent relatively rigid structures (see Chap. II.2.2) responsible for the achievement and maintênance of cell polarity (Inoué 1982) and orientation of organelles and extracytoplasmic cellulose fibrils (e.g. Lloyd and Seagull 1985; Quader 1986). They are involved in mechanical stabilization of the cell cortex in protists (see Chap. II.2), of long cell extensions such as axopodia of Radiolaria and the heliozoan Actinosphaerium (see Chaps. III.1 and III.3.2), filopodia of Foraminifera, or nerve axons (e.g. Bray 1973). In multimicrotubular structures the microtubules are crosslinked with each other by flexible bridges formed by various MAPS, which may also link microtubules to cytoplasmic membranes (Hirokawa 1982). Rigidity of structures formed by crosslinked microtubules is qualitatively demonstrated by mitotic spindles (Hirokawa et al. 1985) which can be moved around in a cell when pushed with a needle or isolated intact (see also Chap. II.3).

Whether microtubules are involved in the production of motive force for anaphase movement of chromosomes, or whether they provide an efficient system to slow down chromosome movement and thus protect the genetic material from being disrupted, is still an open question. The extremely low specific power output of 6 erg s^{-1} cm^{-3} (max. power output per unit weight of the "engine") calculated by Nicklas (1984) is in good agreement with this assumption (Nicklas 1983). For comparison, the specific power output of the cytokinetic furrow was calculated to be 3×10^2, of a flagellum 3 10^5, and of striated muscle 2×10^6 erg s^{-1} cm^{-3} (Nicklas 1984).

The contribution of microtubules to the viscoelastic properties of cytoplasm still needs to be elucidated (cf. Chaps. II.2.3 and II.3). Measurements of the packing volume of sedimented microtubules, either purified or with MAPS, are a first step in clarifying the role of microtubules in the physical compartmentation of cytoplasm (Brown and Berlin 1985). Densely packed microtubules become oriented as parallel rods and behave like liquid crystals imposing little barrier to diffusion of macromolecules through the space between the single microtubules. Due to the low packing density, microtubules and their associated proteins may comprise more than 40–50% of the total cytosolic volume, and constitute a large mechanically irreducible volume without affecting transport by diffusion (Brown and Berlin 1985).

In migrating white blood cells (Schliwa et al. 1982) or endothelial cells (Gotlieb et al. 1981, 1983), centrioles precede the nucleus, suggesting an involvement of microtubule orientation in directionality of locomotion. However, microtubule-free cytoplasts do not differ from entire cells with respect to maintainance of direction of locomotion (Gelfand et al. 1985). A stabilization of the spread state of cells by microtubules seems to be well documented (Domnina et al. 1985).

The role of microtubules in organelle transport and axonal flow is discussed in detail in Chap. II.2 and microtubule "contractility" in Chap- II.3, cilia and flagellae are treated in Chap. II.4, therefore, these topics are not considered in this introductory chapter.

1.5.4 Intermediate Filaments and Related Structures

This group of fibrillar cytoplasmic structures contains five groups, the cytokeratins, desmin, vimentin, neurofilaments, and the glial fibrillar acidic protein (GFAP) (for review see e.g. Weber and Osborn 1982; Traub 1985). The term "intermediate filaments" refers to the diameter (8 – 14 nm), intermediate to that of the actin-containing microfibrils (5 – 6 nm) and microtubules (22 – 25 nm). With a few exceptions (e.g., mammalian germ cells and vertebrate spermatozoa) this type of filament also seems to be ubiquitous among the metazoan cells so far investigated. Under physiological conditions, these fibrils are very stable; there is apparently no pool of monomers in equilibrium with the polymerized state, as is the case for actin and tubulin. Which genes of intermediate filaments are expressed, depends strictly on the cell lineage. Despite this fact, cells in culture may contain more than one type of intermediate filaments.

Cytokeratins are heteropolymers characteristic for cells of epithelial origin. The fibrils (tonofibrils) have a diameter of approx. 8 nm. They are twisted around each other and in culture cells exhibit a curly arrangement. Moreover, in culture cells, cytokeratin fibrils form a loose cage around the nucleus with bundles emerging radially and terminating at their widespread anchor sites at the plasma membrane (Goldman et al. 1985) particularly at the desmosomes. In some cell lines cytokeratins have been found to be connected with nuclear pore complexes (Goldman et al. 1985; Jones et al. 1982). At the desmosomes, the fibrils are crosslinked laterally with the membrane (for review see e.g. Skerrow CJ 1986). In keratinizing tissues, these fibrils are the main differentiation products; furthermore, they are present in many nonkeratinizing stratified epithelia, but they are absent from cells of mesenchymal origin. In epidermal stratum corneum, keratin fibrils are bundled by the associated protein filaggrin. Other factors, however, may also be involved in lateral bundling (Bereiter-Hahn 1971; Skerrow D 1986). In epithelia with a terminal web, cytokeratin fibrils underlie the actin-containing web (Fig. 5). In addition, they may form a fibrillar sheet at the cell periphery extending in large spirals from attachment sites at the basal cell membrane (i.e., hemidesmosomes) toward the apical part of a cell.

From their arrangement, i.e., intertwining and loop formation, the tonofibrils can be assumed to form an elastic meshwork which maintains some tension throughout the cells integrated in an epithelium. The close association with actin fibrils may provide the structural basis for generating tension by active contraction, which is then maintained by the elasticity of the cytokeratin bundles without the need for further energy supply (Kunzenbacher et al. 1982).

Desmin filaments (10 nm diameter) represent the main intermediate filament type of muscle cells. In mature skeletal muscle, desmin is concentrated at Z-disks and in regions of cell – cell contact. In cardiomyocytes desmin is localized in the intercalated disks, while in smooth muscle cells it is the main component of the dense bodies and the interconnecting network and membrane-bound plaques (cited from Schliwa 1986). Desmin fibrils are thought to integrate mechanically all contractile actions of a muscle fiber by linking individual myofibrils laterally at their Z-disks and by linking Z-disks to the plasma membrane and to other membranous organelles (Lazarides 1980).

Vimentin is found in cells of mesenchymal origin. It occurs in a variety of culture cells, in addition to the intermediate filament type that is characteristic of the tissue of origin (Traub 1985). Vimentin extends more or less radially, gently curving as single filaments or loose bundles, through the cytoplasm (Fig. 6). Filaments are concentrated around the nucleus and terminate at the nuclear lamina (Penman et al. 1983). Experiments with isolated nuclei suggest that the nucleus could be an organizing center for vimentin (Wang et al. 1985). Vimentin does not strictly colocalize with other filament types, however, formation of a perinuclear cap on treatment with antimicrotubular drugs reveals a connection with microtubules (Goldman and Knipe 1973; Bloom et al. 1985). Bundling of vimentin filaments was shown to be mediated by an associated protein termed synemin (Lazarides and Granger 1983) or with epinemin which seems to be correlated with senescence processes in human fibroblasts (Wang and Huie 1985). Bundling of vimentin in mitotic cells is accompanied by extensive phosphorylation (Zieve 1985). Turnover of vimentin may be accomplished by its sensitivity to a Ca^{2+} activated protease.

Neurofilaments are intensively interconnected with microtubules and appear to function as a three-dimensional structural lattice that provides tensile strength (Lazarides 1980) or more importantly, stiffness to the axons. This interpretation is based on dissolution of gellike extruded axoplasm by degradation of neurofilaments (Gilbert et al. 1975). Nothing is known of the specific functions of the GFAP.

1.6 Aspects of Cytoplasmic Architecture

Intracellular localization of organelles, the shape of cells, and shape changes during spreading, locomotion, or differentiation all depend on the interaction of cytoskeletal fibers with each other, with organelles, the plasma membrane, and the extracellular space. Any effort to explain cell shape and behavior on the basis of current knowledge of these parameters is necessarily a highly speculative synopsis of experimental findings and the author's imagination. The design of decisive experiments is often very difficult. However, it is this type of wholistic comprehension of cell structure and its dynamics which is needed to understand the role of cytoplasm in growth control and differentiation. The mirror symmetry in migratory paths of mitotic sister cells (Albrecht-Bühler 1978) is an impressive example for the complexity of interaction of cytoskeletal elements and their high information content. In the following a few examples of organelle localization and formation of cell extensions are discussed which by no means can cover the whole range of the field.

1.6.1 Localization of Organelles

The position of the nucleus, more or less in the center of many cells seems to be controlled by the centrosomes and the microtubules emanating from the centrosome region. EM pictures revealed fibrils connecting centrosomes with the nucleus (Nadezhdina et al. 1979), thus both organelles form a mechanical unit. Elongation of microtubules from the centrosome toward the periphery results in the formation of an asterlike structure with considerable rigidity. The aster can be assumed to become centered by repulsory forces developed by the elongating microtubules touching the cell periphery. This mechanism is suggested by the sensitivity of the process to treatment with colchicin (Wang et al. 1982). A similar mechanism has been described for energy distribution in the furrowing eggs of insects (Wolf 1980). Microtubules determine the position of nuclei also in cells without centrioles (Neuhaus-Url and Kiermayer 1982). The intermediate filament arrangement anchoring the nuclei also depends on the existence of properly organized microtubules (Wang et al. 1982). This "framework" of microtubules represents a scaffold that organizes the pattern of the endoplasmic reticulum (Fuhrmann and Bereiter-Hahn, unpubl.); aligns mitochondria, lysosomes and melanosomes; and is responsible for axis determination in eggs (Scharf and Gerhart 1983).

In cells treated with antimicrotubular drugs, location of the nuclei and other organelles seems to be mediated by the actin fibrillar system. F-actin is also the main component responsible for the regulation of macrophage phagosome arrangement (Valberg and Albertini 1985). In higher plant cells chloroplast arrangement is controlled by actin fibrils (e.g. Witztum and Parthasarathy 1985). Actin-based cytoplasmic streaming may move nuclei through the cell. No explanation can be given for the nuclear rotation seen in many plant as well as animal cells. Induction of this rotation in animal cells in culture by viruses may indicate a loss of the intermediate filament anchorage.

In stratified epithelia, as exemplified by the vertebrate epidermis, the preferential orientation of cytofibrils changes markedly when progressing from basal to superficial cells. In the basal epidermal cells, the main direction of fibrils is parallel to the axis of stratification; in superficial cells, the fibrils are oriented parallel to the surface of the epithelium. This arrangement of tonofilaments restricts the possibilities for the orientation of the mitotic spindle. In densely packed basal cells the spindle axis will always be perpendicular to the

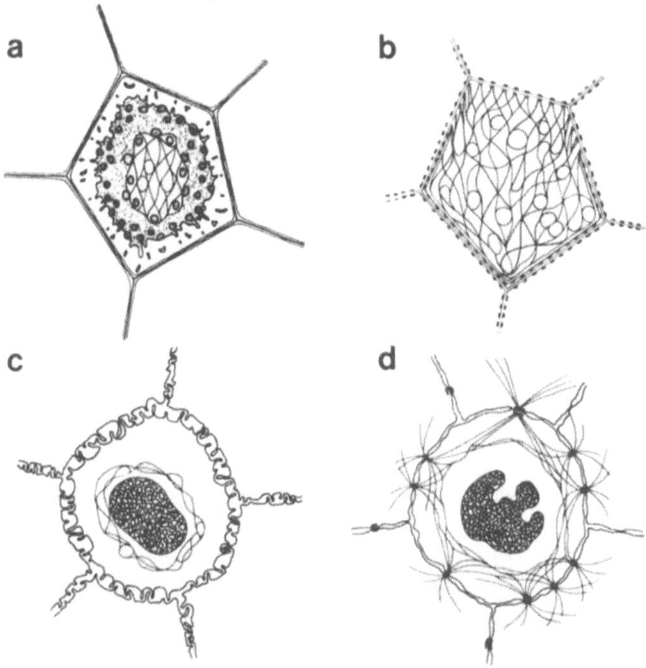

Fig. 7a–d. Reconstruction of fibril organization in superficial cells of *Xenopus* tadpole epidermis in primary culture according to electron-micrographs of serial sections. **a** Surface region at the level of tight and adherens junctions. Mucus granules lay immediately beneath the plasma membrane and are embedded in a meshwork of membrane-associated actin fibrils. **b** Level of desmosome rows which are interconnected by actin bundless running parallel to the rows of desmosomes and by loops of tonofibrils, which seem to form an elastic net. **c** Central part of the cell. No specific membrane junctions, relatively large intercellular spaces, only a few tonofibrils around the nucleus. **d** Basal part of the cell with tonofibrils interconnecting desmosomes, relatively small intercellular spaces. Desmosomal connections are formed not only to lateral neighbors, but also to the underlying cell layer (not shown). (Reconstruction provided by I. Kunzenbacher, unpubl.)

epithelial surface, while in epithelia which are stretched, e.g., due to growth of the enveloped body or by wounding, the spindle axis may orient in the plane of the epithelium. In flat cells near the epithelial surface, mitosis is always oriented in the plane of the epithelium, as can be seen in the nonkeratinizing cells in anuran tadpole epidermis. Figure 7 shows the organization of the tonofibrils in these cells. This principle of orienting mitotic spindles by mechanical forces that orient the cytofibrillar system may well be of general significance (see also Chap. IV.2.3).

1.6.2 Interaction of Cytoskeletal Elements in Generating Cell Shape

In plant cells, shape is determined mainly by cell wall structure and growth which are treated in detail in Chaps. IV.1 and IV.2.

If microtubules of cultured animal cells are destroyed, for example, by colchicine, peripheral cytoplasm flows to the perinuclear region which becomes thickened. The outline of the cell margins is left more or less constant. This behavior provides some insight in the interaction of cytoplasmic fibrillar systems. Microtubules can supposedly form a scaffold resisting the stress exerted by the strained, three-dimensional actomyosin fibrillar meshwork. Strain is relased on disassembly of the rigid microtubules first by elastic recoil, then by further active contractions. Comparable reactions of actomyosin fibrils can be provoked by cytochalasin (Tillmann and Bereiter-Hahn 1986), and by detaching the trailing edge of a migrating cell (Chen 1981). The stressed state results from active contractions requiring metabolic energy. This conclusion is based on the observation that deprivation of cellular ATP inhibits formation of actin-containing patches and of arborization in cytochalasin-treated cells and also prevents rounding of cells on loss of adhesion by treatment with EDTA and trypsin. The abutment for a fibrillar meshwork under stress is provided by the microtubules, the plasma membrane, and extracellular surfaces when cells adhere to an external matrix. Disturbance of contraction control due to loss of tropomyosin (Carley et al. 1985) or lack of adhesion, as is the case in many malignantly transformed cells, results in diminished formation of stress fibers. The cells are more rounded (Heckmann et al. 1985), and as a consequence, they do not arborize properly on cytochalasin treatment (Brett and Godmann 1986). Also, colchicin-treated normal cells do not arborize (Tillmann, pers. comm).

Stress fibers are relatively stable structures emanating from focal contacts, which are stable points of adhesion. These elements stabilize cell shape and therefore are characteristic of nonmoving cells (Badley et al. 1980; Couchman et al. 1983). In thick cytoplasmic strands, such as those of *Physarum* or *Amoeba* actin fibers may traverse the cytoplasmic tubes, thus adding to mechanical stability (Stockem et al. 1983; Kukulis et al. 1985). On the other hand, in loosely adhering culture cells, only actomyosin meshworks develop. If local instabilities of these meshworks occur, cytoplasm becomes displaced, resulting in shape changes and locomotion (see Chap. V.2). Therefore, malignant epithelial cells are considerably more motile than their normal counterparts (Manger and Heckman 1984). The same principle underlies the high blebbing activity of cells in suspension (Fig. 1 in Chap. V.2), which is mostly lost during spreading. A further example of adhesion-dependent formation of actin filament bundles are spreading cells (Fig. 2b–d).

Interaction of plasma membrane and fibrils is also required in the formation of cytotic vesicles, microspikes, filopodia, microvilli, and acrosomal and axonal processes.

Pinocytotic vesicles, for instance, are pulled into the cell by actin fibrils attached to the vesicular membrane (Condeelis 1981) and interwoven with the cytoplasmic fibrillar meshwork. Furthermore, interiorization of small vesicles requires some rigidity of the plasma membrane-cortical fibrillar complex, otherwise a broad area of cell surface would be deflected toward the interior of the cell. The required stiffness could be achieved by the viscoelasticity of the cortical gel and/or by tension in this layer due to contraction, or alternatively by an intracellular hydrostatic pressure (Chap. V.2.7). The formation of a pseudopodiumlike protrusion, concomitant with the internalization of pinocytotic vacuoles, is a strong indication for the involvement of hydrostatic pressure in amoebae (Klein and Stockem 1979). Comparable problems arise for the mechanics underlying the formation of microspikes, filopodia, microvilli, and acrosomal processes. A central bundle of actin filaments extending from the membrane forming the tip through the process is common to all these structures. Mooseker and Tilney (1975) assumed that formation of acrosomal processes and intestinal microvilli is an immediate consequence of actin assembly and crosslinking between the fibrils and the membrane. However, if a hydrostatic pressure exists in living cells, then this pressure would contribute to the extension as has been exemplified for the formation of microridges at the surface of many mucus-bearing epithelia (Bereiter-Hahn et al. 1979). In this case the sites of microvilli ejection could well be defined by local weakenings in the terminal web, which will spontaneously generate a hexagonal pattern when traction and/or fiber density exceed a critical value (Oster et al. 1985). Extensive formation of filipodialike structures in hypotonic media (Hyatt et al. 1984) is also evidence in favor of hydrostatic pressure supporting extension. In any case, the forces generated by actin assembly mediating extension need to be antagonized, e.g., by the stiffness of the cortical fibrillar layer to which the core filaments are crosslinked.

Extension of axonal processes and the large cytoplasmic extensions seen, for instance, in *Reticulomyxa* could result from the pressure generated by the cytoplasmic transport along the interior of the processes (cf. Chap. II.2.2).

Ruffles are folds in the dorsal surface of many cells in culture arising just behind the edge of cytoplasmic lamellae and receding back into the cytoplasm while they move toward the perinuclear area (Abercrombie et al. 1970). The three-dimensional organization of ruffles of human leukocytes was revealed by high voltage electron micrographs (Pryzwansky et al. 1983). A loosely arranged, irregular fibrillar meshwork is found in the interior of the ruffles. Fibrils are oriented along the length of the crest only at the top of the folds. Such a foldlike structure could be generated by excessive incorporation of membrane material at the edge of the lamella. The resulting surplus of membrane would pile up forming a ruffle (Bereiter-Hahn and Strohmeier 1986).

1.6.3 Cytoplasmic Streaming

Cytoplasmic streaming as a mass movement of cytoplasm is a widespread phenomenon among plant cells, and in acellular and cellular slime molds. It is involved in locomotion of amoebae and in many animal cells. Most of these phenomena are discussed in Chaps. II.6 and V.2.3. Therefore, only the rotational streaming of cytoplasm and the vacuolar sap in characean cells (cyclosis) is considered in this introduction. From centrifugation experiments, a driving force of $1-4$ dyn cm^{-2} has been determined for *Chara* internoidal cells. Using this value and a Reynold's number of $R=10^{-4}$ for endoplasm and $R=0.05$ for the vacuolar sap, Nothnagel and Webb (1982) performed calculations for three models to test whether they would be sufficient to drive cyclosis.

The driving force may be provided by (1) myosinlike molecules migrating along actin cables; (2) myosinlike molecules attached to small spherical particles moving along actin cables; and (3) myosinlike molecules which are incorporated into a fibrous or membranous network or a gel extending into the cytoplasm.

Condition 3 would exert enough viscous pull on the endoplasm to account for the observed streaming. Evidence for the generation of cytoplasmic flow by interaction of ectoplasmic actin cables (in *Nitella*) and myosin-covered beads has been provided by the elegant experiments of Sheetz and Spudich (1983).

1.7 Physiological Effects of Mechanical Stresses

1.7.1 Mechanical Aspects of Morphogenesis During Embryo Development

Whenever cytoplasm or extracellular matrix material is subjected to tension, the fibrillar material becomes oriented. In an embryo, forces generated at one position equilibrate virtually instantly over the entire embryo. Thus, mechanical forces are sufficient to coordinate shape changes of large tissue sheets and define migration tracks without the intervention of long-range chemical signaling (Beloussov et al. 1975; Kolega 1986; Murray and Oster 1984 a, b; Oster et al. 1983; Phillips 1984; Waddington 1939; Weiss and Garber 1952, see also Chap. IV.2). Different cell types interact with a supporting collagen layer at different strength (Tucker et al. 1985). A fruitful method for modeling this type of self-organizational event is the finite element theory (see Chaps. I.3 and III.3). Oster and Odell (1984 b) used this approach to describe dermo-epidermal interactions.

1.7.2 Influences of Mechanical Stresses on Cellular Metabolism

Normal cells derived from solid tissues require adhesion to a solid substrate in order to grow. This anchorage-dependent growth is supposed to be related to the degree of spreading (Greenspan and Folkman 1977) which may be one crucial factor in density-dependent control of growth (O'Neill et al. 1986, see also Chap. V.3.3). Experimental evidence is provided by investigations of lens epithelial cells grown on a collagen layer attached to a plastic dish. If the collagen becomes free-floating, the cells contract, and there is a sharp decline in uridine and thymidine incorporation (Iwig et al. 1981). A general decrease of macromolecular synthesis is found in mammary cells growing on free-floating collagen (Emermann and Pitelka 1977). However, metabolite patterns, characteristic for the differentiated state are maintained (Emermann et al. 1981). These experiments point to the influence that cytoskeletal elements have on gene expression and macromolecular synthesis (for review see Ben Ze'ev 1985), as well as energy metabolism (Masters 1984; Clegg 1984; Tillmann and Bereiter-Hahn 1986). Bisell et al. (1982) developed a model for gene expression involving the cytoskeleton which they termed "dynamic reciprocity" between the extracellular matrix (ECM) and the cytoskeleton and the nuclear matrix. It is postulated that "the ECM exerts physical and chemical influences on the geometry and biochemistry of the cell via transmembrane receptors so as to alter the pattern of gene expression by changing the association of the cytoskeleton with the mRNA and the interaction of the chromatin with the nuclear matrix. This in turn would affect the ECM." Such a model could well explain tension-mediated stimulation of growth (e.g., Curtis and Seehar 1978) and differentiation (e.g., Vandenburgh 1983; Yeh and Rodan 1984).

References

Abercrombie M, Heaysman JE, Pregrum SM (1970) The locomotion of fibroblasts in culture. II. "Ruffling". Exp Cell Res 60:437–444

Albrecht-Buehler G (1978) The tracks of moving cells. Sci Am 238:68–76

Albrecht-Buehler G (1985) Is cytoplasm intelligent too? J Muscle Res Cell Motil 6:1–21

Badley RA, Couchman JR, Rees DA (1980) Comparison of the cell cytoskeleton in migratory and stationary chick fibroblasts. J Muscle Res Cell Motil 1:5–14

Beloussov LV, Dorfman JG, Cherdantzev VG (1975) Mechanical stresses and morphological patterns in amphibian embryos. J Embryol Exp Morphol 34:559–574

Ben-Ze'ev A (1985) Cell shape, the complex cellular networks, and gene expression. J Muscle Res Cell Motil 6:23–53

Bereiter-Hahn J (1971) Licht- und elektronenmikroskopische Untersuchungen von Tonofilamenten in den Epidermiszellen von Fischen. Cytobiologie 4:73–102

Bereiter-Hahn J, Strohmeier R (1987) Biophysical aspects of motive force generation in tissue culture cells and protozoa. Fortschr Zool 34: (in press)

Bereiter-Hahn J, Vöth M (1983) Metabolic control of shape and structure of mitochondria in situ. Biol Cell 47:309–322

Bereiter-Hahn J, Osborn M, Weber K, Vöth M (1979) Filament organization and formation of microridges at the surface of fish epidermis. J Ultrastruct Res 69:316–330

Bloom GS, Luca FC, Vallee RB (1985) Cross-linking of intermediate filaments to microtubules by microtubule-associated protein 2. Ann N Y Acad Sci 455:18–31

Bray D (1973) Branching patterns of individual sympathetic neurons in culture. J Cell Biol 56:701–712

Brett JG, Godman GC (1986) Cytoskeletal organization affects cellular responses to cytochalasins: comparison of a normal line and its transformant. Tissue & Cell 18:175–199

Brown PA, Berlin RD (1985) Packing volume of sedimented microtubules: Regulation and potential relationship to an intracellular matrix. J Cell Biol 101:1492–1500

Byers HR, Fujiwara K, Beckerle MC, Porter KR (1980) Microtubule organization and function in chromatophores and scleroblasts in vitro and in situ. In: De Brabander, De Mey (eds) Microtubules and microtubule inhibitors. Elsevier, North Holland

Byers TJ, Branton D (1985) Visualization of the protein associations in the erythrocyte membrane skeleton. Proc Natl Acad Sci USA 82:6153–6157

Cain H, Kraus B (1981) Cytoskeleton in cells of the mononuclear phagocyte system. Virchows Arch B Cell Pathol 36:159–176

Carley WW, Bretscher A, Webb WW (1985) F-actin aggregates in transformed cells contain α-actinin and fibrin but apparently lack tropomyosin. Eur J Cell Biol 39:313–320

Cheer A, Vincent J-P, Nuccitelli R, Oster G (1986) Cortical activity in vertebrate eggs. I.: The activation waves. J Cell Comp Physiol 1:1–9

Clegg JS (1984) Intracellular water and the cytomatrix. J Cell Biol 99:167s–171s

Cohen C, Reinhardt B, Castellani L, Norton P, Štirewalt M (1982) Schistosome surface spines are "crystals" of actin. J Cell Biol 95:987–988

Condeelis J (1981) Microfilament–membrane interactions in cell shape and surface architecture. Int Cell Biol 1980–1981:306–320
Condeelis J (1983) Rheological properties of cytoplasm: Significance for the organization of spatial information and movement. Mod Cell Biol 2:225–240
Couchman JR, Badley RA, Rees DA (1983) Redistribution of microfilament-associated proteins during the formation of focal contacts and adhesions in chick fibroblasts. J Muscle Res Cell Motil 4:647–661
Curtis ASG, Seehar GM (1978) The control of cell division by tension or diffusion. Nature 274:52–53
Dembo M, Harlow F (1986) Cell motion, contractile networks, and the physics of interpenetrating reactive flow. Biophys J 50:109–121
Dembo M, Harris AK (1981) Motion of particles adshering to the leading lamella of crawling cells. J Cell Biol 91:528–536
Dembo M, Maltrud M, Harlow F (1986) Numerical studies of unreactive contractile networks. Biophys J 50:123–137
DeRosier DJ, Tilney LG (1982) How actin filaments pack into bundles. Cold Spring Harbor Symp Quant Biol 46:525–540
Domina LV, Rovensky JA, Vasiliev JM, Gelfand IM (1985) Effect of microtubule-destroying drugs in the spreading and shape of cultured epithelial cells. J Cell Sci 74:267–282
Draber P, Draberova E, Zicconi D, Sellitto C, Vilocky V, Cappuccinelli P (1986) Heterogeneity of microtubules recognized by monoclonal antibodies to alphatubulin. Eur JCell Biol 41:82–88
Drenckhahn D, Mannherz HG (1983) Distribution of actin and the actin-associated proteins myosin, tropomyosin, alpha-actinin, vinculin, and villin in rat and bovine exocrine glands. Eur J Cell Biol 30:167–176
Dustin P (1984) Microtubules, 2nd ed. Springer, Berlin Heidelberg New York
Emerman JT, Pitelka DR (1977) Maintenance and induction of morphological differentiation in dissociated mammary epithelium on floating collagen membranes. In vitro 13:316–328
Euteneuer U, McIntosh JR (1981) Polarity of some motility-related microtubules. Proc Natl Acad Sci USA 78:372–376
Evans L, Mitchison T, Kirschner M (1985) Influence of the centrosome on the structure of nucleated microtubules. J Cell Biol 100:1185–1191
Fung YC (1981) Biomechanics – Mechanical properties of living tissues. Springer, Berlin Heidelberg New York, 433 pp
Fung YC (1984) Structure and stress–strain relationship of soft tissues. Am Zool 24:13–22
Gabella G (1976) The force generated by a visceral smooth muscle. J Physiol 263:199–213
Geiger B, Avnur Z, Kreis TE, Schlessinger J (1984a) The dynamics of cytoskeletal organization in areas of cell contact. In: Shay JW (ed) The Cytoskeleton. Cell and muscle motility. Plenum, New York
Geiger B, Avnur Z, Rinnerthaler G, Hinssen H, Small VJ (1984b) Microfilament-organizing centers in areas of cell contact: cytoskeletal interactions during cell attachment and locomotion. J Cell Biol 99:83s–91s
Gelfand VI, Glushankova NA, Ivanova OY, Mittelman LA, Pletyushkina OY, Vasiliev JM, Gelfand IM (1985) Polarization of cytoplasmic fragments microsurgically detached from mouse fibroblasts. Cell Biol Int Rep 9:883–892
Goldacre RJ (1961) The role of the cell membrane in the locomotion of Amoebae and the source of the motive force and its control by feedback. Exp Cell Res 8:1–16
Goldman RD, Knipe DM (1973) Functions of cytoplasmic fibers in non-muscle cell motility. Cold Spring Harbor Symp Quant Biol 37:523–534
Goldman R, Goldman A, Green K, Jones J, Lieska N, Yang HY (1985) Intermediate filaments: Possible functions as cytoskeletal connecting links between the nucleus and the cell surface. Ann N Y Acad Sci 455:1–17
Gotlieb AI, McBurnie May L, Subrahmanyan L, Kalnins VI (1981) Distribution of microtubule organizing centers in migrating sheets of endothelial cells. J Cell Biol 91:589–594
Gotlieb AI, Subrahmanyan L, Kalnins VI (1983) Microtubule-organizing centers and cell migration: effect of inhibition of migration and microtubule disruption in endothelial cells. J Cell Biol 96:1266–1272

Greenspan HP, Folkman J (1977) Hypotheses on cell adhesion and actin cables. J Theor Biol 65:397–398

Harris AK, Wild P, Stopak D (1980) Silicone rubber substrata: A new wrinkle in the study of cell locomotion. Science 208:177–179

Harvey EN (1931) The tension at the surface of marine eggs, especially those of the sea urchin, arbacia. Biol Bull 61:273–279

Heckman CA, Campbell AE, Wetzel B (1985) Characteristic shape and surface changes in epithelial transformation. Adv Cell Cult 4:1–34

Hill TL (1981) Microfilament or microtubule assembly or disassembly against a force. Proc Natl Acad Sci USA 78:5613–5617

Hiramoto Y (1967) Observations and measurements of sea urchin eggs with a centrifuge microscope. J Cell Physiol 69:219–230

Hirokawa N (1982) Cross-linker system between neurofilaments, microtubules, and membranous organelles in frog axons revealed by the quick-freeze, deep-etching method. J Cell Biol 94:129–142

Hirokawa N, Takemura R, Hisanaga S-I (1985) Cytoskeletal architecture of isolated mitotic spindle with special reference to microtubule-associated proteins and cytoplasmic dynein. J Cell Biol 101:1858–1870

Hoffmann-Berling H (1956) Das kontraktile Eiweiß undifferenzierter Zellen. Biochem Biophys Acta 19:453–463

Hyatt HA, Shure MS, Begg DA (1984) Induction of shape transformation in sea urchin coelomocytes by the calcium ionophore A 32187. Cell Motil 4:57–71

Ingber DE, Jamieson JD (1985) Cells as tensegrity structures: Architecture regulation of histodifferentiation by physical forces transduced over basement membrane. In: Andersson LC, Gahmberg CG, Ekblom P (eds) Gene expression during normal and malignant differentiation. Acad Press, London, pp 13–32

Inoue S (1982) The role of Self-Assembly in the generation of biological form. In: Developmental Order: Its Origin and Regulation. Alan R Liss, New York, pp 35–76

Inouye K (1984) Measurement of the motive force of the migrating slug of Dictyostelium discoideum by a centrifuge method. Protoplasma 121:171–177

Isenberg G, Wohlfarth-Bottermann KE (1976) Transformation of cytoplasmic actin: importance for the organization of the contractile gel reticulum and the contraction-relaxation cycle of cytoplasmic actomyosin. Cell Tissue Res 173:495–528

Iwig M, Glaesser D, Bethge M (1981) Cell shape-mediated growth control of lens epithelial cells grown in culture. Exp Cell Res 131:47–55

Izzard CS, Lochner LR (1980) Formation of cell-to-substrate contacts during fibroblast motility: an interference-reflexion study. J Cell Sci 42:81–116

Izzard CD, Radinsky R, Culp LA (1986) Substratum contacts and cytoskeletal reorganization of BALB/c 3T3 cells on a cell-binding fragment and Heparin-binding fragments of plasma fibronectin. Exp Cell Res 165:320–336

James DW, Taylor JF (1969) The stress developed by sheets of chick fibroblasts in vitro. Exp Cell Res 54:107–110

Jen CJ, McIntire L, Bryan J (1982) The viscoelastic properties of actin solutions. Arch Biochem Biophys 216:126–132

Jones JCR, Goldman AE, Steinert PM, Yuspa S, Goldman RD (1982) Dynamic aspects of the supramolecular organization of intermediate filament networks in cultured epidermal cells. Cell Motil 2:197–213

Kamimura S, Takahashi K (1981) Direct measurement of the force of microtubule sliding in flagella. Nature 293:566–568

Klein HP, Stockem W (1979) Pinocytosis and locomotion of amoebae. 12. Dynamics and motive force generation during induced pinocytosis in A. proteus. Cell Tissue Res 197:263–279

Kolega J (1986) Effects of mechanical tension on protrusive activity and microfilament and intermediate filament organization in anepidermal epithelium moving in culture. J Cell Biol 102:1400–1411

Kreis TE, Birchmeier W (1980) Stress fiber sarcomeres of fibroblasts are contractile. Cell 22:555–561

Kreis TE, Avnur Z, Schlessinger J, Geiger B (1984) Dynamic properties of cytoskeletal proteins in focal contacts. In: Borisy GG, Cleveland DW, Murphy DB (eds) Molecular Biology of the Cytoskeleton. Cold Spring Harbor Lab, pp 45–57

Kristofferson D, Mitchison T, Kirschner M (1986) Direct observation of steady-state microtubule dynamics. J Cell Biol 102:1007–1019

Kucera P, de Ribaupierre Y (1982) In situ recording of the mechanical behaviour of cells in the chick embryo. In: Embryonic Development, Part B. Cellular aspects. Alan R Liss, New York (ed), pp 433–444

Kukulies J, Brix K, Stockem W (1985) Fluorescent analog cytochemistry of the actin system and cell surface morphology in Physarum microplasmodia. Eur J Cell Biol 39:62–69

Kunzenbacher I, Bereiter-Hahn J, Osborn M, Weber K (1982) Dynamics of the cytoskeleton of epidermal cells in situ and in culture. Cell Tissue Res 222:445–457

Lazarides E (1980) Intermediate filaments as mechanical integrators of cellular space. Nature 283:249–256

Lazarides E, Granger BL (1983) Transcytoplasmic integration in avian erythrocytes and striated muscle: the role of intermediate filaments. In: Satir BH (ed) Modern Cell Biology 2:143–162. Alan R. Liss, New York

Lazarides E, Weber K (1974) Actin antibody: The specific visualization of actin filaments in non-muscle cells. Proc Natl Acad Sci USA 71:2268–2272

Lloyd CW, Seagull RW (1985) A new spring for plant cell biology: microtubules as dynamic helices. Trends Biochem Sci 10:476–478

MacLean-Fletcher SD, Pollard TD (1980) Viscometric analysis of the gelation of Acanthamoeba extracts and purification of two gelatin factors. J Cell Biol 85:414–428

Manger RL, Heckman CA (1984) Behavioral anomalies of highly malignant respiratory tract epithelial cells. Cancer Res 44:688–696

Masters C (1984) Interactions between glycolytic enzymes and components of the cytomatrix. J Cell Biol 99:222s–225s

Mastro AM, Keith AD (1984) Diffusion in the aqueous compartment. J Cell Biol 99:180s–187s

Matsumura F, Yoshimoto Y, Kamiya N (1980) Tension generation by actomyosin thread from a non-muscle system. Nature 285:169–170

Mitchison JM, Swann MM (1956) The mechanical properties of the cell surface. I. The cell elastimeter. J Exp Biol 33:524–532

Mitchison T, Kirschner M (1984) Dynamic instability of microtubule growth. Nature 312:237–242

Mooseker MS (1974) Brush border motility: Microvillar contraction in isolated brush border models. J Cell Biol 63:231 a

Mooseker MS, Tilney LG (1975) Organization of an actin filament-membrane complex. J Cell Biol 67:725–743

Murray D, Rosenberg MD (1963) The relative extensibility of cell surfaces. J Cell Biol 17:289–297

Murray JD, Oster GF (1984a) Cell traction models for generating pattern and form in morphogenesis. J Math Biol 19:265–279

Murray JD, Oster GF (1984b) Generation of Biological Pattern and Form IMA. J Math Med Biol 1:51–75

Nadezhdina ES, Fais D, Chentsov YS (1979) On the association of centrioles with the interphase nucleus. Eur J Cell Biol 19:109–115

Neuhaus-Url G, Kiermayer O (1982) Observations of microtubules and microtubule-microfilament associations in osmotically treated cells of Micrasterias denticulata Breb. Eur J Cell Biol 27:206–212

Nicklas RB (1983) Measurements of the force produced by the mitotic spindle in anaphase. J Cell Biol 97:542–548

Nicklas RB (1984) A quantitative comparison of cellular motile systems. Cell Motil 4:1–5

Nothnagel EA, Webb WW (1982) Hydrodynamic models of viscous coupling between motile myosin and endoplasm in characean algae. J Cell Biol 94:444–454

O'Neill C, Jordan P, Ireland G (1986) Evidence for two distinct mechanisms of anchorage stimulation in freshly explanted and 3T3 swiss mouse fibroblasts. Cell 44:489–496

Oster GF (1984) On the crawling of cells. J Embryol Exp Morphol 83:329–364

Oster GF, Odell GM (1984a) The mechanochemistry of cytogels. Physica 12D:333–350
Oster GF, Odell GM (1984b) Mechanics of Cytogels. I. Oscillations in Physarum. Cell Motil 4:469–503
Oster GF, Murray JD, Harris AK (1983) Mechanical aspects of mesenchymal morphogenesis. J Embryol Exp Morphol 78:83–125
Oster GF, Murray JD, Odell GM (1985) The formation of microvilli. In: Molecular Determinants of animal form. Alan R. Liss, New York, pp 365–384
Penman S, Fulton A, Capco D, Zeev AB, Wittelsberger S, Tse CF (1982) Cytoplasmic and nuclear architecture in cells and tissue: form, functions, and mode of assembly. Cold Spring Harbor Symp Quant Biol, Part 2, 61:1013–1028
Penman S, Capco DG, Fey EG, Chatterjee P, Reiter T, Ermish S, Wan K (1983) The three three-dimensional structural networks of cytoplasm and nucleus: Function in cells and tissue. In: Satir BH (ed) Modern Cell Biology, Vol. 2. Alan R. Liss, New York, pp 385–415
Phillips HM (1984) Physical analysis of tissue mechanics in amphibian gastrulation. Am Zool 24:657–672
Pollard EC (1979) Relationship of synthetic processes in the cell to cytoplasmic viscosity. In: Keith AD (ed) The aqueous cytoplasm. Dekker, New York, 230 pp
Porter KR (1984) The cytomatrix: A short history of its study. J Cell Biol 99 (1):3–6
Portzehl H (1951) Muskelkontraktion und Modellkontraktion. Z Naturforsch Teil 6 B:355–361
Pouchelet M, Anteunis A, Gansmuller AG (1986) Correspondence of two nuclear networks observed in situ with the nuclear matrix. Biol Cell 56:107–112
Przywansky KB, Schliwa M, Porter KR (1983) Comparison of the three-dimensional organization of unextracted and Triton-extracted human neutrophilic polymorphonuclear leukocytes. Eur J Cell Biol 30:112–125
Quader H (1986) Cellulose microfibril orientation in oocystis solitaria: proof that microtubules control the alignment of the terminal complexes. J Cell Sci 83:223–234
Rahat M, Parnas I, Nevo A (1969) Extensibility and tensile strength of the stalk "muscle" of Carchesium SP. Exp Cell Res 54:69–76
Rees DA, Lloyd CW, Thom D (1977) Control of grip and stick in cell adhesion through lateral relationships of membrane glycoproteins. Nature 267:124–128
Rockwell MA, Fechheimer M (1984) A comparison of methods used to characterize gelation of actin in vitro. Cell Motil 4:197–213
Sato M, Wong TZ, Allen RD (1983) Rheological properties of living cytoplasm: endoplasm of Physarum plasmodium. J Cell Biol 97:1089–1097
Scharf SR, Gerhart JC (1983) Axis determination in eggs of Xenopus laevis: A critical period before first cleavage, identified by the common effects of cold, pressure and ultraviolet irradiation. Dev Biol 99:75–87
Schliwa M (1986) The cytoskeleton. Springer, Berlin Heidelberg New York, 325 pp
Schliwa M, Bereiter-Hahn J (1975) Pigment movements in fish melanophores: Morphological and physiological studies. 5. Evidence for a microtubule-independent contractile systems. Cell Tissue Res 158:61–73
Schliwa M, Pryzwansky KB, Euteneuer U (1982) Centrosome splitting in neutrophils: an unusual phenomenon related to cell activation and motility. Cell 31:705–717
Schulze E, Kirschner M (1986) Microtubule dynamics in interphase cells. J Cell Biol 102:1020–1031
Sheetz MP, Spudich JA (1983) Movement of myosin-coated fluorescent beads on actin cables in vitro. Nature 303:31–35
Skerrow CJ (1986) Desmosomal proteins. In: Bereiter-Hahn J, Matoltsy AG, Richards KS (eds) Biology of the Integument, Vol 2. Springer, Berlin Heidelberg New York, pp 763–787
Skerrow D (1986) Epidermal a-keratin: structure and chemical composition. In: Bereiter-Hahn J, Matoltsy AG, Richards KS (eds) Biology of the Integument. Springer, Berlin Heidelberg New York, pp 621–643
Spelt JK, Absolom DR, Zingg W, van Oss CJ, Neumann AW (1982) Determination of the surface tension of biological cells using the freezing front technique. Cell Biophys 4:117–131
Stein MB, Gordon R (1982) Epithelia as bubble rafts: A new method for analysis in embryonic and intercellular adhesion in embryonic and other epithelia. J Theor Biol 97:625–639

Steinberg MS (1978) Cell–cell recognition in multicellular assembly. Levels of specificity. In: Curtis ASG (ed) Cell–cell recognition. Symp Soc Exp Biol 52:25–49

Stockem W, Hoffmann H-U, Gruber B (1983) Dynamics of the cytoskeleton in Amoeba proteus. 1. Redistribution of microinjected fluorescein-labeled actin during locomotion, immobilization and phagocytosis. Cell Tissue Res 232:79–96

Stossel TP, Hartwig JH, Yin HL, Zaner KS, Stenddahl OI (1982) Actin gelation and the structure of cortical cytoplasm, part 2. Cold Spring Harbor Symp Quant Biol 61:569–578

Strohmeier R, Bereiter-Hahn J (1984) Control of cell shape and locomotion by external calcium. Exp Cell Res 154:412–420

Taylor EW (1963) Brownian and saltatory movements of cytoplasmic granules and the movement of anaphase chromosomes. Proc Fourth Int Congr Rheol Part 4:175–191

Taylor DL, Condeelis JS, Moore PL, Allen RD (1973) The contractile basis of amoeboid movement. I. The chemical control of motility in isolated cytoplasm. J Cell Biol 59:378–394

Taylor DL, Moore PL, Condeelis JS, Allen RD (1976) The mechanochemical basis of amoeboid movement. I. Ionic requirements for maintaining viscoelasticity and contractility of amoeba cytoplasm. Exp Cell Res 101:127–133

Tillmann U, Bereiter-Hahn J (1986) Relation of actin fibrils to energy metabolism of endothelial cells. Cell Tissue Res 243:579–585

Tilney LG (1977) Actin: Its association with membranes and the regulation of its polymerization. In: Brinkley BR, Porter KR (eds) Int Cell Biol 1976–1977. Rockefeller University Press, New York, pp 388–402

Tranquillo RT, Lauffenburger DA (1986) Consequences of chemosensory phenomena for leukocyte chemotactic orientation. Cell Biophys 8:1–46

Traub P (1985) Intermediate filaments. A review. Springer, Berlin Heidelberg New York, 265 pp

Tucker RP, Edwards BF, Erickson CA (1985) Tension on the culture dish: Microfilament organization and migratory behavior of Quail neural crest cells. Cell Motil 5:225–237

Valberg PA, Albertini DF (1985) Cytoplasmic motions, rheology, and structure probed by a novel magnetic particle method. J Cell Biol 101:130–140

Vandenburgh HH (1982) Dynamic mechanical orientation of skeletal myofibers in vitro. Dev Biol 93:438–443

Volberg T, Geiger B, Kartenbeck J, Franke WW (1986) Changes in membrane-microfilament interaction in intercellular adherens junctions upon removal of extracellular Ca^{2+} ions. J Cell Biol 102:1832–1842

Waddington CH (1939) Order of magnitude of morphogenetic forces. Nature 3649:637

Wang E, Huie P (1985) Changes in the interaction between vimentin and an associated protein, p50, correlated with the formation of vimentin bundles in senescent human fibroblasts. Ann NY Acad Sci 455:715–717

Wang E, Roos DS, Heggeness MH, Choppin PW (1982) Function of cytoplasmic fibers in syncytia. Cold Spring Harbor Symp Quant Biol 46:997–1012

Wang E, Fischman D, Liem RKH, Sun TT (eds) (1985) Intermediate filaments. Ann NY Acad Sci 455:832

Weber HH (1958) The motility of muscle and cells. Harvard University Press, Cambridge, Mass

Weber K, Osborn M (1982) Cytoskeleton: Definition, structure and gene regulation. Pathol Pract 175:128–145

Weiss P, Garber B (1952) Shape and movement of mesenchyme cells as functions of the physical structure of the medium. Contributions to a quantitative morphology. Proc Natl Acad Sci USA 38:264–280

Witztum A, Parthasarathy MV (1985) Role of actin chloroplast clustering and banding in leaves of Egeria, Elodea and Hydrilla. Eur J Cell Biol 39:21–26

Wolf R (1980) Migration and division of cleavage nuclei in the gall midge, Wachtiella persicariae. II. Origin and ultrastructure of the migration cytaster. W Roux's Arch 188:65–73

Yeh C, Rodan GA (1984) Tensile forces enhance prostaglandin E synthesis in osteoblastic cells grown on collagen ribbons. Calcif Tissue Int 36:S67–S71

Zieve GW (1985) Mitosis-specific phosphorylation of vimentin. Ann N Y Acad Sci 455:720–723

I.2 Evaluation of Cytomechanical Properties

YUKIO HIRAMOTO[1,2]

2.1 Introduction

Although a number of investigations have been carried out up to now on the morphological structures of the cell by light and electron microscopy, our knowledge of the mechanical properties of the cell is limited. This is because it is generally difficult to apply conventional methods for measuring mechanical properties of bodies of macroscopic dimensions to measure the properties of living cells of microscopic dimensions.

Many investigations of the mechanical properties of the cell have been carried out using egg cells, because they are relatively large in size, simple in form, and generally free from other cells. Also, a large number of investigations have been carried out in muscle fibers in connection with the generation of forces during activity. In the present chapter, methods and results of the evaluation of the mechanical properties of egg cells are described together with those of some other animal cells. However, those of muscle fibers are excluded; the muscle fiber is particularly differentiated for generating active force, thus the mechanical properties of muscle fibers definitely differ from those of ordinary cells.

For previous reviews on the mechanical properties of the cell, the reader is referred to Heilbrunn (1928, 1958); Harvey (1954) and Hiramoto (1970, 1981).

2.2 Physical Structure of the Cell

The protoplasm is limited by the cell membrane from the surrounding medium. The cell membrane, which is recognized as a membrane several nanometers in thickness by electron microscopy and which acts as a barrier for diffusion of substances into and out of the cell, is thought to have a mechanical tension. The protoplasm surrounded by the cell membrane consists of two parts, the cortex or ectoplasm with a high consistency, and the endoplasm with a low consistency. The difference in consistency between these two parts is shown by the difference in the behavior of endogenous particles when the cell is centrifuged – the particles in the cortex are hardly moved by a centrifugal force capable of translocating the particles in the endoplasm. It is thought that the cortex has a positive tension in a direction tangential to the cell surface.

---

[1] Biological Laboratory, Tokyo Institute of Technology, O-okayama, Meguro-ku, Tokyo 152 and Department of Cell Biology, National Institute for Basic Biology, Okazaki 444, Japan
[2] Present address: The University of the Air, Wakaba, Chiba 260, Japan

Some cells are surrounded by membranes, layers, and/or coats outside the cell membrane, which in this chapter are referred to as "extraneous coats". The stiffness of the extraneous coats varies in different cells: in some cells it is smaller than those of the cell membrane and the cortex, while in other cells it is larger. For example, the cell wall in higher plant cells is extremely stiff in comparison to the cell membrane and the cortex, and withstands a tension far larger than the tensions of the cell membrane and the cortex. In some eggs, the chorion and/or vitelline membrane also generate tensions far larger than the tensions of the cell membrane and the cortex.

2.3 Mechanical Properties of the Cell Surface

2.3.1 Relationship Between the Surface Force and the Internal Pressure of the Cell

The forces in the direction tangential to the cell surface, which are generated by the cell membrane, the cortex, and the extraneous coats, balance the internal pressure of the cell, provided the forces generated by the structures in the endoplasm are negligible as compared with the forces generated by the above structures at the cell surface. Since the tensions of the cortex, the cell membrane, and the extraneous coats cannot be determined separately due to technical difficulties, the sum of the forces generated by them per unit length of the cell surface may be used to characterize the force at the cell surface, and is referred to as "surface force" (Cole 1932) or "tension at the surface" (Harvey 1931). It has the same dimensions as the surface tension of liquids (force divided by length). When the total thickness of the cortex, the cell membrane, and the extraneous coats is very small as compared with the radius of curvature of the cell surface, then

$$T_1/R_1 + T_2/R_2 = P_0, \tag{1}$$

where T_1 and T_2 are surface forces in directions perpendicular to each other, R_1 and R_2 are radii of curvature of the cell surface in the respective directions, and P_0 is the internal pressure of the cell (cf. Fig. 1a). In spherical cells, $T_1 = T_2 = T_0$ and $R_1 = R_2 = R_0$, thus

$$2\,T_0/R_0 = P_0, \tag{2}$$

where T_0 is the surface force and R_0 is the radius of the cell (cf. Fig. 1b). In cells with a form of axial symmetry, e.g., dividing sea urchin egg, the radii of curvature at any points of the cell surface are determined from the form of the cell measured from the direction perpendicular to the axis of symmetry (cf. Fig. 1c). In this case, the surface forces are determined using Eqs. (3) and (4), provided the internal pressure of the cell (P_0) is known (cf. Hiramoto 1968):

$$T_1 = P_0\,R_2/2. \tag{3}$$
$$T_2 = (P_0\,R_2/2)\,(1 - R_2/R_1). \tag{4}$$

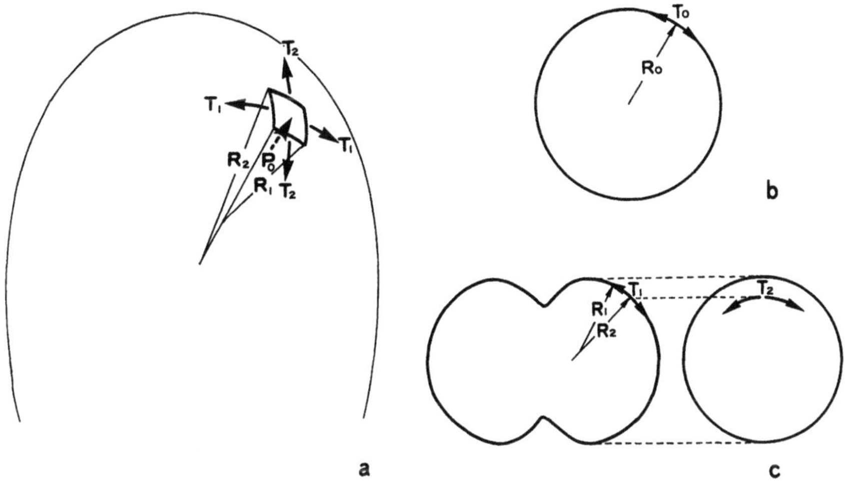

Fig. 1 a – c. Balance between the internal pressure and the surface forces of a cell. P_0 Internal pressure; R_0, R_1, R_2 radii of curvature of the cell surface; T_1, T_2 surface forces. For further explanations see text and equations (1) to (4)

2.3.2 Direct Measurement of the Internal Pressure

The internal pressure of the cell can be determined with a manometer connected to a micropipette inserted into the cell if the pressure is sufficiently large to overcome the technical errors of measurement. The high intracellular pressure reported in some cells (e.g. fish eggs, cf. Kao and Chambers 1954; Kao et al. 1954, and Table 1 in Chap. V.2) may be generated by the tension of the extraneous coats outside the cell proper, as indicated by the report of a very low surface force in the order of 10^{-4} N/m in fish eggs deprived of the chorion (Sakai 1961).

2.3.3 Indirect Measurements of the Surface Force and the Internal Pressure

In cells without or with tender extraneous coats, such as echinoderm eggs, direct measurement of the internal pressure of the cell by the manometric method described above is practically impossible, since the pressure difference at the interface between the protoplasm and the liquid in the micropipette inserted into the cell introduces a serious error in measurement. The low intracellular pressure in these cells may indicate that the surface forces due to the cortex and the cell membrane are generally very small and that the high intracellular pressure occasionally obtained in some other cells is due to the high tensions of the extraneous coats rather than the egg proper.

Internal pressure and surface force provide a cell with stiffness. The resulting mechanical resistivity against externally applied forces is used as a measure for surface force in the compression, the suction, the stretching, and the sessile drop-method described in the following paragraphs.

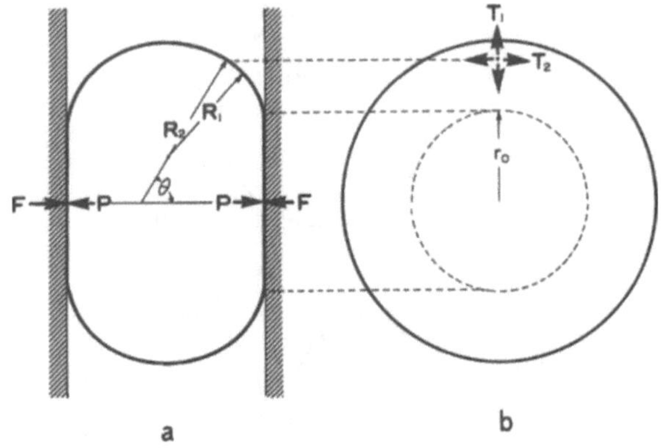

Fig. 2a, b. Compression method. The cell is compressed between two parallel plates. **a** Top view; **b** side view. *F* Applied force; *P* internal pressure of the cell, R_1, R_2 radii of curvature of the cell surface, r_0 radius of the circular area of the cell surface in contact with the compressing plate; θ angle formed by the line normal to the cell surface against the direction of compression

2.3.3.1 Compression Method

When a cell is compressed between two parallel plates, the surface force and the internal pressure of the cell can be calculated from the applied force and the measurements of the cell surface (cf. Fig. 2) using Eqs. (5), (6) and (7). For the balance of forces applied to the cell surface,

$$F = \pi r_0^2 P \tag{5}$$
$$T_1/R_1 + T_2/R_2 = P \tag{6}$$
$$2 \pi R_2 T_1 \sin^2 \theta + F = \pi R_2^2 P \sin^2\theta, \tag{7}$$

where F is the applied force, P is the internal pressure, R_1 and R_2 are radii of curvature of the cell surface as shown in Fig. 2, r_0 is the radius of the area of the cell surface in contact with the plate, T_1 and T_2 are surface forces, and θ is the angle between the normal to the cell surface and the direction of compression as shown in Fig. 2. Equation (7) becomes Eq. (8) where θ equals 90° (Hiramoto 1963), and the calculation becomes simpler if it is assumed that T_1 equals T_2 (cf. Cole 1932):

$$2 \pi R_2 T_1 + F = \pi R_2^2 P \tag{8}$$

Yoneda (1964, 1972, 1986) presented a different method to calculate the surface force assuming that the surface force is uniform over the entire surface of the compressed cell. It was, however, pointed out by Hiramoto (1976) that this assumption may introduce a serious error in the calculation of the surface force.

For practical details of the compression method, see Yoneda (1986).

2.3.3.2 Suction Method

The stiffness (mechanical resistivity) of the cell surface can be determined from its deformation when a negative pressure is applied to a part of the surface

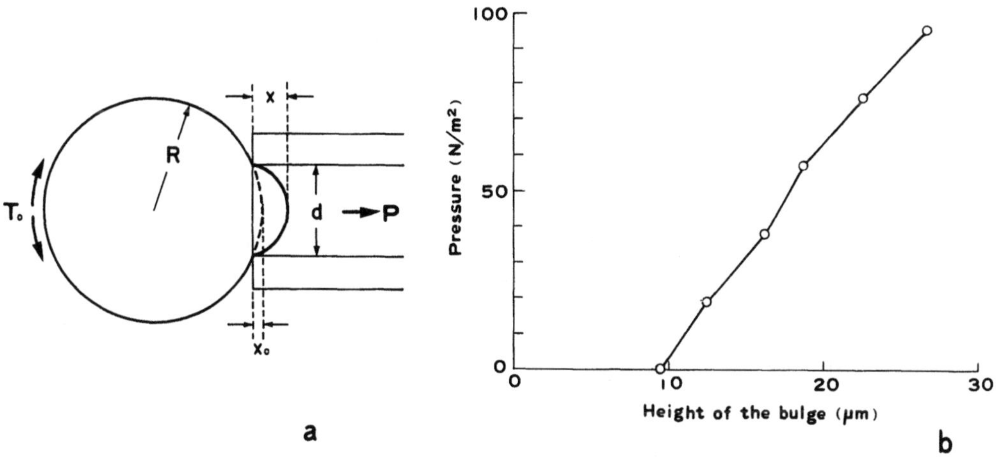

Fig. 3a, b. Suction method. The cell surface is deformed by the negative pressure applied locally with a micropipette. *d* diameter of the micropipette; *P* applied negative pressure; *x* height of the bulge; $x = x_0$ when $P = 0$ (Hiramoto 1986)

through a micropipette whose tip is in close contact with it (cf. Fig. 3 and Chap. V.1). This method was first used by Vlès (1933) on sea urchin eggs and was thereafter improved by Mitchison and Swann (1954). A practical linear relationship between the height of the bulge formed at the cell surface and the negative pressure applied to it was found in various kinds of cells including sea urchin eggs (Mitchison and Swann 1954), starfish eggs (Nakamura and Hiramoto 1978), and amphibian eggs (Selman and Waddington 1955). The surface force can be calculated from the slope of the linear relationship if it is assumed that the surface force is not increased by the deformation of the cell surface (Hiramoto 1970; Nakamura and Hiramoto 1978).

Thus,

$$T = \frac{(x_0^2 + d^2/4)^2}{d^2 - 4x_0^2} S_0 , \tag{9}$$

where d is the inner diameter of the micropipette, x_0 is the height of the bulge when no negative pressure is applied, and S_0 is the slope of the negative pressure-deformation curve (cf. Fig. 3b). Nakamura and Hiramoto (1978) estimated the initial surface force when the cell is not deformed and, by analyzing the force-deformation relationship, the rate of increase in the surface force caused by stretching the cell surface (elasticity value, cf. Hiramoto 1970; Nakamura and Hiramoto 1978).

This method is useful for determining regional differences in the stiffness (and the surface force) in the same cell (cf. Sawai and Yoneda 1974; Ohtsubo and Hiramoto 1985). For practical details of the suction method, see Hiramoto (1986).

2.3.3.3 Stretching Method

Harvey (1931) determined the force required to separate sea urchin eggs into light and heavy halves by centrifuging them in a medium whose density was the same as that of the egg protoplasm. In this case, the protoplasm is stratified as a result of centrifugal migration of heavy components and centripetal migration of light components, until finally the eggs are torn into two halves. Harvey (1931) estimated the surface force at the torn part from the centrifugal acceleration applied to the egg and the volumes and densities of the light and heavy halves. Although notable as the first determination of the surface force of the cell, the calculation involved many assumptions and included an error (cf. Hiramoto 1970).

Norris (1939) determined the force–deformation relationships of urchin eggs and amphibian blood cells impaled on a pair of microneedles by stretching the cell with the needles, one of which was used to determine the applied force from its bending. The surface force could be determined from his measurements, assuming the internal pressure of the cell is negligible.

2.3.3.4 Sessile Drop Method

Harvey and Fankhauser (1933); Harvey (1933) and Sakai (1961) determined the surface forces of salamander, molluscan, and teleost eggs, respectively, from the form of the egg placed on a flat substratum. The egg displays a form similar to a liquid drop placed on a flat plane (sessile drop) (cf. Fig. 5): in both cases, the form is determined by the balance of gravity, tending to flatten (or drop) the egg, and the tension at the surface (surface force in the egg and surface tension in the liquid drop) opposing flattening. They determined surface forces of the eggs from the form of the cell observed from a horizontal direction, the difference in the densities of the protoplasm and the surrounding medium, and the size of the cell, using Dorsey's theoretical equations (1928) for the liquid drop.

Small cells such as sea urchin eggs are not appreciably deformed by gravity, but they display sessile drop forms when they are put in a centrifugal field (Hiramoto 1967). Hiramoto (1967) and Hiramoto and Yoneda (1986) derived a theoretical relationship between the surface tension and the degree of flattening

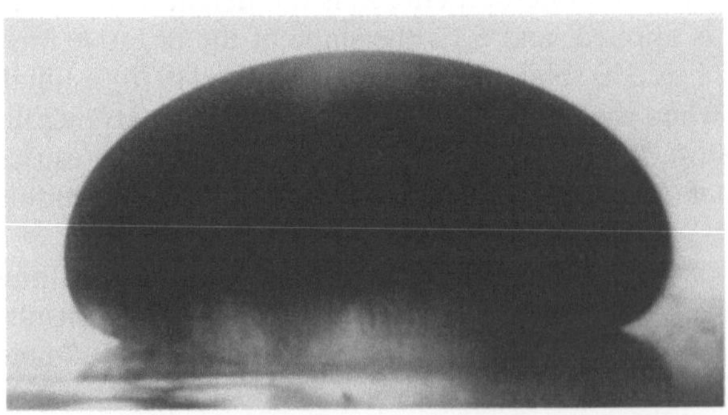

Fig. 4. A fertilized egg of *Xenopus laevis* deprived of the vitelline membrane and placed on a flat substratum

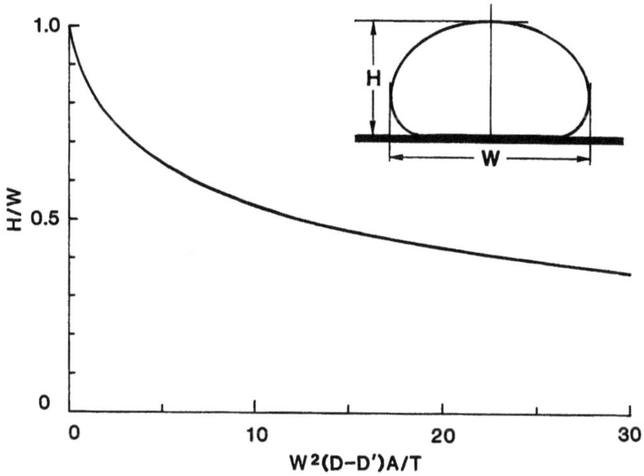

Fig. 5. Theoretical relation among the deformation of a liquid drop placed on a flat substratum, the surface tension, the size of the drop, the density difference between the liquid and the surrounding medium, and the acceleration due to gravity (or centrifugal force). *A* acceleration due to gravity (or centrifugal force); *D* density of the liquid drop; *D'* density of the surrounding medium; *H* height of the drop; *T* surface tension; *W* width of the drop

of a liquid drop under gravity or a centrifugal field. The degree of flattening (H/W) is determined by

$$H/W = W^2 (D - D') A/T, \tag{10}$$

where W is the horizontal and H the vertical diameter of the drop, $D - D'$ is the difference between the densities of the liquid and the surrounding medium, A is the acceleration of the gravitational or centrifugal field, and T is the surface tension of the liquid as shown in Fig. 5. Hiramoto (1967) determined the surface force of the sea urchin egg from the degree of deformation of the egg in a centrifugal field observed with a centrifuge microscope, using the relation between the surface tension and the degree of flattening shown in Fig. 5.

Kamiya and Kuroda (1958b) determined the surface force of a drop of protoplasm extruded from an internodal cell of *Nitella* using Dorsey's equation (1928). They also used an equation for a hanging liquid drop (pendant drop equation).

For practical details of the sessile drop method, see Hiramoto and Yoneda (1986).

2.3.4 Elasticity and Viscoelasticity of the Cell Surface

It is of interest whether or not the surface force increases by stretching the cell surface, in other words, whether or not the cell surface is elastic. By the com-

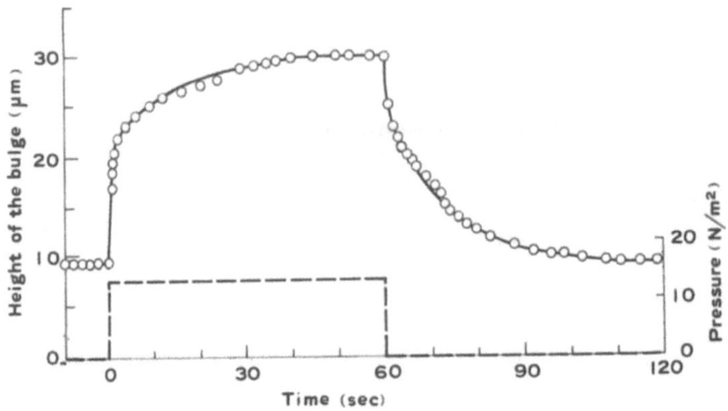

Fig. 6. Change in the height of the bulge (*solid line*) when a constant negative pressure (*broken line*) is suddenly applied to part of the starfish egg surface and then removed (Hiramoto 1986)

pression method, Cole (1932) found that in sea urchin eggs, the surface force increases as the deformation increases. A similar result was obtained in sea urchin eggs by Hiramoto (1963, 1967, 1976) and in starfish eggs by Nakamura and Hiramoto (1978). On the other hand, Yoneda (1964, 1972) concluded that in sea urchin eggs the surface force is almost constant independent of the stretching of the cell surface. The discrepancy in the above results may be due to the inapplicability of the method used by Yoneda for the evaluation of the surface force in sea urchin eggs, as pointed out by Hiramoto (1967, 1976).

Although not described in the preceding section, the deformation of the cell gradually changes even when a constant force is being applied. Figure 6 shows the change in the height of the bulge at a part of the cell surface in starfish eggs when a constant negative pressure is applied through a micropipette whose tip is in close contact with the cell surface, and the change when the pressure is removed after a short time (Nakamura and Hiramoto 1978). The change in the height of the bulge during the application of negative pressure and its recovery after the removal of the pressure indicate that the cell surface is viscoelastic. Viscoelastic properties are generally observed in all kinds of cells. It is thought that the surface force temporarily increases when the cell is suddenly deformed and that it gradually decreases when the deformation of the cell remains unchanged.

2.4 Mechanical Properties of the Endoplasm

2.4.1 Measurements of Mechanical Properties of the Endoplasm

Five different methods have been reported for determining the mechanical properties of the endoplasm: the centrifuge method, magnetic particle method, capillary method, Brownian movement method, and diffusion method.

2.4.1.1 *Centrifuge Method*

When the cell is put in a centrifugal field, particles in the endoplasm move either in a centrifugal or a centripetal direction due to the density difference between the particles and the cytoplasmic matrix surrounding them. The velocity of the particle (v) is given by Stokes' law if the cytoplasmic matrix is regarded as a Newtonian fluid:

$$v = \frac{2a^2(D - D')\,c\,g}{9\eta}, \tag{11}$$

where a is the radius of the particle, $D - D'$ is the density difference between the particle and the surrounding cytoplasmic matrix, c g is the acceleration of the centrifugal field (g is the gravity constant), and η is the viscosity coefficient of the cytoplasmic matrix.

Heilbrunn and his collaborators determined the viscosity of the endoplasm following the above principle in various cells (for reviews, see Heilbrunn 1928, 1958). They estimated the mean velocity of definite particles naturally present in the cell from the duration of the applied centrifugal field and the degree of stratification, which was checked immediately after centrifugation. The radius of the particles (a) was determined with a microscope. The density of the particles and that of the protoplasm were determined by finding the density of liquid in which the particle, or the whole cell neither floated nor sank.

Foreign particles introduced into the cell (e.g., oil droplets or metal particles) can be used instead of natural particles. When the particle is not spherical, an appropriate correction should be made (cf. Crick 1950). The velocity of the particle in a centrifugal field can be determined directly by using a centrifuge microscope (cf. Harvey and Marsland 1932).

2.4.1.2 *Magnetic Particle Method*

Movements of ferromagnetic particles in the magnetic field are influenced by the viscoelastic properties of the surrounding medium as well as the magnitude and gradient of the magnetic field and the magnetic susceptibility of the magnetic particles (cf. Crick 1950). Therefore, it is possible to determine viscoelasticity and yield stress of the protoplasm from the behavior of a ferromagnetic particle in the protoplasm when a magnetic field is applied.

Such an experiment was first carried out by Heilbronn (1922) using slime mold plasmodia and then by Seifriz (1924) using sea urchin eggs. When a magnetic field is applied to a ferromagnetic particle, the particle, if it is not spherical, orients itself in the direction of the magnetic field. In addition, the particle moves along the gradient of the magnetic field if such a gradient exists. The theoretical relationship between the intensity and the gradient of the applied magnetic field, the size, form, and the nature of the magnetic particle, the viscosity of the surrounding medium, and the twisting or translational velocity of the particle have been well documented by Crick (1950).

In cells such as amoebae, the magnetic particle can be introduced by applying it after coating with substances that facilitate phagocytosis (cf. Crick and Hughes 1950; Yagi 1961). In nonphagocytotic cells such as sea urchin eggs, the particle is introduced into the cell by injection with a microneedle or by a strong magnetic force (cf. Hiramoto 1969 a). When a drop of ferrofluid (liquid in which many minute magnetic particles are suspended) is used as a magnetic particle, a conventional microinjection technique can be used for introduction into the cell (cf. Hiramoto 1975).

Figure 7 shows the movements of a spherical iron particle which has been introduced into a sea urchin egg when a magnetic field with a steep gradient is applied by switching the circuit of an electromagnet placed close to the cell on and off. As shown in this figure, when a constant force was suddenly applied, the particle in the protoplasm moved toward the electromagnet fairly fast. The speed gradually decreased, approaching a steady value when the force was maintained, and the particle gradually recoiled when the force was removed. The response of the particle in the protoplasm mentioned above can be explained, assuming the mechanical properties of the protoplasm are in practice represented by the viscoelastic model shown in Fig. 8. In this case, the displacement of the particle at t = 0 when a constant force is suddenly applied is expressed by

$$x = \frac{F}{6\pi a} \frac{1 - \exp(-t/\tau_1)}{G} + \frac{t}{\eta_2} \quad (12)$$

or

$$x = \frac{F}{6\pi a G} 1 - \exp(-t/\tau_1) + \frac{t}{\tau_2}, \quad (13)$$

in which x is the displacement of the particle, F is the applied force, a is the radius of the particle, t is time, and G, η_2, τ_1 (= η_1/G), and τ_2 (= η_2/G) are the viscoelasticity constants of the protoplasm as shown in Fig. 8. This assumes that

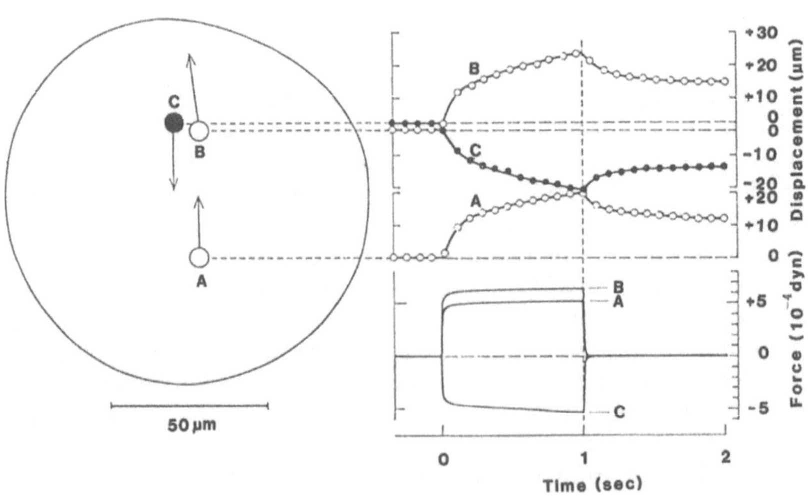

Fig. 7. Movements of a spherical iron particle introduced into a sea urchin egg (see text) (Hiramoto 1969 b)

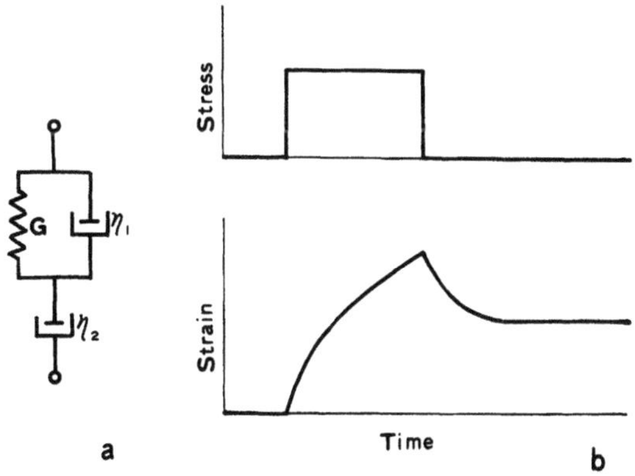

Fig. 8a, b. Three-element viscoelastic model representing the properties of the endoplasm. **a** Viscoelastic model consisting of one elastic element and two viscous elements. **b** The change in the strain when a definite stress is applied to the model and then removed (Hiramoto 1969a)

the force–displacement relationship in the protoplasm is linear, although it was shown experimentally to be not exactly so (Hiramoto 1969a).

Hiramoto (1969a) determined the applied force (F) from the bending of a fine glass fiber, which was fixed in position at one end, and supported the cell containing the magnetic particle at the other end, when the magnetic field was applied by switching on the electromagnet circuit.

The force applied to the particle can be estimated from the magnetic field parameters (i.e., the magnitude and the gradient of the magnetic field) and a constant, which depends on the size, shape, and magnetic properties of the particle (cf. Crick 1950; Crick and Hughes 1950; Sato et al. 1983, 1984; Valberg and Albertini 1985), and which can be determined experimentally using liquids of known viscosity. The viscoelasticity constants of the protoplasm can be determined from the applied force and from the movement of the particle in the protoplasm determined experimentally, using Eqs. (12) and (13).

2.4.1.3 Capillary Method

In the plasmodial strand of the slime mold, the endoplasm flows back and forth with a fairly regular period (cf. Chap. II.6). Kamiya (1950) and Kamiya and Kuroda (1958a) found that the velocity distribution of the endoplasm in the strand is like a parabola, the apex of which is flattened or nearly truncate, as shown in Fig. 9. They also determined the velocity distribution of the protoplasm when a pressure gradient was applied along the strand in order to record the passive flow of the endoplasm when the motive force of the streaming is zero. The velocity profile of this passive endoplasmic flow (Fig. 10b) was similar to that of the endoplasmic flow in normal protoplasmic streaming, suggesting that the protoplasmic streaming in slime mold plasmodia is caused by a pressure difference in the cell and that the endoplasm has to be regarded as a non-Newtonian liquid. Although Kamiya and Kuroda (1958a) did not mention the viscosity value in their paper, an apparent viscosity of 0.3 Pa · s at a shear rate of 2 s^{-1} and of almost infinity at zero shear rate can be determined from their data.

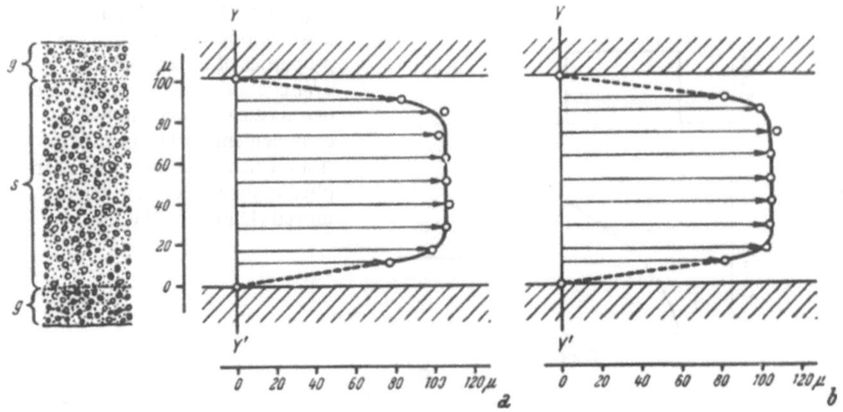

Fig. 9 a, b. Velocity distribution in a plasmodial strand of slime mold. **a** During normal cytoplasmic streaming. **b** During the application of a pressure gradient between the two ends of the strand when the autonomous cytoplasmic streaming is stopped. The velocity is expressed by the displacement of cytoplasmic particles within 3 s (Kamiya and Kuroda 1958)

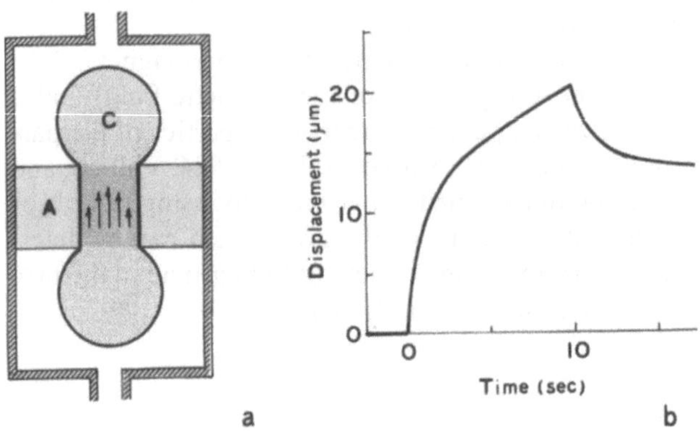

Fig. 10 a, b. Endoplasmic flow caused by an externally applied pressure gradient. **a** Schematic diagram of the experimental setup. The cell (*C*) has been deformed to a dumbbell shape by passing through a cylindrical hole formed in an agar septum (*A*). The pressure difference is applied between two compartments separated by the agar septum. **b** Displacement of the endoplasm at the central axis of the cylindrical part of the cell when a constant pressure difference is applied between the two compartments

Kamiya and Kuroda (1965) determined the viscosity of the endoplasm isolated from *Nitella* cells from the velocity of the endoplasm flowing in a capillary made of agar gel. They showed that the shear rate, and therefore the apparent viscosity, depend on the shear stress.

The capillary method can be applied to the protoplasm of egg cells by modifying the above method. Shôji et al. (1978) determined the velocity of the endoplasm under a pressure gradient applied externally to starfish eggs. Figure 10 shows the movement of the endoplasm at the central axis of the cylindrical part of the cell formed by sucking the cell through a circular tube (cf. Fig. 10a) when a constant pressure difference is applied to both ends of the cylindrical part. It is shown that the protoplasm behaves as a viscoelastic body as is represented by the three-element viscoelastic model shown in Fig. 8a. When the viscoelastic body represented by the above model is forced to move in a capillary with an internal radius R and length L by a pressure difference (P) applied between its two ends, the displacement of the body at the central axis of the capillary (x^0) is represented by

$$x^0 = \frac{P}{L}\frac{R^2}{4}\left\{\frac{1}{G}(1 - e^{-G/\eta_1}) + \frac{t}{\eta_2}\right\}. \tag{14}$$

Equation (14) is valid only for a linear stress/strain relationship.

The viscoelasticity constants can be determined approximately by comparing the record of the movement of the endoplasm in the cylindrical part of the cell with the movement expected from Eq. (14), although, since the stress–strain relationship in the endoplasm is in fact nonlinear, the constants do not exactly represent the mechanical properties of the endoplasm.

2.4.1.4 Brownian Movement Method

The viscosity of the endoplasm was determined by Heilbrunn and collaborators (for reviews, see Heilbrunn 1928, 1958) from the rate of Brownian movement of small particles in the endoplasm using Einstein's formula for Brownian movement (cf. Chap. II.1.2.1).

$$D_x^2 = \frac{R}{N} \cdot \frac{Tt}{3\pi\eta a}, \tag{15}$$

in which D_x is the displacement of a particle along one axis in the time t at the absolute temperature T, R is the gas constant, N the Avogadro number, η the viscosity, and a the radius of the particle. Using a value of 6.02×10^{23} for the Avogadro number and 8.3×10^7 for the value of the gas constant, this formula can be expressed by

$$D_x^2 = 14.7 \cdot 10^{18} \cdot \frac{Tt}{\eta a}. \tag{16}$$

It is, however, problematical whether the movement of particles in the protoplasm is actually "Brownian", because it is thought that meshwork structures exist in the protoplasm (cf. Hiramoto 1967, 1969a, 1970, Chap. I.1.5).

2.4.1.5 Diffusion Method

The viscosity of the aqueous solution can be determined from the rate of diffusion of the substances in it, since the diffusion rate is inversely proportional to the viscosity. The diffusion rate of dyes or other substances labeled with dyes in the protoplasm can be determined by comparing it with the diffusion rate of these substances in a liquid of known viscosity, e.g., pure water, by microphotometry after microinjecting them into the protoplasm (Kaneda and Hiramoto, unpublished) or by photobleaching recovery (Chap. II.1). It seems likely that the viscosity obtained by this method reflects the viscosity of the continuous phase of the protoplasmic matrix rather than that of the cytoskeletal structures.

2.4.2 Relationship Between the Mechanical Properties and Submicroscopic Structure of the Endoplasm

Among the various structures in the protoplasm revealed by electron microscopy, cytoskeletal structures consisting mainly of microfilaments and microtubules are considered to be closely related to the mechanical properties.

The centrifuge microscope observations of Harvey and Marsland (1932) and Allen (1960) showed that particles in amoeba cytoplasm moved, stopped, and moved in a centrifugal field as if they had met invisible obstructions. Similar behavior of particles in a centrifugal field have been observed in echinoderm egg protoplasm (Kaneda, Kamitsubo and Hiramoto, unpublished). These observations are regarded as strong evidence for the existent structures in the protoplasm.

The viscosity of the protoplasm in sea urchin eggs determined by Heilbrunn (1926) from the rate of stratification of the protoplasm is only in the order of centipoises, whereas the viscosity of the protoplasm in eggs of different species of sea urchin determined from the velocity of the nucleus in a centrifugal field by Hiramoto (1967) is about 30 poises, about 1000 times Heilbrunn's viscosity value. This discrepancy may be explained if the protoplasmic particles whose sedimentation rates were used for calculating the protoplasmic viscosity by Heilbrunn (1926) could pass through the interspace of the cytoskeletal meshwork while the nucleus had to move in the protoplasm, pushing the cytoskeletal structures aside. Viscoelastic properties determined by the magnetic particle method (Hiramoto 1969a, b) and the capillary method (Shôji et al. 1978) in echinoderm egg protoplasm may mainly reflect the viscoelastic properties of the cytoskeleton. The low viscosity values (several centipoises) of echinoderm egg protoplasm obtained by the diffusion method (Kaneda and Hiramoto, unpublished) supports the above assumption.

The coincidence of the regions of microtubule structures in the endoplasm and the regions of high protoplasmic consistency in fertilized sea urchin eggs (Hiramoto 1969b) may indicate that the microtubule and microfilament structures contribute to the high consistency of the protoplasm. Viscoelastic properties of the protoplasm in unfertilized sea urchin egg endoplasm, in which no microtubules are observed, may indicate the presence in these cells of meshwork structures different from microtubules.

In conclusion, it is considered that the protoplasm is a complex structure consisting of meshwork structures with viscoelastic properties and a fluid cytoplasmic matrix with a low viscosity. It seems likely that the meshwork structures play an important role in keeping organelles in their positions in the cell, while the fluid cytoplasmic matrix is involved in transporting substances within the cell.

References

Allen RD (1960) The consistency of amoeba cytoplasm and its bearing on the mechanism of amoeboid movement II. The effects of centrifugal acceleration observed in the centrifuge microscope. J Biophys Biochem Cytol 8:379–397

Cole KS (1932) Surface forces of the *Arbacia* egg. J Cell Comp Physiol 1:1–9

Crick FHC (1950) The physical properties of cytoplasm. A study by means of the magnetic particle method part II. Theoretical treatment. Exp Cell Res 1:505–533

Crick FHC, Hughes AFW (1950) The physical properties of cytoplasm. A study by means of magnetic particle method part I. Experimental. Exp Cell Res 1:37–80

Dorsey NE (1928) A new equation for the deformation of surface tension from the form of a sessile drop or bubble. J Wash Acad Sci 18:505–509

Harvey EN (1931) The tension at the surface of marine eggs, especially those of the sea urchin *Arbacia*. Biol Bull 61:273–279

Harvey EN (1933) The flattening of marine eggs under the influence of gravity. J Cell Comp Physiol 4:35–47

Harvey EN (1954) Tension at the cell surface. Protoplasmatologia 2 E5. Springer, Wien

Harvey EN, Fankhauser G (1933) The tension at the surface of the eggs of the salamander, *Tritrus (Diemyctylus) viridescens*. J Cell Comp Physiol 3:463–475

Harvey EN, Marsland DA (1932) The tension at the surface of *Amoeba dubia* with direct observations on the movement of cytoplasmic particles at high centrifugal speeds. J Cell Comp Physiol 2:75–97

Heilbronn A (1922) Eine neue Methode zur Bestimmung der Viskosität lebender Protoplasten. Jahrb Wiss Bot 61:284–338

Heilbrunn LV (1926) The absolute viscosity of protoplasm. J Exp Zool 44:255–278

Heilbrunn LV (1928) The colloid chemistry of protoplasm. Borntraeger, Berlin

Heilbrunn LV (1958) The viscosity of protoplasm. Protoplasmatologia 2 C1. Springer, Wien

Hiramoto Y (1963) Mechanical properties of sea urchin eggs I. Surface force and elastic modulus of the cell membrane. Exp Cell Res 32:59–75

Hiramoto Y (1967) Observations and measurements of sea urchin eggs with a centrifuge microscope. J Cell Physiol 69:216–230

Hiramoto Y (1968) The mechanics and mechanism of cleavage in sea-urchin eggs. In: Aspects of Cell Motility (22nd symposium of the Society for Experimental Biology). Cambridge Univ. Press, London, pp 311–327

Hiramoto Y (1969a) Mechanical properties of the protoplasm of the sea urchin egg I. Unfertilized egg. Exp Cell Res 56:201–208

Hiramoto Y (1969b) Mechanical properties of the protoplasm of the sea urchin egg II. Fertilized egg. Exp Cell Res 56:209–218

Hiramoto Y (1970) Rheological properties of sea urchin eggs. Biorheology 6:201–234

Hiramoto Y (1975) Force exerted by the cleavage furrow of sea urchin eggs. Dev Growth Differ 17:27–38

Hiramoto Y (1976) Mechanical properties of sea urchin eggs III. Visco-elasticity of the cell surface. Dev Growth Differ 18:379–388

Hiramoto Y (1981) Mechanical properties of dividing cells. In: Zimmerman AM, Forer A (eds) Mitosis/Cytokinesis. Academic Press, London, pp 397–418

Hiramoto Y (1986) Determination of mechanical properties of the egg surface by elastimetry. Methods Cell Biol 27:435–442

Hiramoto Y, Yoneda M (1986) Determination of mechanical properties of the egg by the sessile drop method. Methods Cell Biol 27:443–456

Kamiya N (1950) The rate of the protoplasmic flow in the myxomycete plasmodium I. Cytologia 15:183–193

Kamiya N, Kuroda K (1958a) Studies on the velocity distribution of the protoplasmic streaming in the myxomycete plasmodium. Protoplasma 49:1–4

Kamiya N, Kuroda K (1958b) Cell operation in *Nitella* IV. Tension at the surface of the effused endoplasmic drops. Proc Jpn Acad 34:435–438

Kamiya N, Kuroda K (1965) Rotational protoplasmic streaming in *Nitella* and some physical properties of the endoplasm. Proceedings of the 4th International Congress in Biorheology. Aug 26–30, 1963. Wiley, New York

Kao C-Y, Chambers R (1954) Internal hydrostatic pressure of the *Fundulus* egg I. The activated egg. J Exp Biol 31:139–149

Kao C-Y, Chambers R, Chambers EL (1954) Internal hydrostatic pressure of the fundulus egg II. Permeability of the chorion. J Cell Comp Physiol 44:447–462

Mitchison JM, Swann MM (1954) The mechanical properties of the cell surface I. The cell elastimeter. J Exp Biol 31:443–460

Nakamura S, Hiramoto Y (1978) Mechanical properties of the cell surface in starfish eggs. Dev Growth Differ 20:317–327

Norris CH (1939) The tension at the surface and other physical properties of the nucleated erythrocyte. J Cell Comp Physiol 14:117–133

Ohtsubo M, Hiramoto Y (1985) Regional difference in mechanical properties of the cell surface in dividing echinoderm eggs. Dev Growth Differ 27:371–383

Sakai Y (1961) Methód for removal of chorion and fertilization of the naked egg in *Oryzias latipes*. Embryologia 5:357–368

Sato M, Wong TZ, Allen RD (1983) Rheological properties of living cytoplasm: Endoplasm of *Physarum* plasmodium. J Cell Biol 97:1089–1097

Sato M, Wong TZ, Brown DT, Allen RD (1984) Rheological properties of living cytoplasm: A preliminary investigation of squid axoplasm (*Loligo pealei*). Cell Motil 4:7–23

Sawai T, Yoneda M (1974) Wave of stiffness propagating along the surface of the newt egg during cleavage. J Cell Biol 60:1–7

Seifriz W (1924) An elastic value of protoplasm, with further observations on the viscosity of protoplasm. Brit J Exp Biol 2:1–11

Selman GG, Waddington CH (1955) The mechanism of cell division in the cleavage of the newt's egg. J Exp Biol 32:700–733

Shôji Y, Hamaguchi MS, Hiramoto Y (1978) Mechanical properties of the endoplasm in starfish oocytes. Exp Cell Res 117:79–87

Valberg PA, Albertini DF (1985) Cytoplasmic motions, rheology, and structure probed by a novel magnetic particle method. J Cell Biol 101:130–140

Vlès F (1933) Sur une déformation méchanique des œufs d'oursin. Arch de Zool Exp Gen 75:425–463

Yagi K (1961) The mechanical and colloidal properties of *Amoeba* protoplasm and their relations to the mechanism of amoeboid movement. Comp Biochem Physiol 3:73–91

Yoneda M (1964) Tension at the surface of sea-urchin eggs. A critical examination of Cole's experiment. J Exp Biol 41:893–906

Yoneda M (1972) Tension at the surface of sea-urchin eggs on the basis of "liquid-drop" concept. Adv Biophys 4:153–190

Yoneda M (1986) The compression method for determining the surface force. Methods Cell Biol 27:421–434

I.3 Use of Finite Element Methods in Cytomechanics: Study of the Mechanical Stability of the Skeletal Basal Plate of *Callimitra*, a Biomineralizing Protozoan

GUY L. STEUCEK[1,4], ROBERT R. ARCHER[2], and O. ROGER ANDERSON[3]

3.1 Introduction

The morphogenesis and efficient adaptation of organisms depend upon a variety of cytomechanical factors. Biophysical principles play important roles in developmental processes (Green 1980). For example, plant growth is a biophysical phenomenon where the compliance of the cell wall under turgor pressure determines the direction and extent of growth (Taiz 1984; Cosgrove 1986). Growth/remodeling of bones is modulated by the mechanical loads impressed on the bone-depositing cells (Frost 1982). Other examples of cytomechanical problems amenable to finite element method (FEM) analysis are presented in this publication including growth analyses of extracellular skeletons (see Chap. III.3.3).

In this chapter, some analyses of the mechanical integrity of extracellular skeletons are presented toward elucidation of broader questions in cytomechanics that can be answered by FEM analyses. What mechanical loads do cells experience? What is the distribution of mechanical stress within a cell and how do extracellular walls and cytoskeletons bear mechanical loads? What are the consequences of construction mistakes in cellular structures? Is the utilization of cellular structural material optimal in terms of mechanical integrity? How well do cells conserve structural material in view of the metabolic costs incurred in synthesis and fabrication? Is there an alternative architecture (perhaps nonexistent in nature) that is more appropriate for a particular cell type within a specified environmental milieu? One can employ a variety of techniques to answer these questions, including the following: (1) direct observations of deformations within cellular structures exposed to loads, (2) construction and observation of physical models of cellular structures, and (3) creation of theoretical models of cellular structures and computer assessment of the mechanical integrity of these models.

[1] Department of Civil Engineering, University of Massachusetts, Amherst, Massachusetts 01002, USA
[2] Department of Civil Engineering, University of Massachusetts, Amherst, Massachusetts 01002, USA
[3] Biological Oceanography, Lamont-Doherty Geological Observatory of Columbia University, Palisades, New York 10964, USA
[4] Present Adress: Department of Biology, Millersville University Millersville, Pennsylvania 17551, USA

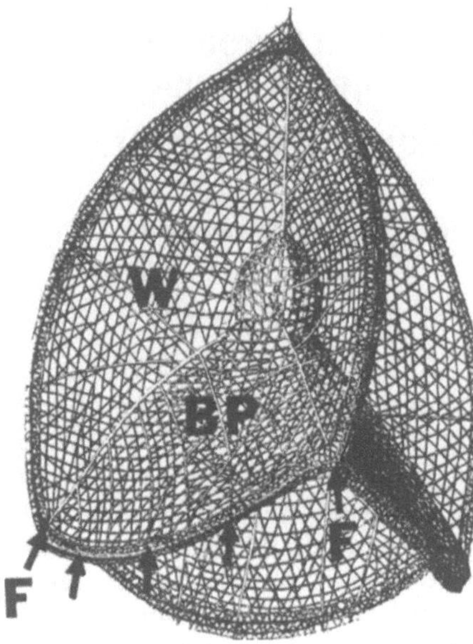

Fig. 1. Drawing of a *Callimitra* skeleton by Haeckel (1881); the cell diameter is 0.15 mm. The basal plate (*BP*) mesh of spars and fibers is bounded by feet which extend from the central cephalic region distally to the *arrows* labeled *F*. *Unlabeled arrows* denote the distal ends of one set of spars. The wings (*W*) are also composed of spars and fibers

The application of the finite element method to mechanics has permitted engineers to study the mechanical integrity of complex structures over the past few decades; these techniques are now being applied to the study of cytomechanics. In contrast to physical models, very complex theoretical models may be constructed relatively easily and they may be readily modified. In addition, a number of software packages are available for finite element analysis; some have been adapted for use on personal computers (Falk and Beardsley 1985).

The purpose of this work is to illustrate how one may use a finite element method to investigate the mechanical integrity of a cellular structure; i.e., the basal plate (BP, Fig. 1) of the siliceous, latticed, extracellular skeleton produced by *Callimitra*, a radiolarian occurring in marine plankton.

3.2 *Callimitra* Architecture

Haeckel's (1881) drawing depicts the siliceous skeleton of *Callimitra* (Fig. 1). Three winglike plates with arcuate margins and three basal plates (BP), composed of a meshwork, join the central capsular skeleton (central latticed helmet-shaped structure) to form the structure of a spherical tetrahedron. The detail of this intricate architecture is illustrated well with scanning electron microscopy (Anderson 1983; Robinson and Goll 1978). The central capsular skeleton, surrounding the cell body, is ca. 50 μm in diameter and the wings and basal plates, supporting rhizopodial strands in the living organism, extend approximately 100 μm beyond the cell body. Three feet attached to the central capsular region

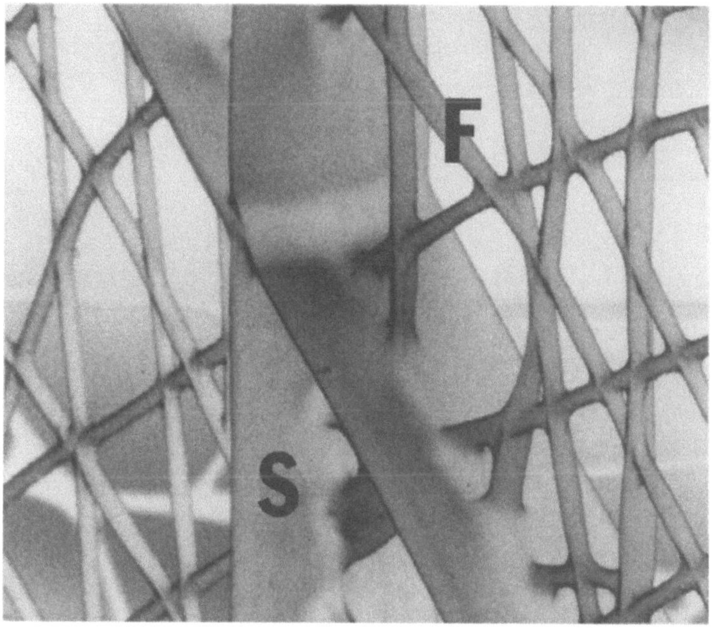

Fig. 2. Scanning electron micrograph of a portion of the *Callimitra* basal plate illustrating the intersection of two spars (*S*) and fibers (*F*); spars are 4 μ in depth

form a tripod and a BP is produced by the mesh supported between two of the feet; hence, two feet and the BP are all in one plane (Figs. 1 and 3a).

The feet supporting the BP form an angle of ca. 60° between their longitudinal axes. Mesh fibers suspended between the feet have a circular cross-sectional diameter of 0.5 μm and extend ca. 5.5 μm from one intersection in the mesh to the next one, where either another fiber, a spar, or a foot intersects (Fig. 2). There are three sets of fibers that compose the BP. One set of fibers is parallel to one foot, the second is parallel to the other foot, and the third set of fibers extend from one foot to the other, joining the feet at equal distances from the cell body (Fig. 3a). These three sets of fibers generate a mesh with a pattern of equilateral triangles. Approximately 26 fibers comprise each set within a BP.

Within the BP meshwork, three sets of spars (thickened strands appearing as bars arranged parallel to the feet) occur at locations 9, 16, and 22 fibers from the central cell body (Figs. 1 and 3a). Spars have a rectangular cross-section (0.5 μ × 4.0 μm) with the broad side oriented normal to the plane of the BP (Fig. 2).

3.3 Finite Element Approach

The problem addressed by FEM is: how do the spars influence the distribution of stress in a BP when a normal load is applied? We could write equations to

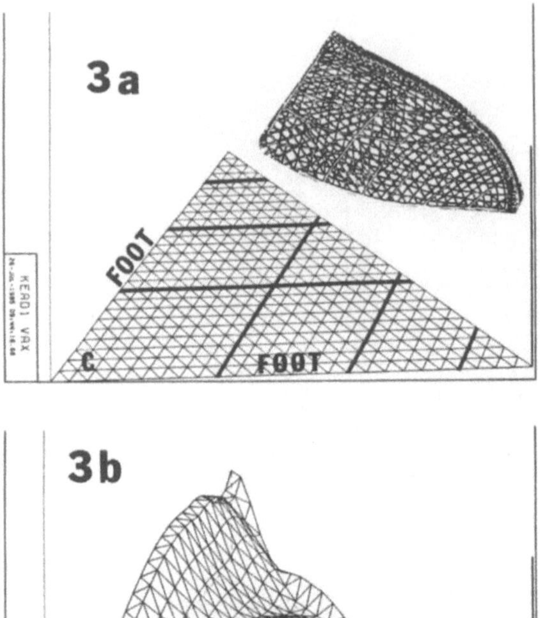

Fig. 3a. A view of a FEM model of the *Callimitra* BP in 3-D space. The cephalic region (*C*) and feet bounding the BP are labeled; the *inset* at the *upper right* is a portion of the drawing from Fig. 1 labeled BP, it is included for orientation. BP spars are represented by *heavy lines* and fibers by *thin lines;* all elements of the BP are in one plane. Each line segment represents an element of the FEM model; line segments, finite elements, join at nodes. The maximum stress at a node is plotted on the vertical axis and is zero for all nodes in this figure since the BP was loaded. The vertical axis is plotted with its origin at the apex of the cephalic (*C*) region. The *heavy bar* at the *right* represents 200 MPa. The distribution of maximal stress at each node can be ascertained by comparing this figure with **b** and **c. b** Maximum stress (Z-axis) at each node is plotted as a function of nodal location (X-Y plane) in a BP constructed of fibers only. A load of 500 Pa was applied normal to the plane of the BP by applying a uniform load to each node of the BP. The *heavy bar* at the *right* represents a stress of 200 MPa. **c** Maximum stress (Z-axis) at each node is plotted as a function of nodal location (X-Y plane) in a BP constructed of fibers and spars as depicted in **a.** A load of 500 Pa was applied normal to the plane of the BP by applying a uniform load to each node of the BP. The *heavy bar* at the *right* represents a stress of 200 MPa

describe the distribution of stress within the BP of *Callimitra*; however, their solution by exact methods would be a formidable task. Through the process of discretization, the rather complex structure of the BP is divided into smaller, manageable units: a finite number of elements. These can be segments of linear components within a structure (e.g., segments of skeletal fibers and spars as occurs in the BP of *Callimitra*) or two- or three-dimensional geometric forms identified within the structure as explained in Chapter III.3.3. Material properties, equations for elements, and boundary conditions are brought into a global assemblage; solutions for primary unknowns, such as nodal displacements, are then obtained. Elemental stresses can subsequently be determined. Introductions to finite element methods have been published by Becker et al. (1981), Desai (1979), and Reddy (1984).

While commercial FEM programs are able to generate node locations for a great variety of structures, we used simple FORTRAN programs to define node locations and constraints, elements, and loads. This approach permitted flexibility in the study of modifications of the BP; i.e., spars could be converted to fibers, fibers and/or spars could be deleted from the BP structure, etc. Elements (1053) were defined as segments of spars and fibers and were 5.5 μ in length. Nodes (378) were defined as locations where fibers/spars intersected as illustrated in Fig. 3a. The line segments in Fig. 3a represent the elements, either fibers as thin lines or spars as thick lines. The BP was loaded by applying a 500 Pa force normal to the plane of BP with appropriate forces on the nodes.

The commercial program ANSYS (DeSalvo and Swanson, 1985) was used to conduct finite element analysis with spar and fiber segments defined as three-dimensional beams. ANSYS computed components of stress for all elements and these were then used to compute the maximum stress over the set of elements at each node. The maximum stress for each set of elements at each node was then plotted as a function of nodal location for BPs without spars (Fig. 3b) and with spars (Fig. 3c).

Locations of high stress represent regions where one would expect failure within the BP structure. When loaded normal to the plane of the BP, the maximal stresses are found in elements located along the leg toward the periphery of the BP (Fig. 3b). Major stress is indicated by the displacement of the peripheral edge of the triangular mesh exhibiting an undulating margin, while the region at the vertex (representing the union of the two feet) is less stressed. Note also the deviation from a planar configuration of the unloaded model (Fig. 3a). These stresses are reduced considerably with the introduction of spars (Fig. 3c), resulting in a form more closely approximating the planar organization in Fig. 3a. Therefore, the spar architecture contributes to the mechanical integrity of the BP. This effect is more obvious when, for illustrative purposes, buoyancy forces of an attached photosynthetic symbiont (algal cell held in the rhizopodial net) is simulated with a concentrated load in the central region of the BP. These data clearly illustrate the insights obtained by FEM analysis coupled with model alterations to determine the architectural significance of components within a biological structure. Other detailed analyses of the cytomechanical significance of *Callimitra* skeletal architecture are currently under investigation. Some biometric data for the organization of fiber patterns in the mesh of wings of *Callimitra* and the possible ordering role for the spars during skeletal morphogenesis have been presented by Anderson (1986).

3.4 Further Applications of FEM and Their Implications

The complexity of phenomena in cell biology and structural engineering along with the relatively recent development of finite element techniques and software has provided little opportunity for substantial applications of FEM to cytomechanics. Among some of the current applications, Sugihara (1985) has studied the deformation of red blood cells in a shear flow. Pitt and Davis (1984) have modeled aggregates of parenchyma cells in fruits and vegetables. In addition, Cooke et al. (1976) studied the mechanics of guard cells, which regulate gas exchange between a plant and the atmosphere. More recently, applications have been made .to the growth of cells, tissues, and mineralized products (Skalak et al. 1982; Moss et al. 1985; Chap.III.3.3). The well-developed theoretical base for FEM analyses and the clear acceptance of the validity of the method in engineering analyses provides a sound background for applications of these techniques to cytomechanical problems. The availability of commercial packages for finite element analysis will greatly assist the biologist. However, it takes time to become familiar with the packages and much effort will be devoted to the definition of the cytomechanical problem; help from engineers is invaluable here. While default procedures enhance the utility of commercial packages for the experienced user, they could prove frustrating to the novice. Graphic displays aid comprehension of models with many elements and one may discover that these facilities are not available with an FEM package.

The definition of realistic loads represents a problem with this approach in the investigation of cytomechanics. Where direct observations do not exist, assumptions must be made in order to proceed with an analysis. However, it is a simple matter to modify loads and study the influence of a variety of loading situations with this modeling method. Some critics may note that this approach is used to design large structures such as buildings and therefore question its utility in dealing with cellular structures. While the analytical technique is not influenced by scaling, the definition of elements could influence results and should receive careful consideration.

As analytical facilities become more available, the finite element method will receive more attention when dealing with cytomechanical problems. In addition, this approach will open the way to other aspects of cellular engineering such as the study of transport within and between cells.

Acknowledgments. Technical assistance of Kai Bletzinger, Dee Breger, and Doris Peterson was invaluable. Support by SFB 230, NSF OCE 82-07495, and Millersville University was greatly appreciated. Availability of ANSYS at the University of Massachusetts and the Universität Stuttgart is gratefully acknowledged.

References

Anderson OR (1983) Radiolaria. Springer, Berlin Heidelberg New York

Anderson OR (1986) Biomineralization in Radiolaria: Silica Deposition and Ontogenetic Origins of Form. In: Leadbeater BSC, Riding R (eds) Biomineralization in lower plants and animals. Oxford Univ Press, pp 375–391

Becker EB, Carey GF, Oden JT (1981) Finite Elements: an introduction, vol 1. Prentice-Hall, Englewood Cliffs, N.J.

Cooke RJ, DeBaerdemacker JG, Pond RH, Mang HA (1976) A finite element shell analysis of guard cell deformations. Trans ASAE 19:1107–1121

Cosgrove D (1986) Biophysical control of plant growth. Ann Rev Plant Physiol 37:377–405

Desai CS (1979) Elementary finite element method. Prentice-Hall, Englewood Cliffs, NJ

DeSalvo GJ, Swanson JA (1985) ANSYS engineering analysis system user's manual. Swanson Analysis Systems, Houston, PA

Falk H, Beardsley CW (1985) Finite element analysis packages for personal computers. Mech Eng 107:54–71

Frost HM (1982) Mechanical determinants of bone modeling. Metab Bone Dis Relat Res 4:217–229

Green PB (1980) Organogenesis – a biophysical view. Ann Rev Plant Physiol 31:51–82

Haeckel E (1881) Entwurf eines Radiolaren-Systems auf Grund von Studien der Challenger-Radiolarian. Jenaische Z Naturwiss Ser 15:418–472

Moss ML, Skalak R, Patel H, Sen K, Moss-Salentijn L, Shinozuka M, Vilmann H (1985) Finite element method modeling of cranio-facial growth. Am J Orthod 87:453–472

Pitt RE, Davis DC (1984) Finite element analysis of fluid-filled cell response to external loading. Trans ASAE 27:1976–1983

Reddy JN (1984) An introduction to the finite element method. McGraw-Hill, New York

Robinson WJ, Goll RM (1978) Fine skeletal structure of the radiolarian *Callimitra carolotae* Haeckel. J Micropaleontol 24:432–439

Skalak R, Dasgupta G, Moss M, Otten E, Dullemeijer P, Vilmann H (1982) Analytical description of growth. J Theor Biol 94:555–577

Sugihara M (1985) Motion and deformation of a red blood cell in a shear flow: a 2-dimensional simulation of the wall effect. Biorheology 22:1–20

Taiz L (1984) Plant cell expansion: regulation of cell wall mechanical properties. Ann Rev Plant Physiol 35:585–657

I.4 Mechanics and Hydrodynamics of Rotating Filaments[1]

ROBERT JAROSCH[2]

4.1 The Molecular Basis of Filament Rotation

Microtubules (∅ 25 nm), microfilaments (7 nm), and bacterial flagellin filaments (12 to 19 nm) are the most prominent fibers involved in cell motility. Their motile function relates to the interaction with associated proteins. Many of them are minifilaments (2 to 4 nm; e.g., spectrin, filamin, actin-binding protein, myosin), and have a high α-helical content. Also, intermediate filaments consist of coiled-coils twisted into ropes (∅ 10 nm). Minifilaments seem to interconnect all cytoskeletal elements forming a "cytoplasmic matrix" (for review see e.g. Schliwa 1986). Model experiments with elastic helices that show mechanical properties similar to those supposed for protein helices are important for the comprehension of the mechanical characteristics of these filaments. Models allow the simulation of many cell motility phenomena on a macroscopic scale, for example, torsional force production by a single α-helix (Fig. 1) by changing the pitch height (Jarosch 1963, 1964a, 1969; Fraser and MacRae 1973). The tubular helix model shown in Fig. 2 demonstrates impressively the torsional rotation (arrow c) after a pitch decrease. Pitch changes of helices are induced, for instance, by pH changes close to the pK due to the appearance or disappearance of intrahelical electrostatic repulsion between the charged side chains. The repulsion forces may be strong enough to break α-helical H-bonds, e.g., of polyglutamic acid, if the pH >2. The torsional force produced by the mot proteins of bacterial flagella is likewise proton controlled (Macnab 1983). A conformational change of an α-helix 52-turns-long (ca. 28.3 nm) can produce one torsional rotation. Therefore, an α-helical filament of 1-μm length may easily produce 30 rotations (Jarosch 1969).

Certain proteins forming the stable coiled-coil configuration show the ability of transformation of chemical to mechanical energy on addition (or removal) of Ca^{2+}. Figure 3 shows a model of energy transformation by a conformational change-mediated torsion of a coiled coil. The striking length change as with the single α-helix (Fig. 2) does not occur (for details, see Jarosch 1979, 1985a). Additional Ca-binding proteins like calmodulin may function as a Ca^{2+} trap or buffer and may support the regulation by a specific conformational change. The number of rotations n performed by the free end of a coiled coil, if the other end is fixed should be:

---

[1] Dedicated to my late friend Bob Allen (1927–1986). The results of his video enhanced microscopy encouraged the author to write this chapter.

[2] Institut für Pflanzenphysiologie der Universität Salzburg, Austria

$n = l/p_1 - l/p_2$, where l is the length of the coiled coil, p_1 is its smaller, p_2 its larger pitch. As a consequence of the torsion, the filaments twist and may self-intertwine ("telephone cord phenomenon", e.g. typical for spectrin) or may intertwine with other filaments as in the formation of intermediate filaments. Twisted filaments associated with the protofilaments of microtubules or with microfilaments may rotate these coarser cytoskeletal elements (arrow CC) by winding and unwinding as shown in Fig. 4. This mechanism, and not a power stroke of crossbridges, is supposed to be the action of the ATP-dependent mechanoenzymes (Jarosch and Foissner 1982; Jarosch 1986b). However, force production and ATP splitting may also be separated and a passive winding of associated minifilaments may occur along the cytoskeletal elements that couples other enzymatic work energetically with the ATP splitting. Thermody-

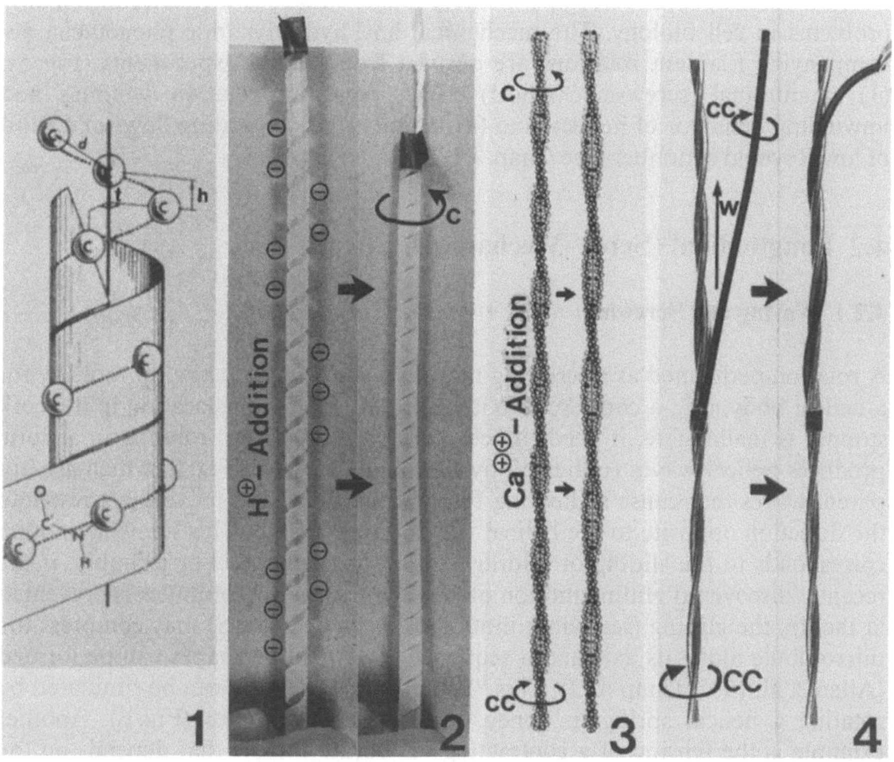

Fig. 1. Simplified tubular α-helix (Fraser and McRae 1973). Every change of the unit height *h* is connected with a change of the unit twist *t*

Fig. 2. A pitch decrease of a helical plastic tube causes a torsional rotation as indicated by the *arrow c*. This corresponds to the conformational change of a negatively charged α-helix after a decrease of the pH value

Fig. 3. A conformational change is shown and torsional rotations (*arrows c* and *cc*) may occur at a coiled-coil filament after Ca^{2+} addition

Fig. 4. The torsional rotation (*arrow cc*) of a twisted rubber string turns (*arrow CC*) a double helix by winding. The double helix represents a microfilament

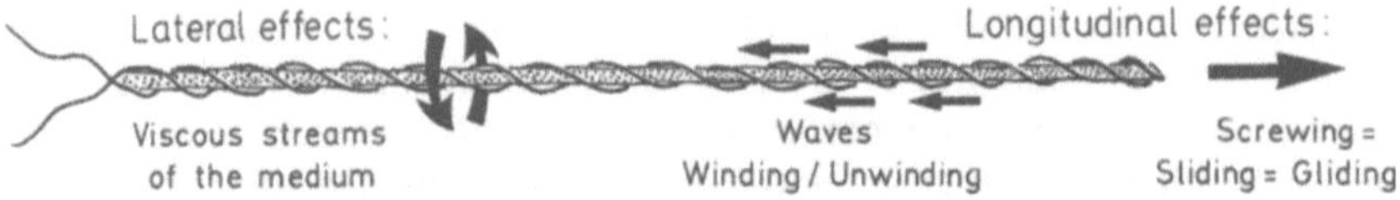

Fig. 5. Diagram of the main effects produced by the rotation of a helical filament. For details, see text

namic fluctuations and their high number statistics, which are important for binding kinetics, are not directly involved here, because the rotating helical molecules move regularly and directionally like threads in a spinning mill. Determinated molecular motions of this kind were already postulated for living systems by Schrödinger (1944).

The filament rotation model is like a magic wand for the solution of many problems in cell biology. The mechanical and hydrodynamic phenomena accompanying filament rotations are derived from model experiments (Fig. 5): (1) longitudinal (screw-mechanical) effects regarding rotation, winding, and unwinding behavior of helices, and (2) lateral effects regarding flows of a fluid of low Reynold's number (see Chap. I.1.2).

4.2 Longitudinal (Screw-Mechanical) Effects

4.2.1 Waving and Screwing

A rotation performed as a screwing motion is the adequate kind of motility for a helical body, e.g., a corkscrew. Without rotations its translocation in the cork stopper is inadequate, it needs much more force. A screw, rotating in a fluid, produces helical waves (indicated by the small arrows in Fig. 5) which are apparent waves that cause a shearing force in the fluid, and move the screw into the direction opposite to the helical waves (large arrow). This screwing motion corresponds to the sliding or gliding motion of filaments. For example, if the recently discovered gliding motion of isolated axonal microtubules is prevented at the tip, the gliding (screwing) motion of its end (arrow S) may compress the microtubule along its axis and a sequence of serpentine shapes can be formed (Allen et al. 1985, Chap. II.2). This "fishtailing" (arrow F) can be simulated by rotating a helical spring in honey or between two sheets (Fig. 6). Another example is the tension of a contracting vertebrate muscle that depends on the shearing force produced by the waves of the thin filament helices that are directed to the Z-disks if the muscle is kept isometrically. The isotonic contraction corresponds to the counterdirected screwing motion of these filaments (Jarosch 1979, 1986a). The high number of myosin crossbridges that contact the thin filaments of an activated muscle forms an abutment for the thin filament double helices preventing slipping. If they slip, the crossbridges have to cycle passively (Jarosch and Foissner 1982).

On the other hand, displacement of a screw by pushing or pulling it through a solid or viscous medium is connected with passive rotations. A simple conse-

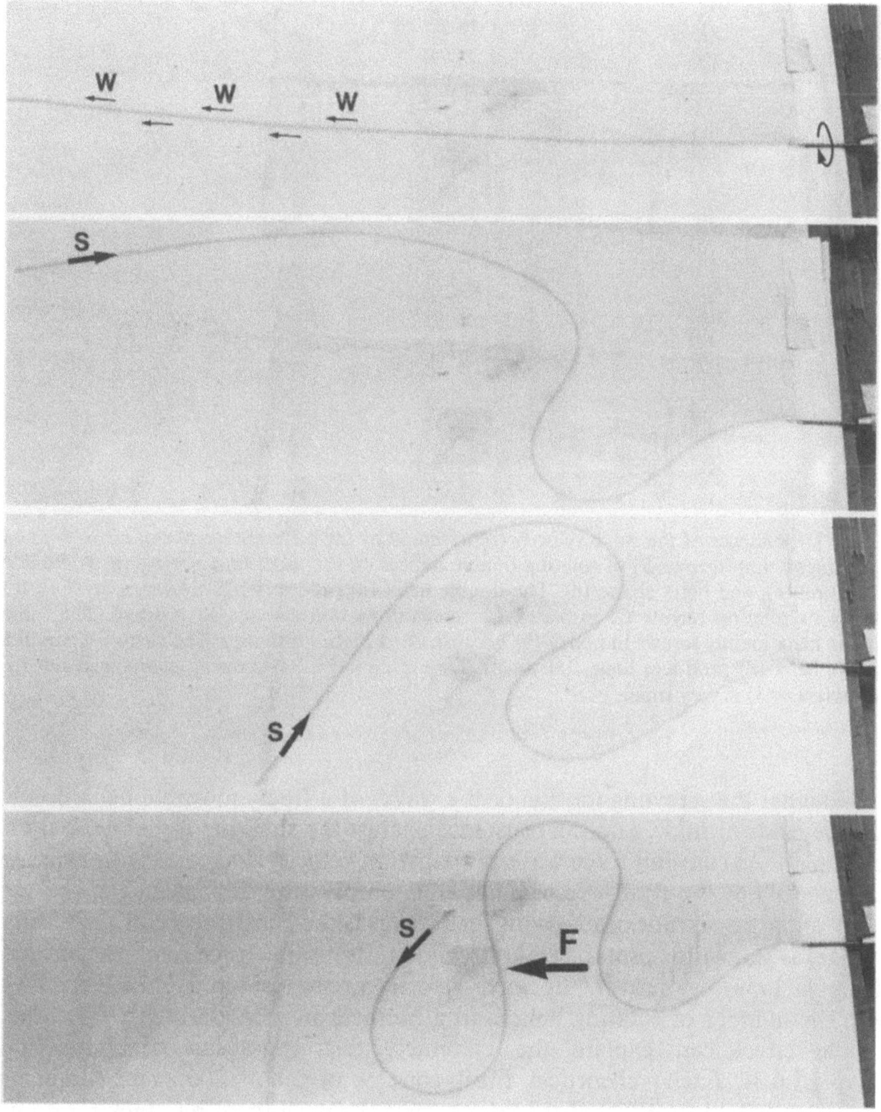

Fig. 6. An electromotor-driven helical spring rotates between a plastic sheet and a cloth substratum. It simulates the "fishtailing" motion (*arrow F*) described for single axonal microtubules when their gliding motion is prevented by an obstacle. The *arrows W* indicate the direction of the helical waves, *arrow S* the resulting screwing motion of the free helix end

quence is the stretching activation of a muscle. Every stretch produces passive rotations of the thin filaments. Since these are anchored unipolar to the Z-disks, the passive rotations cause an increase of their torsional force ("force enhancement") and a change in the structure of the Z-disks by winding the connecting α-actinin filaments (Jarosch 1985b, 1986a).

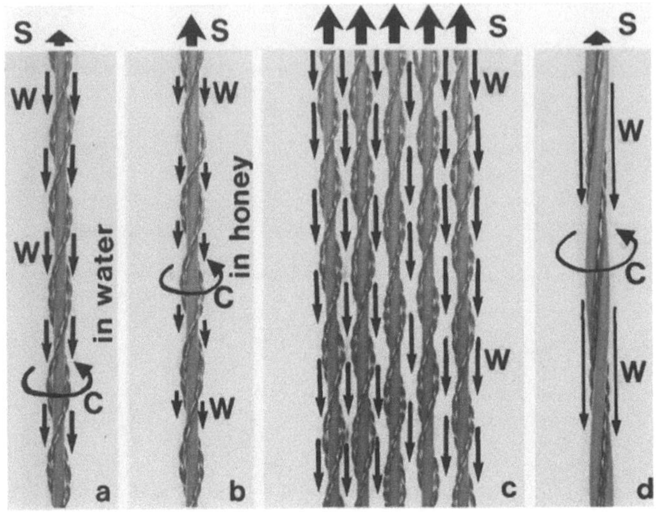

Fig. 7. Dependence of the wave velocity (represented by the different lengths of *arrow W*) and screwing motion (*arrows S*) of rotating double helices on the medium viscosity (**a, b**), bundle formation (**c**), and helix shape (**d**). The double helix (**a**) produces mainly waves (*arrows W*) during its rotation (*arrow C*) in water. Its screwing motion (*arrows S*) is small. The same double helix mainly screws in honey (**b**) because of its higher viscosity. The steep F-actin-like double helix (**d**) produces long, flat waves (*long arrows W*). Its screwing component per rotation (*arrow S*) is very small

Whether the screwing motion or the waves of a freely movable helix dominate depends primarily on the resistance against the shearing force in the fluid (cf. Fig. 7). At constant force wave propagation velocity decreases with increasing viscosity of the fluid. Accordingly, the temperature dependence of the velocity of particles translocated with such waves is the consequence of a viscosity change, as shown for protoplasmic streaming. Its motive force remains constant (Kamiya 1959). Moreover, the wave velocity (arrows W in Fig. 7c) increases with the number of rotating helices in a filament bundle (Jarosch 1972). This bundle effect can explain the extremely fast characean streaming (cf. Chap. I.1.6.3). Each subcortical fibril consists of 50 to 200 actin filaments (Nagai and Rebhun 1966) that may produce shearing force by quick rotations (up to 200 s^{-1}, Foissner and Jarosch 1981).

In addition, bundles of rotating filaments can produce waves visible by light microscopy, when for example the filaments form superhelices that rotate in phase, e.g., bacterial flagella (Jarosch 1972), actin filaments in filopodia of certain amoebae (Bovee 1964; Jarosch 1971), and the "false discharge" of the *Limulus* sperm acrosomal process (Tilney 1975), or if straight bundles form angles or bendings, e.g., the F-actin rings and polygons of characean cytoplasm (Jarosch 1956, 1976; Kuroda 1964; Foissner and Jarosch 1981; Foissner 1985) and microtubule bundles of axostyles (Mc Intosh 1973; Langford and Inoué 1979). As postulated (Jarosch 1968) the doublet microtubules of eukaryotic flagella turned out to be superhelices (Costello 1973; Zobel 1973) that produce the helical or planar bending waves, most probably by torsional rotations (Jarosch 1986c).

4.2.2 The Oscillation

The quick rotation of helices (and tubules) produces undampened oscillations and standing waves of a frequency that corresponds to the rotation number (Jarosch 1964b, 1966). This causes a dynamic self-stabilization that may prevent the Brownian motion of rotating filaments and their attached particles. Similar standing waves occur along quickly rotating bacterial flagella and spirochetes and can be demonstrated by dark-field microscopy (Jarosch 1967). A mechanical contact results in harmonic oscillations propagating very quickly in both directions. Thus, the oscillations provide a mechanical basis of irritability.

4.2.3 Control of Polymerization and Depolymerization

A noncontrolled polymerization of microtubules and microfilaments is possible if the concentration of subunits is high. However, their in vivo elongation is limited and can be explained by winding up of associated proteins that carry subunits (Jarosch and Foissner 1982). The in vivo elongation of microtubules is either center-controlled (Brinkley et al. 1981) and may occur by addition of subunits in the center, or it is tip-controlled and may occur by the addition of subunits at the distal ends, e.g., of ciliary microtubules when "pluglike" structures are present (Witman 1975). Both cases are simulated by the model of Fig. 8 for three growing protofilaments of a microtubule. The motion of the rods corresponds to the tubulin flux (arrows T_{FL}) or treadmilling. The electron dense material seen in cell centers may correspond to the MAP tubulin material not yet wound.

The controlled protofilament elongation of microtubules will stop when the winding supply ceases. During center-directed translocations (arrows W_A in Figs. 9 and 10) anchorage at the center (A) prevents the screwing-out motion of microtubules, thus the anchoring filaments have to change their twist. This is important during the poleward transport in asters, during aster migration (cf. Bajer and Molè-Bajer 1982), during pigment aggregation in chromatophores, and during feeding in suctorian tentacles (Foissner and Jarosch 1985; Jarosch and Foissner 1985). The anchorage is also very important during the anaphase A motion of chromosomes that is simulated in the model of Fig. 10. Here, the controlled depolymerization of the kinetochore microtubules is assumed to occur at their kinetochore ends (Gruzdev 1972; Forer 1976, Gorbsky et al. 1986) by unwinding of tubulin-complexed helices (MAPs) from the protofilaments. Their rotations (arrows C) cause the poleward-directed motion (arrows W_A) of chromosomes and of particles situated between the chromosome and the pole. The opposite process is shown in Fig. 8c.

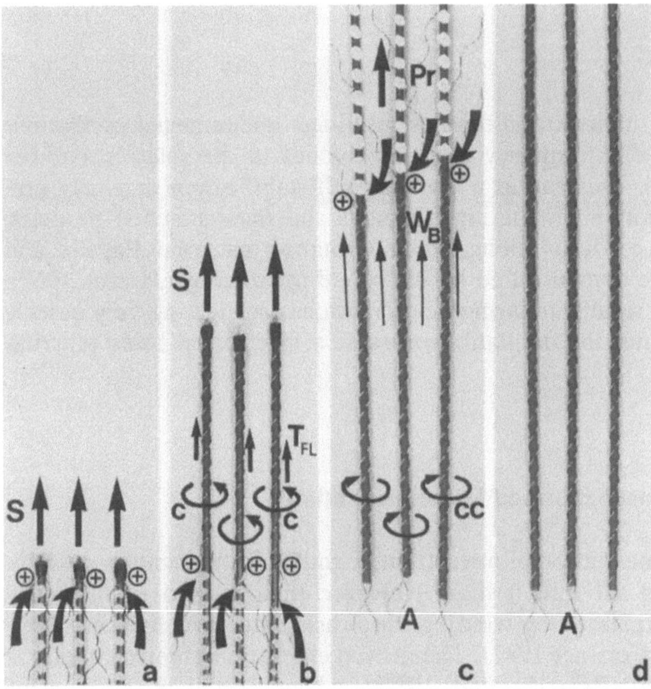

Fig. 8a–d. Model for the translocational polymerization of a microtubule. The elongation of three protofilaments of a growing microtubule is simulated by winding (*curved arrows*) of two helices per rod in the center (**a, b**) or at the tips (**c, d**). The helices represent tubulin-complexed MAPs. The protofilament rotation (*arrows C* and *CC*) causes either screwing into the cytoplasm (*arrows S*), if the screwing component is distinct as with the helix of Fig. 7b, or progression of the tips (*arrow Pr*) if the screwing component is small (cf. Fig. 7d). The center may also represent an obstacle for the center-directed screwing motion. *A* indicates an anchorage at the center. The part of the rods not yet wrapped is punctuated

Less controlled polymerization and depolymerization may occur also along the whole length of the microtubules (Inoué and Ritter 1975) by lateral winding and unwinding of tubulin-complexed MAPs, e.g., via crossbridges.

The progressive motion (arrow Pr in Fig. 8c) is strongly promoted if the screwing component of the rotating filament is relatively small as with the F-actin double helix (Fig. 7d). The winding up of membrane-associated minifilaments like spectrin or α-actinin (cf. Bereiter-Hahn et al. 1981) that carry actin subunits or oligomeric actin would result here in a progressive motion of the cell membrane (arrow Pr) as it is shown in the model of Fig. 11, since the winding along the F-actin grooves proceeds without a corresponding screwing motion. Such a cell progression on the basis of a "translocation polymerization" of actin filaments was already proposed and discussed in detail by Lindberg et al. (1981) who wrote: "The double helical nature of the actin filaments would be

9
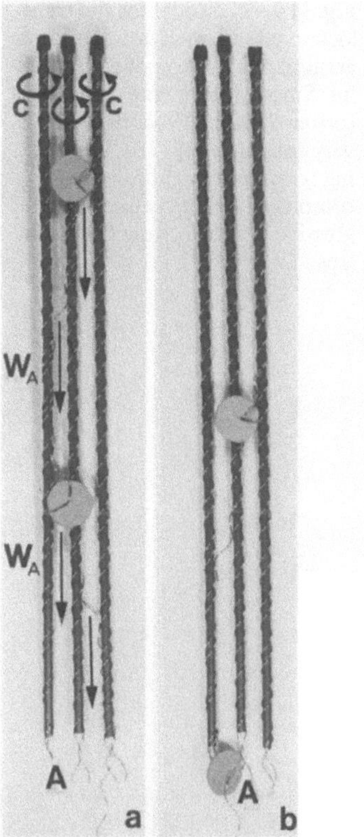

10
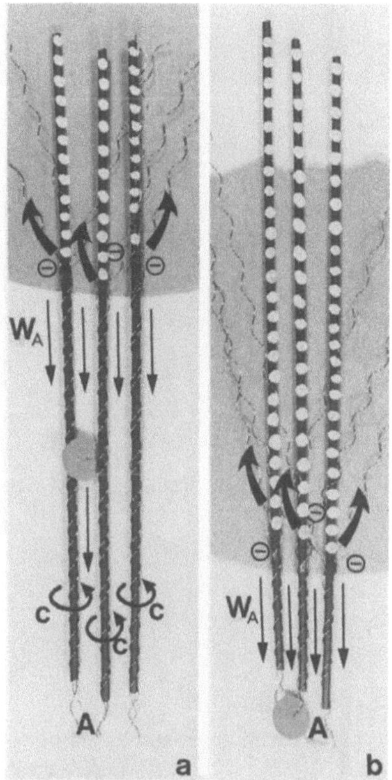

Fig. 9a, b. Simulation of the center(poleward)-directed transport of particles along protofilaments of a spindle microtubule

Fig. 10a, b. Simulation of the anaphase A: Unwinding (*curved arrows*) of tubulin-complexed helices (MAPs) from protofilaments of a kinetochore microtubule *at the kinetochore end* causes their depolymerization (indicated by *punctuation*) at the kinetochore end and poleward transport (*arrows* W_A) of the chromosome and a particle

crucial to the polymerization mechanism." Note that the direction of the waves (arrows W_A in Fig. 11) that may transport subunits to the tip (cf. Tilney and Kallenbach 1979) and the direction of the progression (arrow Pr) is the same. This model simulates the elongation and polymerization of actin filaments from the membrane-associated ends, e.g., in differentiating *Limulus* spermatids (Tilney et al. 1981), in the acrosomal process of *Thyone* sperm (Tilney and Kallenbach 1979; Tilney and Inoué 1982), and in microvilli (Mooseker et al. 1982). It simulates the elongation of filopodia, microspikes , and the motion of the leading edge (leading lamella) of glia cells, fibroblasts (Abercrombie et al. 1970; Lindberg et al. 1981; Small and Langanger 1981; Wang 1985), and epithelial cells (Strohmeier and Bereiter-Hahn 1984). Figure 12 shows a simplified functional model for a creeping monopodial amoeba (cf. Fig. 11). It is assumed here that associated minifilaments polymerize the actin filaments by

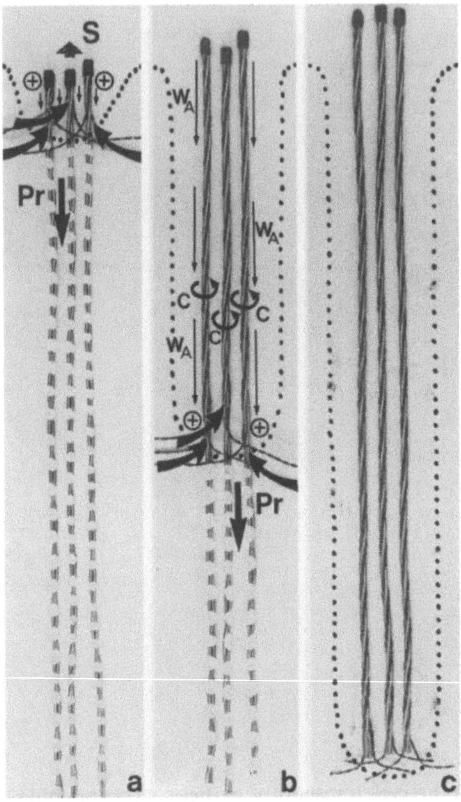

Fig. 11 a – c. Model for the translocational polymerization of three actin filaments, e.g., of a microvillus. Since the screwing component (*arrow S*) of the F-actin helices is very small (cf. Fig. 7 d), the winding (*curved arrows*) of G-actin-complexed helices causes the progressive motion (*arrow Pr*) of the tips

winding at the plus ends in the front and depolymerize them by unwinding at the minus ends in the tail. It seems that the rotation of the actin filaments is mainly propelled by myosin, since it is found in the cell body (e.g. Allen and Taylor 1975; Strohmeier and Bereiter-Hahn 1984). The permanent unwinding may produce a collapse or a solation of the actin filaments in the tail and, therefore, its continuous "contraction." Many observations in the literature are consistent with that model, e.g., the demonstration of the rotational component in the motion of amoebae (Goldacre 1956; Bovee 1964; Jarosch 1971; Nowakowska 1978; Bynum and Allen 1980) or the "rolling motion" of their membrane (Abé 1962, cf. the locations of the numbered particles in Fig. 12).

4.2.4 The Translocation of Particles

A helix rotating in a fluid moves a layer of the surrounding fluid with small particles in helical paths (Jarosch 1966). This fluid motion – also around a

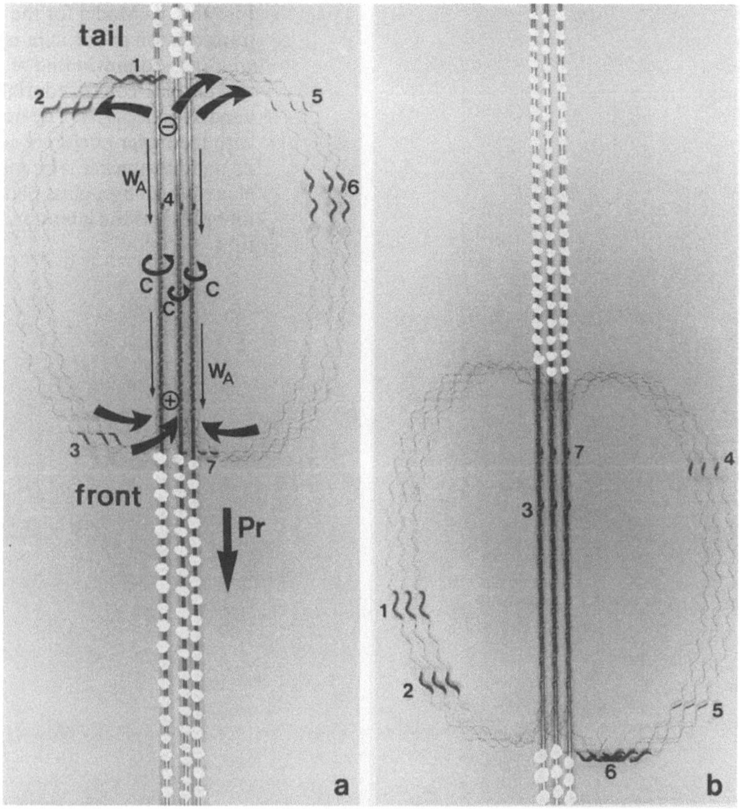

Fig. 12 a, b. Simplified model for a creeping amoeba. The translocational polymerization of actin filaments by winding of helices in the front or leading edge (cf. Fig. 11) is combined with unwinding or depolymerization in the tail. For more details, see text

bundle of rotating helical filaments – may correspond to the microstreams postulated in axoplasmatic transport along microtubules (Weiss and Gross 1982). The lateral motion of a particle crossbridged to one filament of the bundle is prevented by the other filaments of the bundle. Its translocation results in a motion parallel to the fibril. The translocation occurs by winding or unwinding of the connecting helices (Fig. 13). Actin filaments obviously support particle transport by winding and unwinding of myosin (cf. Sheetz and Spudich 1983; Shimmen and Yano 1984; Kachar 1985; Higashi-Fujime 1985), microtubules, probably their protofilaments, by dynein, kinesin (cf. Vale et al. 1985) or other associated proteins. It was surprising that particles can move simultaneously in both directions on a single microtubule and can pass each other without apparent collision (Hayden and Allen 1984; Allen et al. 1985; Schnapp et al. 1985; Schroer and Kelly 1985; Koonce and Schliwa 1985). The counterdirected motion is simulated by using right(R)- and left(L)-handed helices (Fig. 14), but passing and overtaking of particles is not possible on a single rotating element. Therefore, I have suggested that the particle-connected helices may contact and

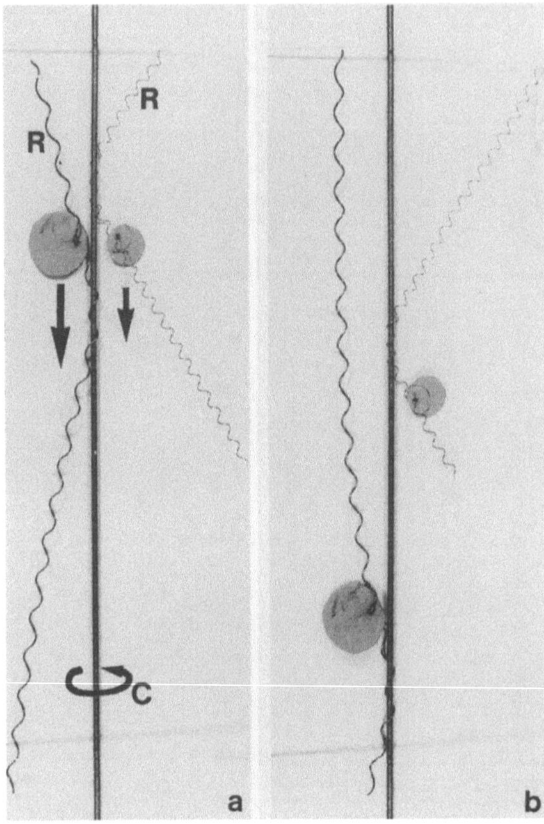

Fig. 13 a, b. Model for the translocation of particles by winding and unwinding of two particle-connected, right-handed helices (*R*). The helix with the larger pitch moves its particle quicker. The model is covered by a glass plate that prevents the lateral motion

rotate different protofilaments of a microtubule (Jarosch 1986 b, Fig. 15). Accordingly, a change in the direction of translocation is possible either by changing the rotational direction of the protofilaments or by translocases that coil in opposite directions.

Isolated axonal microtubules can glide on glass surfaces and move particles preferentially into the direction opposite to gliding. In aged preparations particles frequently move in the same direction (Allen et al. 1985). This is simulated (Fig. 16) by a left-handed "translocase" (L) that moves a particle along a single protofilament in the same foreward direction (arrow W_B) that is performed as a screwing motion (arrow S) by the right-handed helices of most protofilaments.

The MT concept as a bundle of rotating protofilaments also includes the possibility of a medium and small particle transport within the lumen (cf. Burton 1984).

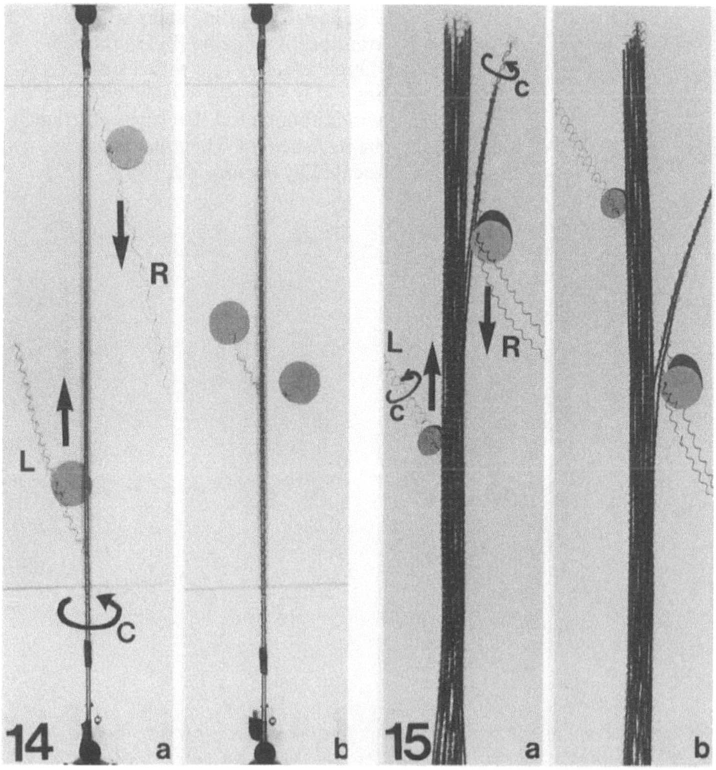

Fig. 14 a, b. Particles, connected with left-handed (*L*) and right-handed (*R*) helices that are wound by the rotating rod, move in opposite directions (*arrows*), but cannot pass each other

Fig. 15 a, b. Bidirectional translocation of particles (*arrows*) that pass each other is possible along a single microtubule, if *L*-helices and *R*-helices that carry particles each wind about different protofilaments

4.2.5 Crossbridges

Crossbridges are believed to break and rebuild during motility events, e.g., between translocated particles and microtubules (Miller and Lasek 1985). However, after quick freezing and deep etching, all crossbridges are attached to filaments and microtubules (e.g. Hirokawa et al. 1984, 1985). This is consistent with the concept that the associated proteins are wound and unwound. According to the models, all crossbridges and side arms move with the helical waves.

The mode of winding depends on the polarity of the filaments and determines the direction of sliding (Jarosch and Foissner 1982). The helices of oblique crossbridges crossing two filaments (Fig. 9) wind directly between near-neighboring filaments that rotate in the same direction. They cause parallel sliding in the same direction. Such crossbridges connect dendritic neurofilaments (Hirokawa et al. 1984). Compound crossbridges as shown in Figs. 17 and

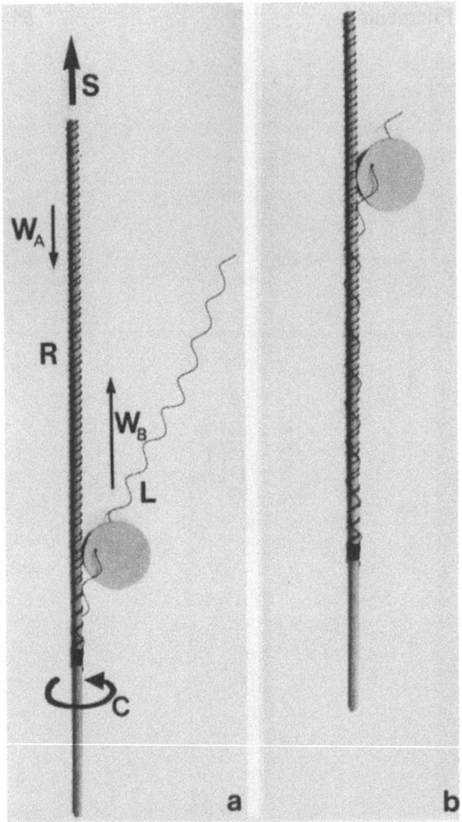

Fig. 16 a, b. Gliding and particle translocation (*arrow* W_B) in the same direction is possible with a single rotating rod if a particle-connected *L*-helix winds about a rod that screws (*arrow S*) by its *R*-helices. Their waves are indicated by the *arrow* W_A

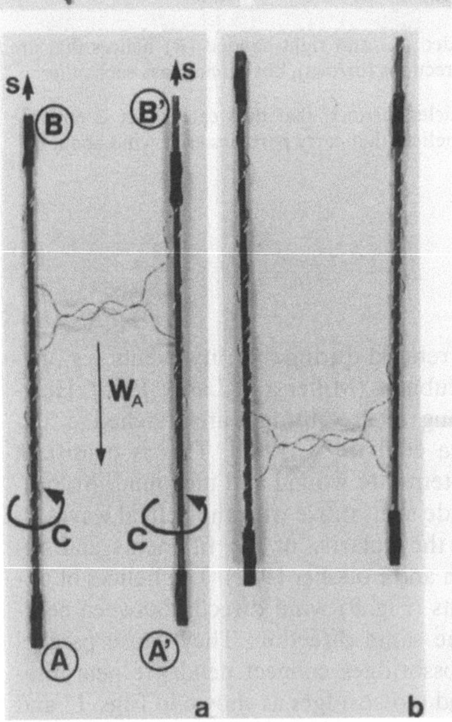

Fig. 17 a, b. Translocation (*arrow* W_A) of a compound crossbridge along two rods rotating in the same directions (*arrows C*)

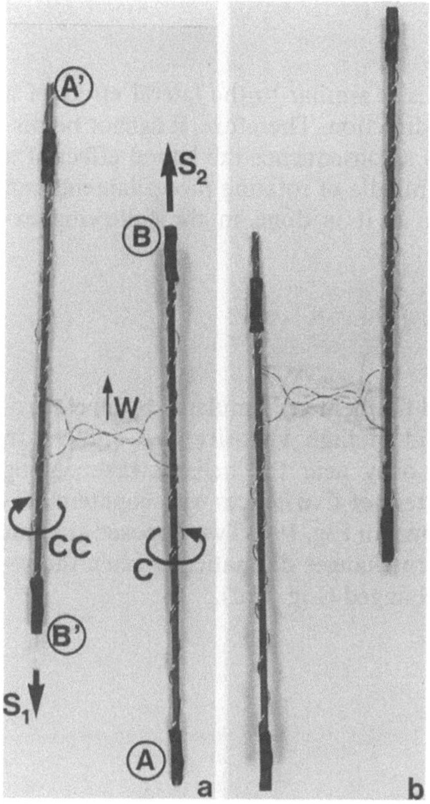

Fig. 18 a, b. Antiparallel sliding (*arrow* S_1 and *arrow* S_2) of two counterrotating rods (*arrow CC* and *arrow C*) that are connected by a compound crossbridge. For more details see the text

18 are perpendicular to the filaments that are spaced farther apart. They are similar to the crossbridges between neurofilaments and neurotubules of axons (cf. Jarosch 1986b). They may cause parallel or antiparallel sliding. Both filaments rotate (arrows C) and slide (arrows S) in the same direction when the helices are wound from A to B′ or from A′ to B (Fig. 17). The crossbridge moves here in the direction of arrow W_A. In Fig. 18 the filaments rotate (arrow C and arrow CC) and slide in opposite directions (arrow S_2 and arrow S_1) because the helices wind from B′ to A or from A′ to B. The waves (arrow W) always run here in the sliding direction of the quicker filament (arrow S_2). These two velocities (V_W and V_{S_2}) may correspond to the two different velocities of particles described along the subcortical characean fibrils (Jarosch 1960; Kamitsubo 1981; Jarosch and Foissner 1983; Kachar 1985) and to the cytoplasmic streaming and the much quicker motion of the "headed streaming bands" in the stalk of *Acetabularia* (Heunert 1978; Koop and Kiermayer 1980).

4.3 Lateral (Hydrodynamic) Effects

The lateral effect of a single rotating helix is similar to the lateral effect of a bundle of helices that rotate in the same direction. Therefore, it cannot be discriminated between these two models. As a consequence the lateral effect of a microtubule which may be assumed as a bundle of rotating protofilaments can be simulated by a single rotating helix as it is done in the following experiments.

4.3.1 Pattern of Flows

Hydrodynamic effects of a rotating helical filament are simulated by an electromotor-driven helix that rotates in a fluid of high viscosity (see Table 1 in Chap. I.1), like honey (Fig. 19a). The honey near the helix is taken along (circular arrow in Fig. 19b). The flow pattern of five helices with counterclockwise rotation arranged in one plane is shown in Fig. 19c. Two opposed streams are produced (arrows St). The flow pattern changes dramatically when the rotational direction of each second helix is changed (Fig. 19d).

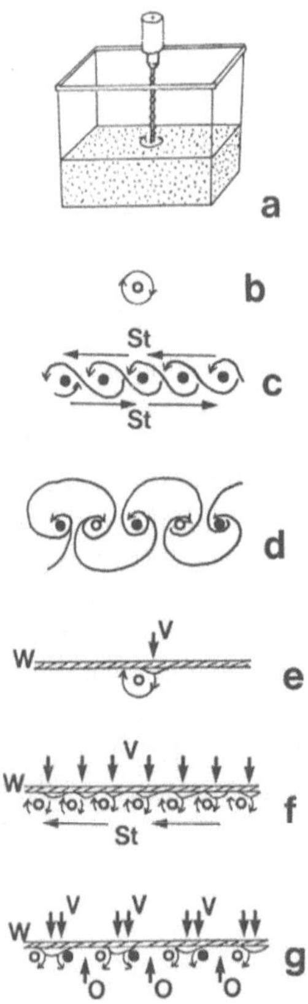

Fig. 19. The lateral flows produced by electromotor-driven helices that rotate in honey are investigated as shown in **a.** **b, c, d** Flow pattern without an adjacent wall. **e, f, g** Flows adjacent to a wall (*W*). The cross-sections of helices that rotate in a clockwise direction are designated in *white,* counterclockwise rotations in *black. Arrows St* streams; *arrows V* sites of low pressure; *arrows O* sites of overpressure. For more details, see text

4.3.2 Flows Adjacent to a Wall

A low pressure (arrow V in Fig. 19e) arises between the rotating helix and a wall (W). The viscous medium is lifted off from the wall. A few helices that rotate in the same direction (Fig. 19f) enhance and enlarge the region of the low pressure and produce a single stream (arrows St). Alternating counterrotation of helices (Fig. 19g) produces a pattern with alternating regions of low pressure (arrows V) and overpressure (arrows O). This streaming behavior may explain the pattern of wall thickening in young plant cells, e.g., of the secondary wall material beneath clusters of microtubules in differentiating xylem elements (Pickett-Heaps and Northcote 1966) or a similar pattern of surface reinforcement in cysts of certain protists (Fig. 20). Schnepf (1973) has suggested for developing *Sphagnum* cells that the plasmalemma between the microtubules and the wall is lifted off against the turgor pressure to form an extraplasmatic space in which wall material is accumulated. That accumulation may also occur during a local reduction of the turgor pressure (cf. Chap. IV.2.3.4).

The model shown in Fig. 21 simulates the function of the preprophase band of microtubules that predicts where the nucleus is situated during its division and where the new cell wall of a dividing plant cell will join the old one.

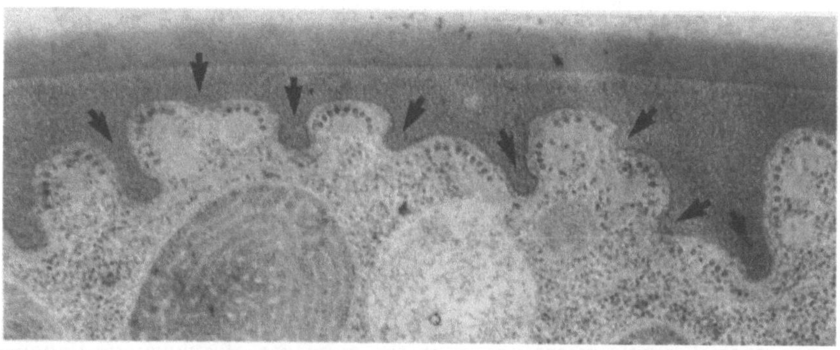

Fig. 20. Effect of cortical microtubules on the wall thickenings of a protozoan cyst (*Kahliella simplex*). The *arrows* indicate possible sites of local underpressure, compare Fig. 19g (Electron micrograph from Foissner and Foissner 1986)

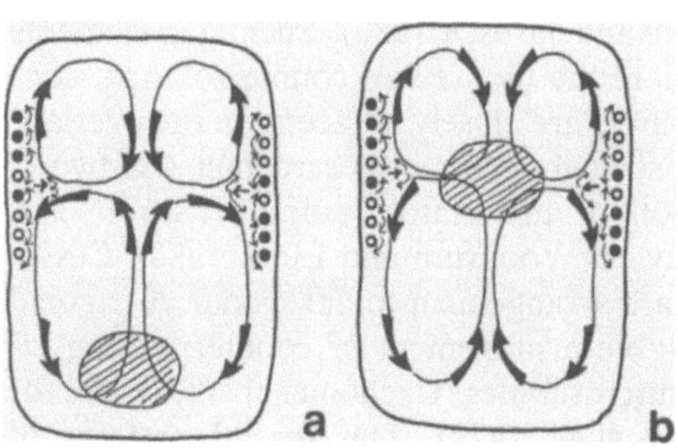

Fig. 21 a, b. Model for the function of the preprophase band of microtubules in young plant cells. Two groups of counterrotating helices produce four large circulating streams (*arrows*) (cf. Fig. 19f). The streams move the "nucleus" excentrically situated (**a**) into a stabilized position in the centre of the preprophase band (**b**)

4.3.3 Flows and Molding of an Adjacent Liquid Surface

Rotating helices close to a liquid surface (Fig. 22a) give rise to an assymetrical rampart and a furrow on the surface (Fig. 22b) which increase when two or more helices rotate in the same direction (Fig. 22c, d). Two counterrotating helices produce either a large, smoothly shaped symmetrical rampart with a stationary surface (Fig. 22e) or a "double roll" with a very thin, fast flowing surface (Fig. 22f). The surface rests if the helices counterrotate side by side (Fig. 22e), or streams if the helices are arranged one over the other (Fig. 22g). Several helices rotating in the same direction produce a deep furrow (Fig. 22h) and may thus simulate the formation of a single furrow of an euglenoid pellicle (cf. e.g. Gallo and Schrével 1982). An unidirectional surface flow occurs around three or more helices which rotate in the same direction (Fig. 22i), so that they may become enveloped, e.g., by a paper ribbon, like enveloping of axons by the plasma membrane of a Schwann cell (Fig. 22k).

4.3.4 Rolling Motions and Self-Arrangements

Nonfixed helices rotating in honey are laterally displaced by themselves and show rolling instead of fluid motions. These increase close to a wall and may explain the increased rotational motion of amoeba cells that are flattened between slides (Goldacre 1956; Czarska and Grebecki 1965).

Two helices which rotate in the same direction close to one another (Fig. 23a) tend to roll over each other (arrows R). A bundle of helices rotating in the same direction is not static, it rolls (arrows R, Fig. 23b). It may simulate the lateral motions of spindle microtubules (Bajer and Molé-Bajer 1975) in telophase after the disappearance of the stabilizing, overlapping region of antipolar microtubules. Lateral motions of microtubules have also been found in the reticulopodial network of foraminifera (Travis et al. 1983). The spectacular rotational motion of axostyles (Tamm 1976, 1979) may be related. Two adjacent counterrotating helices (Fig. 23c) do not roll. A possible representative are the two central, single microtubules in the axoneme of eukaryotic flagella which may rotate around each other (Fig. 23a) in unicellular organisms, but do not move in the cilia of metazoa (Tamm and Tamm 1981).

Counterrotating helices can form a stable ribbon (Fig. 23d). A bundle of counterrotating helices in tetragonal or square arrangement (Fig. 23e) is stable and achieves minimum friction (Schulz and Jarosch 1980). Such arrangements are found in the Z-disks of muscle where the two sets of counterrotating, connecting filaments of neighboring sarcomers are closely adjacent, in the overlapping region of the thin filaments in a strongly contracted sarcomer (Eastwood et al. 1978), in the overlapping region of the diatom spindle (Schulz and Jarosch 1980), and sometimes in axostyles (Woodrum and Linck 1980). Cessation of rotation should change the square arrangement to hexagonal. A trigonal arrangement (Fig. 23f) is a further stable arrangement of counterrotating elements. This is verified for axopodial microtubules (Cachon et al. 1977) and for myosin subfilaments in muscle (Pepe et al. 1979). The special patterns of

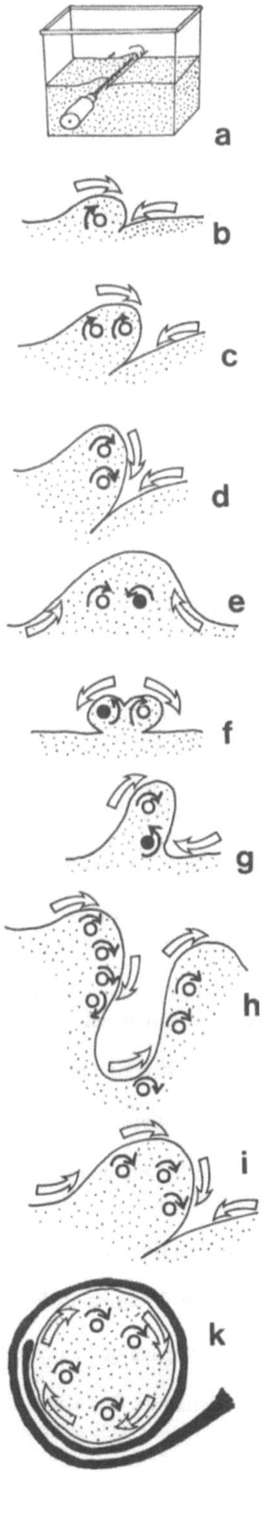

◀ **Fig. 22.** Flows (*white arrows*) and molding of a honey surface (that may respresent a free cell surface) produced by electromotor-driven helices arranged as shown in **a.** The shape of the surface depends on the positions of the helices and their rotational direction (counterclockwise rotations: *black*). The enveloping of a young axon by the membrane of a Schwann's cell is simulated in **i** and **k.** For more details, see text

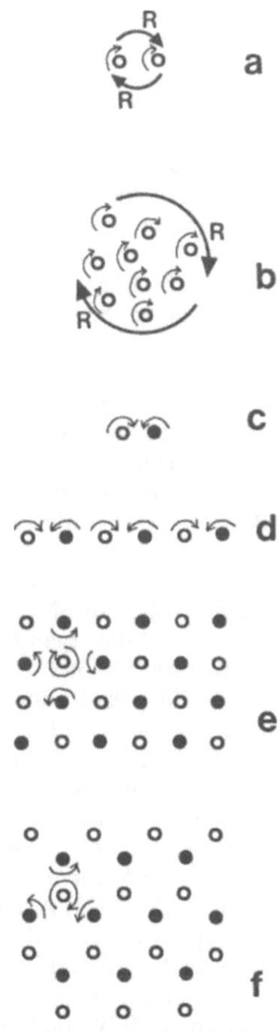

Fig. 23. **a, b** Rolling motions (*arrows R*) of nonfixed helices that rotate in the same direction. **c, d, e, f** Stable arrangements of counterrotating helices (clockwise rotation: *white;* counterclockwise rotation: *black*). For details, see text

microtubule arrangement may be determined by the hydrodynamic effects described and in addition by the associated filaments (crossbridges) that seem to connect single protofilaments of neighboring microtubules. Thus, hydrodynamic and screw-mechanic effects may be combined (cf. Chap. II.2.2).

Acknowledgments. I would like to thank I. Foissner for valuable discussions, D. Weiss for the demonstration of the recent results of AVEC microscopy on gliding axonal microtubules, and J. Bereiter-Hahn for important discussions and improvements of the manuscript.

References

Abé TH (1962) Morpho-physiological study of ameboid movement. II. Ameboid movement and the organization pattern in striata ameba. Cytologia (Tokyo) 27:111–139

Abercrombie M, Heaysman JEM, Pegrum SM (1970) The locomotion of fibroblasts in culture. I. Movements of the leading edge. Exp Cell Res 59:393–398

Allen RD, Taylor DL (1975) The molecular basis of amoeboid movement. Inoué S, Stephens RE (eds) Molecules and cell movement. Raven, New York, pp 239–258

Allen RD, Weiss DG, Hayden JH, Brown DT, Fujiwake H, Simpson M (1985) Gliding movement of and bidirectional transport along single native microtubules from squid axoplasm: Evidence for an active role of microtubules in cytoplasmic transport. J Cell Biol 100:1736–1752

Bajer A, Molé-Bajer J (1975) Lateral movements in the spindle and the mechanism of mitosis. In: Inoué S, Stephens RE (eds) Molecules and Cell Movement. Raven, New York, pp 77–96

Bajer A, Molé-Bajer J (1982) Asters, poles, and transport properties within spindlelike microtubule arrays. In: Organization of the cytoplasm. Cold Spring Harbor Symp Quant Biol vol. 46. Cold Spring Harbor Lab, pp 263–283

Bereiter-Hahn J, Strohmeier R, Kunzenbacher I, Beck K, Vöth M (1981) Locomotion of *Xenopus* epidermis cells in primary culture. J Cell Sci 52:289–311

Bovee EC (1964) Morphological differences among pseudopodia of various small amebae and their functional significance. In: Allen RD, Kamiya N (eds) Primitive motile systems in cell biology. Academic Press, New York, pp 189–219

Brinkley BR, Cox SM, Pepper DA, Wible L, Brenner SL, Pardue RL (1981) Tubulin assembly sites and the organization of cytoplasmic microtubules in cultured mammalian cells. J Cell Biol 90:554–562

Burton PR (1984) Luminal material in microtubules of frog olfactory axons: structure and distribution. J Cell Biol 99:520–528

Bynum RD, Allen RD (1980) Torsional movements in the ameba, *Chaos carolinensis,* suggests a helical cytoskeletal organization. J Protozool 27, 420–423

Cachon J, Cachon M, Tilney LG, Tilney MS (1977) Movement generated by interactions between the dense material at the ends of microtubules and non-actin-containing microfilaments in *Sticholonche zanclea.* J Cell Biol 72:314–338

Costello DP (1973) A new theory on the mechanics of ciliary and flagellar motility. I. Supporting observations. Biol Bull 145:279–291

Czarska L, Grebecki A (1965) Rotary movement in Amoeba proteus. In: Progress in Protozoology, Second Int Conf Protozool, London, Abstr. 327

Eastwood AB, Wood DS, Reuben JP (1978) Unusual thick and thin filament packing in a crustacean muscle. J Cell Biol 77:48–58

Foissner I (1985) The rotation model as a basis for *Nitella* filament – dynamics. In: Alia EE, Arena N, Russo MA (eds) Contractile proteins in muscle and non-muscle cell systems. Praeger, New York, pp 213–218

Foissner I, Foissner W (1986) *Ciliomyces spectabilis,* nov gen, nov spec, a zoosporic fungus which parasitizes cysts of the ciliate *Kahliella simplex.* I. Infection, vegetative growth and sexual reproduction. Z Parasitenkd 72:29 – 41
Foissner I, Jarosch R (1981) The motion mechanics of *Nitella* filaments (cytoplasmic streaming): their imitation in detail by screw-mechanical models. Cell Motility 1:371 – 385
Foissner I, Jarosch R (1985) The rotation model applied to melano(erytro)phore motility. In: Alia EE, Arena N, Russo MA (eds) Contractile proteins in muscle and non-muscle cell systems. Praeger, New York, pp 219 – 228
Forer A (1976) Actin filaments and birefringent spindle fibers during chromosome movements. In: Goldman R, Pollard T, Rosenbaum J (eds) Cell motility. Cold Spring Harbor Lab, pp 1273 – 1293
Fraser RDB, MacRae TP (1973) Conformation in fibrous proteins. Academic Press, London
Gallo J-M, Schrével J (1982) Euglenoid movement in *Distigma proteus.* 1. Cortical rotational motion. Biol Cell 44:139 – 148
Goldacre R (1956) The regulation of movement and polar organization in Amoeba by intracellular feedback. In: Abstr Proceed 1st Int Congress Cybernetics. Namur, pp 715 – 725
Gorbsky GJ, Sammak PJ, Borisy GG (1986) Kinetochore microtubules remain stationary and disassamble from the kinetochore end during anaphase movement of chromosomes. J Cell Biol 103 (5/2) Nr. 1613, 432a
Gruzdev AD (1972) Critical review of some hypotheses concerning anaphase chromosome movements. Tsitologiya 14:141 – 149 (in Russian, English translation in: NRC Technical Translation 1758, National Res Council of Canada, Ottawa)
Hayden JH, Allen RD (1984) Detection of single microtubules in living cells: particle transport can occur in both directions along the same microtubule. J Cell Biol 99:1785 – 1793
Heunert HH (1978) Beobachtungen über die Plasmaströmung der Grünalge *Acetabularia.* Protoplasma 93:477 – 479
Higashi-Fujime S (1985) Unidirected sliding of myosin filaments along bundles of Factin filaments spontaneously formed during superprecipitation. J Cell Biol 101:2335 – 2344
Hirokawa N, Glicksman MA, Willard MB (1984) Organization of mammalian neurofilament polypeptides within the neuronal cytoskeleton. J Cell Biol 98:1523 – 1536
Hirokawa N, Bloom GS, Vallee RB (1985) Cytoskeletal architecture and immunocytochemical localization of microtubule-associated proteins in regions of axons associated with rapid axonal transport: The β,β'-iminodipropionitrile-intoxicated axons as a model system. J Cell Biol 101:227 – 239
Inoué S, Ritter H Jr (1975) Dynamics of mitotic spindle organization and function. In: Inoué S, Stephens RE (eds) Molecules and cell movement. Raven, New York, pp 3 – 30
Jarosch R (1956) Plasmaströmung und Chloroplastenrotation bei Characeen. Phyton (Argentina) 6:87 – 107
Jarosch R (1960) Die Dynamik im Characeen-Protoplasma. Phyton (Argentina) 15:43 – 66
Jarosch R (1963) Grundlagen einer Schrauben-Mechanik des Protoplasmas. Protoplasma 57:448 – 500
Jarosch R (1964a) Screw-mechanical basis of protoplasmic movement. In: Allen RD, Kamiya N (eds) Primitive motile systems in cell biology. Academic Press, New York, pp 599 – 622
Jarosch R (1964b) Zur Mechanik schnell rotierender elastischer Schrauben. Biorheology 2:37 – 53
Jarosch R (1966) On the behavior of rotating helices. In: Warren KB (ed) Intracellular transport. Academic Press, New York, pp 275 – 300
Jarosch R (1967) Studien zur Bewegungsmechanik der Bakterien und Spirochäten des Hochmoores. Österr Bot Z 114:255 – 306
Jarosch R (1968) Rotationen von Superschrauben in viskosen Lösungen. Modellversuche zur Entstehung der Transversalwellen an Geißeln und Protoplasmafibrillen. Biorheology 5:303 – 311
Jarosch R (1969) Das sterische Verhalten der Alpha-Helix. Z Naturforsch Teil 24 B:672 – 680
Jarosch R (1971) Vergleichende Studien zur amöboiden Beweglichkeit. Protoplasma 72:79 – 100
Jarosch R (1972) The participation of rotating fibrils in biological movements. Acta Protozool 11:23 – 36

Jarosch R (1976) Dynamisches Verhalten der Octinfibrillen von *Nitella* auf Grund schneller Filament-Rotation. Biochem Physiol Pflanzen 170:111–131

Jarosch R (1979) The torsional movement of tropomyosin and the molecular mechanism of the thin filament motion. In: Hatano S, Ishikawa H, Sato H (eds) Cell motility: molecules and organization. University of Tokyo Press, pp 291–319

Jarosch R (1985a) The filament rotation model: molecular and screw-mechanical details. In: Alia EE, Arena N, Russo MA (eds) Contractile proteins in muscle and non-muscle cell systems. Praeger, New York, pp 239–251

Jarosch R (1985b) A simple explanation of "force enhancement" during stretch. J Muscle Res Cell Mot: 6:105

Jarosch R (1986a) The thin filament rotation model of muscle contraction. In: Frey M, Freilinger G (eds) 2nd Vienna muscle symposium proceedings. Facultas Universität, Wien, pp 1–7

Jarosch R (1986b) A model for the molecular basis of filament contractility and sliding as demonstrated by helix models. Cell Motility Cytoskeleton 6:229–236

Jarosch R (1986c) The mechanical behavior of doublet microtubules simulated by helical models. Cell Motility Cytoskeleton 6:209–216

Jarosch R, Foissner I (1982) A rotation model for microtubule and filament sliding. Eur J Cell Biol 26:295–302

Jarosch R, Foissner I (1983) The rotation model for filament sliding as applied to the cytoplasmic streaming. In: Earnshaw, Steer MW (eds) The application of laser light scattering to the study of biological motion. Plenum A59, New York, pp 545–558

Jarosch R, Foissner I (1985) Filament winding and unwinding as a basis for microtubule-behavior: a theoretical analysis with helix-models. In: Alia EE, Arena N, Russo MA (eds) Contractile proteins in muscle and non-muscle cell systems. Praeger, New York, pp 229–237

Kachar B (1985) Direct visualization of organelle movement along actin filaments dissociated from characean algae. Science 227:1355–1357

Kamitsubo E (1981) Effect of supraoptimal temperatures on the function of the subcortical fibrils and an endoplasmic factor in *Nitella* internodes. Protoplasma 109:3–12

Kamiya N (1959) Protoplasmic streaming. In: Heilbrunn LV, Weber F (eds) Protoplasmatologia VIII. Springer, Berlin Heidelberg New York

Koonce MP, Schliwa M (1985) Bidirectional organelle transport can occur in cell processes that contain single microtubules. J Cell Biol 100:322–326

Koop HU, Kiermayer O (1980) Protoplasmic streaming in the giant unicellular green algae *Acetabularia mediterranea*. I. Formation of intracellular transport systems in the course of cell differentiation. Protoplasma 102:147–166

Kuroda K (1964) Behavior of naked cytoplasmic drops isolated from plant cells. In: Allen RD, Kamiya N (eds) Primitive motile systems in cell biology. Academic Press, London New York, pp 31–41

Langford GM, Inoué S (1979) Motility of the microtubular axostyle in *Phrysonympha*. J Cell Biol 80:521–538

Lindberg U, Höglund AS, Karlsson R (1981) On the ultrastructural organization of the microfilament system and the possible role of profilactin. Biochimie (Paris) 63:307–323

Macnab RM (1983) An entropy-driven engine – the bacterial flagellar motor. In: Oplatka A, Balaban M (eds) Biological structures and coupled flows. Academic Press, London, pp 147–160

McIntosh JR (1973) The axostyle of *Saccinobaculus*. II. Motion of the microtubule bundle and a structural comparison of straight and bent axostyles. J Cell Biol 56:324–339

Miller RH, Lasek RJ (1985) Cross-bridges mediate anterograde and retrograde vesicle transport along microtubules in squid axoplasm. J Cell Biol 101:2181–2193

Mooseker MS, Pollard TD, Wharton KA (1982) Nucleated polymerization of actin from the membrane-associated ends of microvillar filaments in the intestinal brush border. J Cell Biol 95:223–233

Nagai R, Rebhun LI (1966) Cytoplasmic microfilaments in streaming *Nitella* cells. J Ultrastruct Res 14:571–589

Nowakowska G (1978) Twisting of suspended monotactic specimens of *Amoeba proteus*. Acta Protozool 17:347–352

Pepe FA, Sanger JW, Nachmias VT (1979) The myosin filament. In: Motility in cell function. Academic Press, London, pp 103–116

Pickett-Heaps JD, Northcote DH (1966) Relationship of cellular organelles to the formation and development of the plant cell wall. J Exp Bot 17:20–26

Schliwa M (1986) The cytoskeleton. Cell Biology Monographs, Vol. 13. Springer, Berlin Heidelberg New York

Schnapp BJ, Vale RD, Sheetz MP, Reese TS (1985) Single microtubules from squid axoplasm support bidirectional movement of organelles. Cell 40:455–462

Schnepf E (1973) Mikrotubulus-Anordnung und -Umordnung, Wandbildung und Zellmorphogenese in jungen *Sphagnum*-Blättchen. Protoplasma 78:145–173

Schrödinger E (1944) What is Life? Cambridge University Press

Schroeder GA, Kelly RB (1985) In vitro translocation of organelles along microtubules. Cell 40:729–730

Schulz D, Jarosch R (1980) Rotating microtubules as a basis for anaphase spindle elongation in diatoms. Eur J Cell Biol 20:249–253

Sheetz MP, Spudich JA (1983) Movement of myosin-coated fluorescent beads on actin cables in vitro. Nature 303:31–35

Shimmen T, Yano M (1984) Active sliding movement of latex beads coated with skeletal muscle myosin on *Chara* actin bundles. Protoplasma 121:132–137

Small JV, Langanger G (1981) Organization of actin in the leading edge of cultured cells: influence of osmium tetroxide and dehydration on the ultrastructure of actin meshworks. J Cell Biol 91:695–705

Strohmeier R, Bereiter-Hahn J (1984) Control of cell shape and locomotion by external calcium. Exp Cell Res 154:412–420

Tamm SL (1976) Properties of a rotary motor in eucaryotic cells. In: Goldman R, Pollard T, Rosenbaum J (eds) Cell motility, part C. Cold Spring Harbor Lab, pp 949–967

Tamm SL (1979) Membrane movements and fluidity during rotational motility of a termite flagellate. J Cell Biol 80:141–149

Tamm SL, Tamm ST (1981) Ciliary reversal without rotation of axonemal structures in ctenophore comb plates. J Cell Biol 89:495–509

Tilney LG (1975) Actin filaments in the acrosomal reaction of *Limulus* sperm. J Cell Biol 65:289–310

Tilney LG, Inoué S (1982) Acrosomal reaction of *Thyone* sperm. TI. The kinetics and possible mechanism of acrosomal process elongation. J Cell Biol 93:820–827

Tilney LG, Kallenbach N (1979) The polymerization of actin. VI. The polarity of the actin filaments in the acrosomal process and how it might be determined. J Cell Biol 81:608–623

Tilney LG, Bonder EM, DeRosier DJ (1981) Actin filaments elongate from their membrane associated ends. J Cell Biol 90:485–495

Travis JL, Kenealy JFX, Allen RD (1983) Studies on the motility of the foraminifera. II. The dynamic microtubular cytoskeleton of the reticulopodial network of *Allogromia laticollaris*. J Cell Biol 97:1668–1676

Vale RD, Reese TS, Sheetz MP (1985) Identification of a novel force-generating protein, kinesin, involved in microtubule-based motility. Cell 52:39–50

Wang Y-L (1985) Exchange of actin subunits at the leading edge of living fibroblasts: possible role of treadmilling. J Cell Biol 101:597–602

Weiss DG, Gross GW (1982) The microstream hypothesis of axoplasmic transport: Characteristica, predictions and compatibility with data. In: Weiss DG (ed) Axoplasmic transport. Springer, Berlin Heidelberg New York, pp 362–383

Witman GB (1975) The site of in vivo assembly of flagellar microtubules. Ann N Y Acad Sci 253:178–191

Woodrum DT, Linck RW (1980) Structural basis of motility in the microtubular axostyle: implications for cytoplasmic microtubule structure and function. J Cell Biol 87:404–414

Zobel CR (1973) Effect of solution composition and proteolysis on the conformation of axonemal components. J Cell Biol 59:573–594

II. The Supramolecular Level

II.1 Mechanical Concepts of Membrane Dynamics: Diffusion and Phase Separation in Two Dimensions

KONRAD BECK[1]

1.1 Introduction

Biological membranes are continuously involved in a variety of processes. They separate functional units from the environment and thereby delimit reaction compartments while at the same time allowing a controlled exchange of substances across these boundaries. For maintenance of shape they need sufficient mechanical rigidity, yet this must not interfere with continuous turnover of their components, with cell motility, cell migration, or cytokinesis. Many enzymatically catalyzed reactions, especially multienzymatic processes with high turnover rates take place at membranes. The structural basis which meets all these requirements seems to consist of a lipid bilayer matrix with proteins embedded in it. For eukaryotic cell membranes this simple model is complicated by the fact that they cannot be regarded as a single physical entity: they are connected in an up-to-now not well-understood manner to the inner architecture of the cell (cytoskeleton and/or membrane skeleton) as well as on the outer side to other cells (e.g., tight junctions) or the extracellular matrix (e.g., basal lamina).

In order to obtain an understanding of the relationships of structure and function of membranes from experimental data, models have been proposed. In principle, the simplicity of such models allows a simple grasp of key elements that determine both structure and functional relationships. In practice, however, the theoretical treatment of even simple models is difficult and only rudimentary predictions can be based on them. In this chapter the diffusional two-dimensional transport in membranes will be discussed from the points of view of statistical mechanics and hydromechanics. These concepts are applied to explain lateral diffusion in artificial, well-characterized systems and the results are compared with measurements on biological membranes. Finally, diffusion in a fluid membrane with impermeable domains is treated by the effective-medium and percolation theory and the relevance of phase separation for cell membranes is discussed.

[1] Department of Biophysical Chemistry, Biozentrum, University of Basel, Klingelbergstrasse 70, CH-4056 Basel, Switzerland

1.2 Translational Diffusion in Fluid Phase Membranes

1.2.1 Net Transport by Diffusion: The Einstein-Smoluchowski Equation

Assume an infinite two-dimensional plane of molecules in a fluid phase, macroscopically characterized by a translational diffusion coefficient D; then, if one waits long enough, a single particle reaches any point of this lattice by a random walk. The vectorial sum of all displacements x of the particle over an infinite time t, called the mean displacement $\langle x \rangle$, is zero. If we consider a group of particles located at $x = 0$ at the time $t =: t_0$ (see left side of Fig. 1), and we assume only one-dimensional movement, after a finite time $t > t_0$ they will be found at $x =: x_t$ where, in general, $x_t \neq x_0$. On average, half of the observed particles can be found in the range $x \geqq 0$, the other half in $x \leqq 0$. The mean of the square displacements $\langle x^2 \rangle$ is given by

$$\langle x^2 \rangle = \int_0^{\infty} \frac{x^2 \exp(-x^2/4Dt)}{\sqrt{\pi D t}} dx = 2Dt, \tag{1}$$

which is the well-known Einstein-Smoluchowski equation generally written as

$$D = \langle x^2 \rangle / 2t. \tag{2}$$

This equation gives the probability of finding the particles observed at time $t = 0$ in a region limited by $x < |\sqrt{2Dt}|$ at time t:

$$\int_{-\sqrt{2Dt}}^{+\sqrt{2Dt}} \frac{\exp(-x^2/4Dt)}{2\sqrt{\pi D t}} dx = \frac{2}{\sqrt{\pi}} \int_0^{1/\sqrt{2}} \exp(-t^2)\, dt = \operatorname{erf}(1/\sqrt{2}) \tag{3}$$

with erf(x) denoting the probability integral. Thus, at the time t 68% of the observed particles are inside the interval $x < |\sqrt{2Dt}|$ while the rest is outside (right side of Fig. 1). If the particle is allowed to migrate in both dimensions,

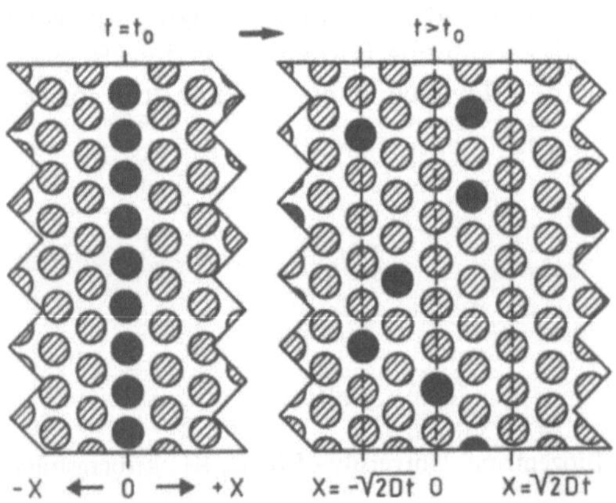

Fig. 1. Transport by diffusion. Assuming only one-dimensional diffusion the *black* particles on the *left side* will be found randomly distributed in the plane at $t > t_0$ (*right side*). According to the Einstein-Smoluchowski equation about two-thirds of these particles located at $t = t_0$ at $x = 0$ can be found inside the interval $x = \pm\sqrt{2Dt}$, whereas the rest has moved further

x and y with $|r| = \sqrt{x^2 + y^2}$, Eq. (2) will change to

$$D = \langle r^2 \rangle / 4t \,. \tag{4}$$

It should be noted that the mean displacements are proportional to the square root of time. Consider a given molecule in a homogeneous lipid bilayer surrounding a sphere of 10 μm diameter. If the diffusion coefficient is 1 $\mu m^2 s^{-1}$, then it will require on average about 1 min for the molecule to reach a diametrically opposite location on the sphere by a random walk.

At the beginning it was mentioned that a molecule moving in a two-dimensional array reaches any point after an infinite time. However, this is not obvious, and is only true for one and two dimensions. In three dimensions there is a good chance that a molecule will never find its target (Polya 1921). This theorem on the probability of encounter together with the realization that a ratio of the two- to the three-dimensional diffusion coefficients larger than 0.01 would speed up reaction velocities under certain conditions (Adam and Delbrück 1968) led to the suggestion that localization of enzymatic processes at membranes should be generally advantageous. For the case of a molecule which is to encounter a receptor this assumption is only valid if the number of receptor sites in a membrane is small and the two-dimensional mobility of the receptor is not smaller than 0.01 of its mobility in three dimensions. It should be noted that none of the translational diffusion coefficients that have been measured for both lipids and proteins since Adam and Delbrück stated their hypothesis, conclusively support the concept of an evolutionary optimization of biochemical reaction processes. A critical survey of this problem destroying some teleological hopes for establishing nature as an optimization generator was recently published by McCloskey and Poo (1986).

1.2.2 Diffusion Modeled as a Stochastic Random Walk: The Free Volume Model

Lateral transport has been treated theoretically from several points of view. Up to now no model has described the whole range of diffusion correctly. Deficiencies of current models are that they are only based on phenomenological parameters or that they contain experimentally inaccessible quantities (for review, see Clegg and Vaz 1985). Here, only two concepts and their validity are discussed.

The basic idea of the free volume model (with respect to two dimensions better called "free area model") is shown schematically in Fig. 2. A membrane is modeled as a plane filled with hard rods of cross-section a_0 and mass m representing the molecules. If all molecules are in direct contact with each other, which mimics a crystalline phase, no independent motion of a single molecule is possible. At a higher temperature the total area increases and on average the molecules are surrounded by a free area a_f. The sum of a_0 and a_f is the mean molecular area $\bar{a}$. This fluid state allows a vibrational motion which is limited by the neighboring molecules. Within this cage the molecules are mov-

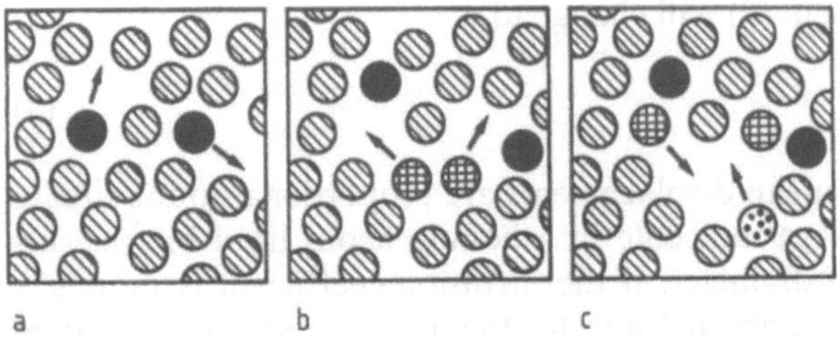

Fig. 2a–c. Diffusion mechanism assumed in the free volume model. Randomly distributed particles in a fluid two-dimenstional array can jump from their location into an adjacent hole, if the hole is large enough (**a**). The abandoned location can be occupied by another molecule (**b**) or the initially observed particle jumps back (**c**)

ing with the two-dimensional gas kinetic velocity $v = \sqrt{2\,kT/m}$ (T = absolute temperature; k = Boltzmann constant). Density fluctuations statistically create holes large enough for adjacent molecules to jump into this new position (Fig. 2a). The hole left by one molecule on leaving can be used by another molecule (Fig. 2b) or the initial one can jump back (Fig. 2c).

This concept has been worked out quantitatively for three-dimensional crystals and its formulation for liquids and glasses by Cohen and Turnbull (1959), adopted to two dimensions, was first applied by Galla et al. (1979). Neglecting any interactions of membrane particles with the adjacent media, the diffusion coefficient for a liquid phase results in

$$D = g \cdot l \cdot \sqrt{2kT/m} \cdot \exp(-c \cdot a^*/a_f)\,, \tag{5}$$

where g = 1/4 is a geometric factor, l is the average length of free travel of the molecule in its cage, c with $0.5 < c < 1$ accounts for overlapping of free areas a_f, and a^* is a critical free area necessary for a molecule to regard it as a hole. In deriving this formula it is assumed that the mean molecular area $\bar{a}$ is smaller than twice the minimum area a_0. To account for the interaction of membrane molecules in the midplain of the bilayer and with the aqueous phase, Vaz et al. (1985a) mainly replaced the preexponential factor of Eq. (5) by kT/f where the friction coefficient f includes the viscosities of the corresponding fluids multiplied by experimental factors.

The phenomenological character of the free volume model is obvious and well demonstrated by the factors g or f, respectively. A thermodynamically more justified approximation, based on the same diffusion mechanism, has been worked out by Blank and Britten (1965) and adopted to a liquid phase by Agrawal and Neuman (1987)

$$D = \frac{(\gamma_0 - \gamma)\,(\gamma - E)\,(a_f + a_0)^2}{2kT\sqrt{E \cdot m}} \exp\left(\frac{-a_0(\gamma - E)}{kT}\right). \tag{6}$$

The main progress in this evaluation is the lack of physically unexplained parameters: γ_0 and γ are the surface tensions of the pure and membrane-covered solvent (e.g., water), E is the energy of extension related to the entropy of extension S by $E = \gamma - TS$. As long as all surface pressure values $\Pi := \gamma_0 - \gamma$ for bilayer membranes are questionable [the extreme values, which are in discussion are 12.5 mN/m (Albrecht et al. 1978) and 92 mN/m (Fulford and Peel 1980), respectively] this approach is of minor practical relevance.

Although the free volume model fails to give an absolute measure of the translational diffusion coefficient, its dependence on the free area can be used to check the validity of the assumed jumping mechanism. Substituting the preexponential term of Eq. (5) by $A := g \cdot l \cdot \sqrt{kT/m}$ and the exponential one by $B := c \cdot a^*$ gives

$$\ln D = \ln A - B/a_f \tag{7}$$

(Peters and Beck 1983). A promising model system for verifying this logarithmic dependence of D on the reciprocal free area is provided by lipid monolayers at an air/water interface. They allow an isothermal change of a_f over a wide range and all physically independent variables can also be changed independently in the experiment, assuring that A and B remain constant. Figure 3a gives a semilogarithmic plot of the diffusion coefficients for two analog phospholipids spread as monolayers onto a clean water subphase. The direct measures in the monolayer technique are the surface pressure Π and the mean molecular area $\bar{a}$. To find $a_f = \bar{a} - a_0$, the measured data were optimized by regression analysis (Beck and Peters 1985), which results in $a_0 = 0.425\ nm^2$. This corresponds well with the theoretical cross-section of two aliphatic chains and with structural data of crystalline mono- and bilayers. From the slope ln D vs $1/a_f$ the critical free area results as $0.26 < a^* < 0.52\ nm^2$, thus giving a value of 0.61 for the ratio $c\,a^*/a_0$. This is significantly lower than those values (0.66–0.81) given for glasses (Cohen and Turnbull 1959) which might reflect the lack of conformational flexibility in three-dimensional liquid systems and the cavity-stabilizing effect of surfactant molecules (O'Leary 1987). The deviation from the predicted linear relationship for $\bar{a} > 2\,a_0$, which corresponds to $a_f > a_0$ is inherent in the derivation of Eq. (5). Qualitatively this can be understood from the supposition that one molecule has only one hole in his neighbourhood (Fig. 2). In this region, however, two molecules instead of one can change position simultaneously or, equivalently, the probability for one molecule to migrate is doubled. This results in a drastic increase of diffusion and is in agreement with this model.

Recently O'Leary (1987) extended the free volume approach using concepts from scaled particle theory (Lebowitz et al. 1965), which considers the work necessary for adding an additional hard rod to an environment as illustrated in Fig. 2. Although the phenomenological restrictions of the free volume model could not be omitted this hybrid model explains the different slopes of the ln D vs. $1/a_f$ curves (Fig. 3) for different homologous phospholipids. The diffusion coefficient can be assumed to be sensitive to chain-length differences and van der Waals interactions between chains, which is expressed by the variable c in Eq. (5). A pre-exponential factor resulting from the scaled particle theory

$$D \sim (a_f/\bar{a}) \exp(-c\,a^*/a_f) \tag{5'}$$

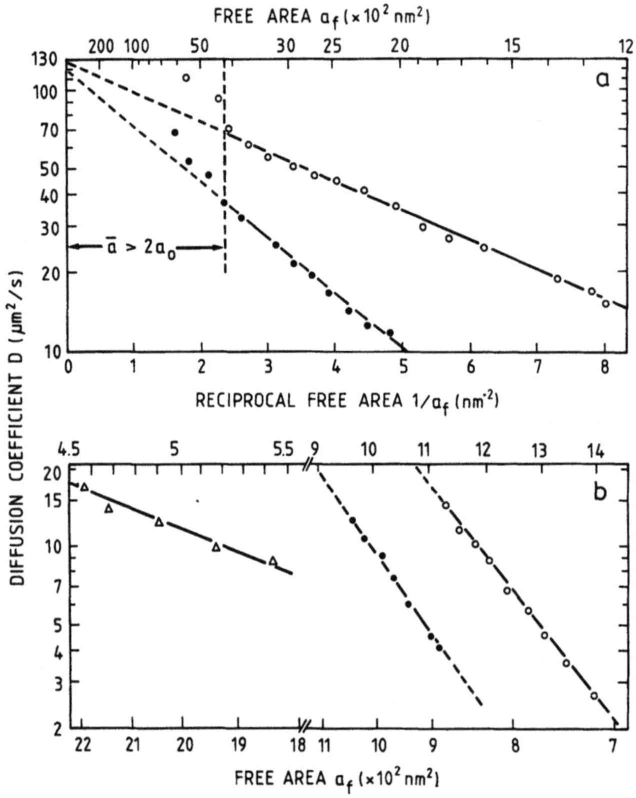

Fig. 3. Lipid translational diffusion in artifical monolayers (**a**) and bilayers (**b**) according to the free volume model. **a** Monolayers: The graphs represent the logarithmic dependence of the translational diffusion coefficient D from the reciprocal free area $1/a_f$ (Eq. 7) for two phospholipids differing only in their aliphatic chain length by two CH_2 groups (*open dots:* dodecanoic chains; *filled dots:* tetradecanoic chains). If the free area a_f is larger than the minimum area a_0 of the molecule, the probability for a diffusional jump increases and the model assumptions are not further valid (data according to Beck and Peters 1983 and Peters and Beck 1983). **b** Bilayers: The temperature-induced decrease of the reciprocal free area also results in an exponential increase of diffusion. The lipids are the same as those used in the monolayer experiments shown in **a.** In addition, the data of the analog lipid with hexadecanoic chains are shown (*triangles*). In comparison to the monolayer technique, bilayers only allow a limited range of a_f to check the model predictions

gives a substantially better agreement of this model with the experimental data (Fig. 3a) than the classical linear dependence according to Eq. (7) in the region $\bar{a} > 2\,a_0$. Furthermore this model provides a reasonable description of the reduction of the lateral diffusion of lipids if proteins are incorporated into the membrane.

In contrast to monolayers, a free area change is very limited in lipid bilayer systems and is always coupled with the change of other parameters. However, the temperature-induced area change (Fig. 3b), as well as the isothermal change induced by different degrees of hydration (McCown et al. 1981), can be interpreted well by the free volume model (Vaz et al. 1984). Absolute diffusion coefficients in lipid monolayers are significantly greater than those of bilayers at the same molecular packing density and temperature: Multilamellar vesicles with a free area of 0.125 nm^2 for the lipid show a diffusion coefficient of $5-8\ \mu m^2/s$ (Peters and Beck 1983; Vaz and Hallmann 1983), whereas the correspondent monolayer at the same temperature and free area exhibits $D = 15\ \mu m^2/s$ (see Fig. 3). This is due to the different adjacent media: whereas in bilayers the corresponding monolayers show some interaction and possibly even a substantial degree of interdigitation (Zaccai et al. 1979), in monolayers any frictional forces resulting from the air interphase are neglectible.

As long as the lateral pressure in bilayers is unknown, the energy-barrier model represented by Eq. (6) is applicable only to monolayer studies. The entropy of extension $S = -\partial\gamma/\partial T$ can be calculated from appropriate $\Pi/\bar{a}$-isotherms from which the energy of extension E results. Because of the exponential dependence of D on E and the experimental errors in deriving E, no decisive critical test of the validity-of this approach can be made. In general the absolute diffusion coefficients resulting from Eq. (6) are about one order of magnitude smaller than the reliable measurements thus far published (Beck and Peters 1983, 1985; Peters and Beck 1983; Agrawal and Neuman 1987; Hänggi-Mojtabai 1985). A straightforward extension of Eq. (6) (Blank and Britten 1965; Agrawal and Neuman 1987) predicts surface viscosities which are several orders of magnitude smaller ($\sim 10^{-12}$ g/s) than those reported in the literature ($< 10^{-5}$ g/s). It should be mentioned, however, that the experimental evaluation of surface viscosities seems doubtful.

1.2.3 Diffusion Modeled by Continuum Hydromechanics: The Saffman-Delbrück Model

Whereas the diffusion of a particle embedded in a fluid matrix of similarly sized particles is described by the free volume model, this approach fails to explain the diffusion of larger molecules like proteins within a liquid lipid bilayer: with increasing area of the diffusant in relation to its adjacent molecules the diffusion coefficient decreases rapidly. Actual values of protein diffusion in artificial lipid bilayers are discussed by Vaz et al. (1982) and tabulated in Clegg and Vaz (1985): these values are much greater than predicted from the statistical approach. Qualitatively this can be understood in relation to the diffusion mechanism (Fig. 2): if a particle with a radius double of that of its neighbors is assumed, it is unlikely that four independent neighboring molecules will show an ordered displacement which would allow the large molecule to jump into this hole.

With respect to a large molecule the adjacent lipid bilayer matrix can be characterized as a continuum simply determined by its thickness h and the (two-dimensional) viscosity η, whereas in the free volume model all molecules were regarded as individual entities. Because the surrounding media embeds the particle in three dimensions, the viscosities η_1' and η_2' of the neighboring media (inside and outside of the enclosed volume, respectively) have to be considered. Such a problem corresponds to the low Reynold's number limit of the Navier-Stokes equation. Restricting the general case to a protein modeled as a cylinder of radius r and the height equal to the thickness h of the bilayer, an absolute solution has been worked out by Hughes et al. (1981; Eq. (3.68)). The total system is characterized by the dimensionless constant

$$\varepsilon = \left(\frac{r}{h}\right)\left(\frac{\eta_1' + \eta_2'}{\eta}\right). \tag{7}$$

A biologically relevant approximation with $\varepsilon \lesssim 1$ gives for the translational diffusion coefficient:

$$D = \frac{kT\varepsilon}{4\pi r(\eta_1' + \eta_2')}\left[\ln\left(\frac{2}{\varepsilon}\right) - C + \frac{4\varepsilon}{\pi} - \frac{\varepsilon^2}{2}\ln\left(\frac{2}{\varepsilon}\right)\right], \tag{8a}$$

where C represents Euler's constant with $C = 0.57722$. For most biological relevant parameters $\varepsilon \lesssim 0.1$ might be realistic and a more restricted approximation was formerly worked out by Saffman and Delbrück (1975) consisting of only the first two terms of the sum:

$$D = \frac{kT}{4\pi\eta h}\left[\ln\frac{2\eta h}{r(\eta_1' + \eta_2')} - C\right]. \tag{8b}$$

Besides the lateral displacements the molecules in a liquid membrane show a rotational diffusion, i.e., they turn around an axis. This motion is characterized by the rotational diffusion coefficient D_R. Assuming the same boundary conditions as above, an approximation for $\varepsilon \lesssim 1$ is given by:

$$D_R = \frac{kT\varepsilon}{4\pi r(\eta_1' + \eta_2')} \cdot \frac{1}{r^2} \cdot \frac{3\pi}{8\varepsilon + 3\pi} \tag{9a}$$

[Hughes et al. 1981; the exact solution can be found there as Eq. (5.36)]. The Saffman-Delbrück approach for $\varepsilon \lesssim 0.1$ results again in the first two factors:

$$D_R = \frac{kT}{4\pi\eta h} \cdot \frac{1}{r^2}. \tag{9b}$$

The practical use of this continuum hydrodynamical model critically depends on the protein structure and the determination of ε. Many membrane-spanning proteins show a significant portion immersed in at least one of the adjacent media. Furthermore, they are frequently connected to intra- and/or extracellular components, but binding constants for such interactions are mostly unknown. Thus, theoretical considerations on this problem are useless. With respect to ε the most discussed parameter is the surface viscosity η (its reciprocal $1/\eta$ is often designated as "membrane fluidity"): some assumptions were established by fluorescence anisotropy decay measurements of indicator molecules, but these values are very uncertain. The concept of a microviscosity itself is not generally accepted. The viscosities η_i' of the surrounding media are normally assumed as that of water.

To circumvent some of the problems in determining ε, Hughes et al. (1981) favored the concept of measuring both translational and rotational diffusion. The ratio of the more general results given by Eqs. (8 a) and (9 a) agrees with the exact solution for $\varepsilon \lesssim 0.4$. The Saffman-Delbrück model yields the more obvious equation:

$$\frac{D}{D_R} = r^2 \left(\ln \frac{kT}{2\pi(\eta_1' + \eta_2')\, r^3 D_R} - C \right), \tag{10}$$

which is independent of the membrane viscosity η and valid for $\varepsilon \lesssim 0.05$.

An experimental test of this model was performed by Peters and Cherry (1982). They measured the translation and rotational diffusion of bacteriorhodopsin, the light-driven proton pump of *Halobacterium halobium*, reconstituted in liposomes. This protein is of ideal choice because of its well-known structure: X-ray and electron diffraction at high resolution give the widest cross-sections as 3.5 nm, thus the hydrodynamic radius is 1.75 nm. Furthermore, it packs nearly ideally into the lipophilic region; thus the height can be assumed as that of the bilayer thickness. The hydrophobic part consisting of seven α-helices is relatively rigid and reconstitution assays for this protein are well established. In contrast to the natural membrane where bacteriorhodopsin is arranged as a trimer, in vesicles it is monomeric in the fluid phase. Using the measured values for D and D_R, Eq. (10) results in 1.89 to 2.39 nm for the protein radius, depending on temperature and the lipid/protein ratio. This slightly greater radius is probably due to the fact that a real protein is not a stiff cylinder with smooth surfaces, as the model assumes. Thus, lipid/protein interaction cannot be neglected. Furthermore, the authors used their measurements to evaluate the membrane viscosity by Eq. (8 b). For low lipid/protein ratios, as assumed by the model, this results in 1.1 to 1.8 P. Using these values the predictions of Eq. (9 a) are in excellent agreement with the measured diffusion constants obtained by varying the viscosity of the environment. Clegg and Vaz (1985), however, argued that this viscosity change could interfere with the order in the lipid phase.

Translational diffusion coefficients of several reconstituted proteins with radii up to 6 nm can be fitted by the Saffman-Delbrück model if the membrane viscosity is assumed as 0.75 to 2.0 P (Clegg and Vaz 1985; Vaz et al. 1984). In contrast to these studies, Kleinfeld et al. (1981) failed to find a modulation of protein diffusion by the lipid viscosity. Whereas the free volume model failed to explain the diffusion of large proteins in lipid bilayers, the continuum hydrodynamic model is in disagreement with the fact that lipid diffusion is independent of acyl chain lengths (Vaz and Hallmann 1983). For three-dimensional diffusion an analogous discrepancy of models for small and large diffusants has been established by Nir and Stein (1971). Thus, the limits of the two models have to be investigated.

Archaebacteria differ in their membrane lipid composition: phytanyl chains are ether linked to a glycerol backbone and the membranes of the methanogens and thermo-acidophiles mainly consist of a monolayer of bipolar lipids. Recently, Vaz et al. (1985b) have measured the mobility of such a membrane-spanning lipid of *Sulfolobus solfataricus* embedded in a phosphatidylcholine bilayer. These diffusion coefficients agree well with both a free volume model and a continuum hydromechanical approach (Nadler et al. 1985).

1.2.4 Diffusion in Biological Membranes

For lipidic probes translational diffusion coefficients in fluid artificial bilayer systems generally range from 1 to 10 μm²/s and proteins reconstituted in phospholipid bilayers show diffusion coefficients from about 0.7 to 8 μm²/s with only a weak dependence on their molecular size. Some of these results can be understood on the basis of the above described models. Biological membranes are of much more complicated composition: the "simple" erythrocyte membrane contains well over a 100 different lipid species. The reason for this lipid diversity is unknown. Model studies show that, for example, a small portion of cholesterol effects a broadening of the phase transition region: above the phase transition temperature diffusion is reduced, whereas below this diffusion is enhanced. An analogous effect results by decreasing the lipid/protein ratio in reconstituted liposomes.

Some diffusion coefficients for lipids and proteins in biomembranes are listed in Table 1. In general, lipid diffusion is three to ten times slower than in bilayer vesicles. Liposomes of plasma membrane lipids exhibit a diffusion coefficient of 2.5 and 6.3 μm²/s at 20° and 37° C, respectively, whereas the same combination of lipids in living cells show 0.8 and 1.6 μm²/s (Jacobson et al. 1981). Protein diffusion is generally much more restricted. A single exception is the value $D \simeq 0.3$ μm²/s, which was measured for rhodopsin in the photoreceptor membrane (e.g., Wey and Cone 1981). Such a high diffusion coefficient is in agreement with the "fluid mosaic" membrane model (Singer and Nicolson 1972), but no comparable high values could be found for other membrane proteins. This might be due to the lack of structural elements adjacent to this membrane.

The mobility of the anion transport protein band 3 (which is assumed to be aggregated mostly in tetramers) in erythrocyte ghosts is increased by a factor of 50 if the underlying spectrin network is chemically weakened (Golan and Veatch 1980) or genetically removed (Koppel et al. 1981). Reconstituted in lipid vesicles, band 3 shows a diffusion coefficient of 1.6 μm²/s at 30° C or taking into account the slower erythrocyte lipid diffusion. Chang et al. (1981) computed $D \simeq 0.73$ μm²/s for free diffusible band 3 in erythrocyte liposomes. Assuming the applicability of the Saffman-Delbrück model, Eq. (8b) results in a membrane viscosity of 20–25 poise for spectrin depleted ghosts (estimated with $r = 3$ nm, $h = 4.5$ nm, $\eta'_1 = \eta'_2 = 0.01$ P). In contrast to the assumption in the model, the N-terminal 40-kD fragment, responsible for nearly half of the total protein molecular mass ($M_r = 93$ kD), extends into the cytoplasm. Thus, this should be regarded as an additional frictional force. Furthermore, lipid diffusion is slower than in liposomes: a reevaluation yields $D \simeq 0.8$ μm²/s at 25° C

Table 1. Lateral diffusion coefficients of lipids (D_L) and proteins (D_P) in cell membranes measured by fluorescence photobleaching recovery

System	T [°C]	D_L [μm²/s]	D_P [× 10^{-2} μm²/s]	Reference
Plasma membrane lipids in human fibroblasts	20	0.8	–	Jacobson et al. 1981
	37	1.6	–	
Rhodopsin in frog rod outer segments	22	–	30	Wey et al. 1981
Mouse erythrocyte ghosts				
normal cells	23	1.4 ±0.5	0.45±0.08	Koppel et al. 1981 [a]
genetically spectrin-depleted cells		1.5 ±0.5	25 ±6	
Fibronectin on 3T3 mouse cells and chicken fibroblasts	37	0.7 −0.9	<0.05	Schlessinger et al. 1977
Concanavalin A on mouse fibroblasts				
parallel to stress fibers	22−23	–	0.14	Smith et al. 1979 [b]
perpendicular to stress fibers		–	0.038	
Mouse neuroblastoma cells				
during mitosis	20−24	0.18±0.02	1.3 ±0.2	de Laat et al. 1980
in early G_1		0.38±0.02	3.1 ±0.3	
during DNA synthesis		0.61±0.004	1.8 ±0.1	
Sea urchin egg				
unfertilized	20−23	0.56±0.12	3.5 ±0.7	Peters and Richter 1981
~20′ after ferilization		0.27±0.12	0.7 ±0.15	
3T3 mouse cells				
untransformed cells	?	0.90±0.08	2.4 ±0.5	Eldrige et al. 1980
SV-40 transformed cells		0.87±0.13	1.3 ±0.5	

[a] A critical reevaluation with special regard to radiation-induced artifacts results in $D_L = 0.8$ μm²/s (25 °) for intact erythrocytes (Bloom and Webb 1983).
[b] A part of this difference for D_P might result from inadequate data evaluation (Aizenbud and Gershon 1985).

for the lipid diffusibility in intact erythrocyte membranes. Qualitatively these unconsidered details would result in a decrease of diffusion thus suggesting that a continuum hydromechanical approach for describing protein diffusion might be justified.

The next tabulated protein, fibronectin in fibrous form, is linked to the extracellular cell surface via a membrane-embedded receptor protein and it also interacts strongly with some membrane glycolipids. This could be responsible for the low mobility measured by Schlessinger et al. (1977).

As long as no conclusive model has been worked out to explain the two-dimensional mobility in biological membranes, biologists prefer to correlate changes of this parameter with cytological events: Smith et al. (1979) reported that lateral transport parallel to stress fibers is about four times faster than perpendicular to those; thus they assumed a regulatory function of the cytoskeleton for membrane diffusion. A part of this anisotropy can result from neglecting the nonplanar form of the membrane (Aizenbud and Gershon 1985). Development biologists look to variations of diffusion during the cell cycle: de Laat et al.

(1980) reported an independent change of lipid and protein diffusion, which reaches a maximum in the S-phase or the early G_1-phase, respectively. Peters and Richter (1981) observed that fertilization of sea urchin eggs results in a significant decrease of the translational diffusion of both lipids and proteins. With respect to the basic relevance of diffusion (see Sect. 1.2.1) this could be essential for allowing pattern formation necessary for the differentiation of germ layers: if the cell cycle time is shorter than the time a particle needs for mixing with other germ layers, differentiation could be regarded as a diffusion controlled process. Consequently, local changes of diffusion in the animal and vegetal half of eggs depending on fertilization were investigated (e.g. Dictus et al. 1984).

Much hope and effort have been invested in the search for mobility changes induced by malignant transformation. Looking to the "microviscosity" indicated by fluorescence anisotropy changes some phenomenological observations were reported (for review, see Shinitzky 1984), but its physiological implications are not understood. Translational diffusion is nearly unaffected by transformation of 3T3 mouse cells: the slight decrease of protein diffusion (Eldrige et al. 1980) does not seem very significant.

1.3 Fluid-Solid Phase Separation in Two Dimensions

1.3.1 Effective Medium and Percolation Theory

The hydromechanical model of diffusion treated above assumes a protein, modeled as a stiff cylinder, embedded in an infinite lipid bilayer. Biological membranes, however, often exhibit a high protein packing density: about 50% of the inner mitochondrial membrane area consists of proteins. Monte Carlo simulations predict only a slight dependence of membrane mobility from protein packing density: a membrane lattice covered to 82% with impermeable domains should diminish the diffusion coefficient only by a factor of 20 in comparison to the zero concentration limit (Pink 1985). Besides the low diffusion coefficients of membrane proteins, the significant large portions of immobile protein fractions are inherent with the above mentioned models: in general, only 20 to 70% of the membrane proteins exhibit wide-range diffusion (for review, see e.g. McCloskey and Poo 1984), whereas the diffusion coefficient of the rest is assumed as several orders of magnitude smaller. In reconstituted liposomes in the fluid phase, however, nearly 100% of proteins are mobile with a uniform diffusion coefficient.

One approach for explaining these differences has been worked out by Saxton (1982) and is based on classical theories for evaluating the conductivity of composite materials. He combined effective medium and percolation theories for describing "lateral diffusion in an archipelago." The effective medium theory assumes an infinite array of conducting cylinders with a random distribution of islands characterized by a lower conductivity. Replacing the conductivities by the diffusion coefficients for fluid and solid domains, namely D_F and D_S, respectively, the formulation of Bruggeman (1935) and Landauer

(1952) for two dimensions results in

$$D = D_F(\Phi_F - \tfrac{1}{2})\left(1 - \frac{D_S}{D_F}\right) + D_F\sqrt{(\Phi_F - \tfrac{1}{2})^2\left(1 - \frac{D_S}{D_F}\right)^2 + \frac{D_S}{D_F}}, \quad (11)$$

where D denotes the average wide-range diffusion coefficient, generally measured by fluorescence, absorption, and excimer techniques, and Φ_F equals the area fraction of molecules diffusing with the higher (fluid phase) diffusion coefficient D_F. With $D_S \equiv D_F$, we obtain the expected result $D = D_F$. For a membrane with completely insulating islands, characterized by $D_S = 0$, Eq. (11) is reduced to the linear relationship:

$$D = \begin{cases} 2\Phi_F - 1 & \text{if} \quad \Phi_F > 0.5 \\ 0 & \text{if} \quad \Phi_F < 0.5 \end{cases}. \quad (12)$$

In this case wide-range diffusion should be completely blocked at a critical area fraction $\Phi_F = 0.5$. Figure 4 demonstrates the quantitative relationship of D from Φ_F according to Eq. (11) for some D_S/D_F ratios (solid lines).

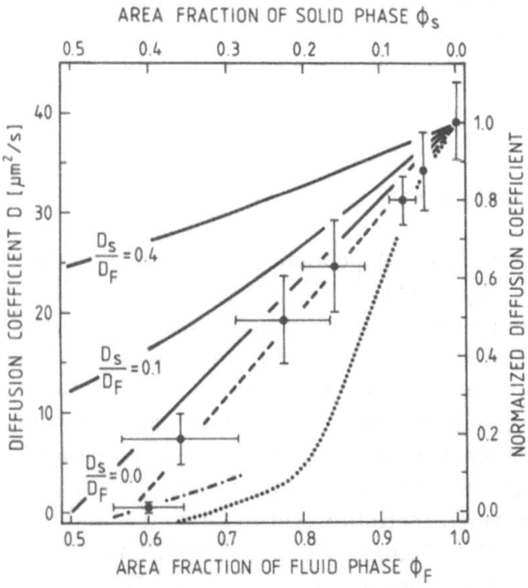

Fig. 4. Phase separation according to the effective medium and percolation theory. The *solid lines* represent the predicted increase of the average translational diffusion D with increasing the area fraction Φ_F of the fluid phase according to the effective medium theory of Bruggeman and Landauer for different ratios D_S/D_F (Eqs. 11 and 12, respectively). The *dots* indicate the diffusion coefficients D and the area fraction of the fluid lipid of the lipid monolayer also discussed in Fig. 5a–c and Fig. 6; *error bars* show standard deviations. The normalized diffusion coefficient correlates to the maximum packing density of the homogeneous fluid monolayer (cf. Fig. 6). The *dashed line* results from linear regression which is justified if $D_S/D_F = 0$ is assumed. The *dashed-dotted* line fits D for a fluid area fraction $\Phi_F = 0.6$ assuming percolation theory (Eq. 13) with $\Phi_c = 0.58$. The *dotted line* represents the cubic polynomal of Saxton (1982, Eq. 7) which connects effective medium and percolation theory with $\Phi_c = 0.668$ (for detailed discussion, see text)

If the conductivity of the insulating phase is very low in comparison to the adjacent one and Φ_F decreases, effective medium theory fails to explain conductivity measurements. Here, percolation theory seems more appropriate: whereas for $D_S > 0$ effective medium theory results in a linear decrease of D with D_S/D_F at $\Phi_F = 0$, in percolation theory a critical area fraction Φ_C is assumed around D which shows a power law dependence:

$$D \cong \begin{cases} D_F(\Phi_F - \Phi_C)^{1.34} & \text{if} \quad \Phi_F > \Phi_C \\ D_S(\Phi_C - \Phi_F)^{-1.1} & \text{if} \quad \Phi_F < \Phi_C \end{cases} . \tag{13}$$

In three dimensions a percolation threshold was found at $\Phi_F = 0.578$. A reduction of dimensionality results in a increase of the fraction of conducting sites necessary for conduction. Different approaches result in $0.64 < \Phi_C < 0.69$ for two dimensions, if $D_S/D_F \ll 10^{-3}$.

In contrast to effective medium theory, which is useful for large ratios of D_S/D_F, percolation theory concentrates on aggregation phenomena and has been successfully applied also to patching effects in immunology (e.g. DeLisi and Perelson 1976). For matching both, Saxton (1982; Eq. 7) constructed a cubic polynomal, which fits Eq. (11) at $\Phi_F = 1$ and Eq. (13) with $\Phi_C = 0.668$ at $\Phi_C + 0.1$. This somewhat arbitrary function is shown in Fig. 4 as a dotted line.

1.3.2 Phase Separation in Lipid Monolayers

An experimental test of the phase-separation dependence of diffusion was performed by the lipid monolayer technique (Beck 1984). The flattened region in surface-pressure/area-isotherms can be interpreted as a first-order phase transition region: by increasing the packing density of the lipid molecules, the fluid phase is transformed to the solid phase part by part. In this process, small islands of lipid in the solid phase are built up, the area fraction increases up to the limit at which the whole monolayer is in the solid phase. Such a creation of "large" lipid shells was observed already by Zocher and Stiebel (1930) using dark-field microscopy and was confirmed by fluorescence (e.g. Peters and Beck 1983; Beck 1984; Beck and Peters, 1985; Hänggi-Mojtabai 1985) and electron microscopy (e.g. Neuman et al. 1984). By choosing an appropriate fluorescence probe, which partitions equally in the solid and fluid phase and indicates its environment by a different fluorescence response (Fig. 5), it is possible to quantify the area fraction of the fluid phase in the intermediate monolayer state (Beck 1984). The pure fluid and solid phase appear homogeneous in the fluorescence microscope (Fig. 5a) with the difference that the intensity per unit area is double for the solid phase. In the intermediate region, islands with such a high intensity can be recognized (Fig. 5b) which are embedded in an environment with a lower fluorescence intensity. After bleaching all dye molecules irreversibly, fluorescence recovers only between the islands, which remain dark for hours (Fig. 5c) thus indicating them as in solid phase. In this

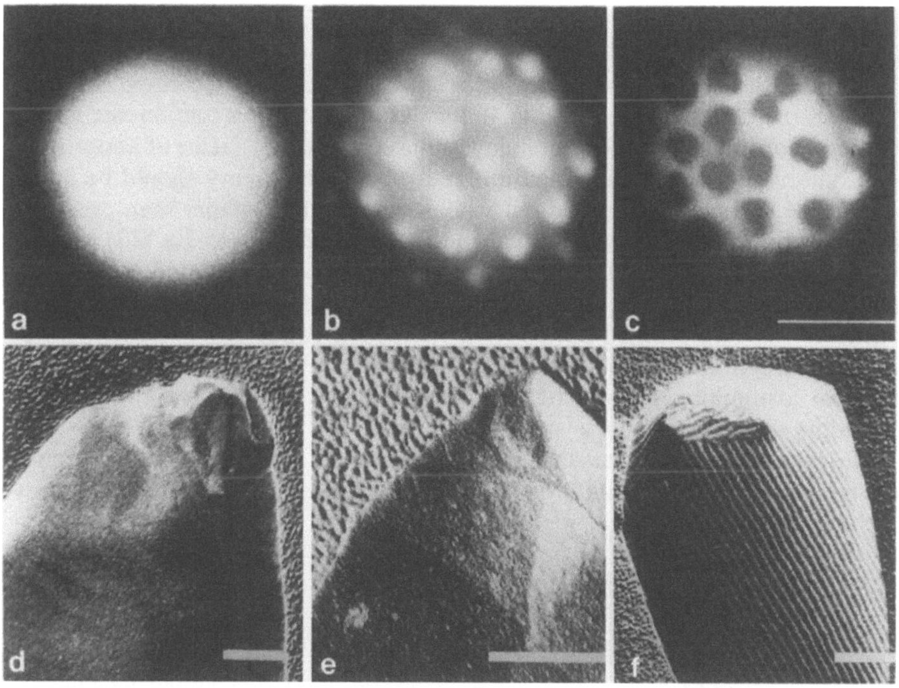

Fig. 5. Visualization of phase separation in lipid monolayers (**a – c**) and bilayers (**d – f**). **a – c** The surface-pressure/area-isotherm of 1,2-dipalmitoyl-*sn*-glycero-3-phosphocholine exhibits a flattened intermediate region between the pure fluid and solid phase at room temperature (the results presented in Fig. 4a – c and Fig. 6 also refer to this monolayer). Doped with an appropriate probe this monolayer appears homogeneously fluorescent in the fluid phase (**a**). If the monolayer is compressed, at the beginning of the intermediate region small bright fluorescent islands become visible which grow with decreasing mean molecular area (**b**). Reaching the pure solid phase the whole monolayer is transformed and it again appears homogeneous as in **a**, but the fluorescence intensity is as large as that of the islands, whereas the fluorescence intensity of the pure fluid phase corresponds to that of the surrounding of the islands. 10 min after bleaching all dye molecules in the observed area irreversibly, fluorescence recovers only in the channels but the islands remain dark for hours (**c**), thus indicating them as in solid phase (*bar* = 20 μm). **d – f** In electron micrographs of the internal membrane surface of freeze-fractured, mixed lecithin/ganglioside (4:1 mol/mol) vesicles the surfaces appear smooth if the vesicles are rapidly cooled from temperatures above the main phase transition (**d**) or below the pretransition temperature. If the aliphatic chain length of the ganglioside significantly differs from that of the host lipid, small lipidic domains appear which consist of about 100 – 200 molecules (**e**). If the vesicles are cooled from a temperature between the main and pretransition temperature, electron micrographs represent a wavy pattern ("ripple structure") with a 27-nm periodicity (**f**) (vesicles were generously prepared by P. End). The pictures are presented in inversed contrast: shadows appear dark (*bars* = 200 nm)

simple case the fluid area fraction Φ_F is described by:

$$\Phi_F = \frac{a_{fl}}{\bar{a}} \cdot \frac{a_s - \bar{a}}{a_s - a_{fl}} \tag{14}$$

(Beck 1984, p. 61), where a_{fl} and a_s denote the molecular area in the fluid and solid phase, respectively. With respect to the mean molecular area $\bar{a}$, for a_{fl} an area expansion results, if a_s is assumed to be constant. This area difference in the phase separation region might be the reason for the finite slope of the corresponding surface pressure/area isotherm. A simultaneous recording of D and Φ_F gives the data indicated in Fig. 4. Because the diffusion coefficients for the homogeneous fluid and solid monolayer phase differ by a factor of about 1000, the combination of effective medium and percolation theory should be applicable. The average of all measured values is somewhat smaller than predicted by effective medium theory but can be fitted in analogy to Eq. (12) linearly (dashed line in Fig. 4). The lowest values, $\Phi_F > 0.65$ can be fitted by both the effective medium theory as well as by percolation theory. The latter one is demonstrated by the dashed-dotted line. The somewhat low assumed value $\Phi_c = 0.58$ might be due to a too large D_s/D_F ratio. To decide on the validity of the two assumptions, more delicate techniques (e.g., low-light-level TV-microscopy) are necessary; this work is in progress.

For thermodynamic reasons, phase separation is accompanied by a change of the lateral compressibility. Figure 6 demonstrates the monolayer compres-

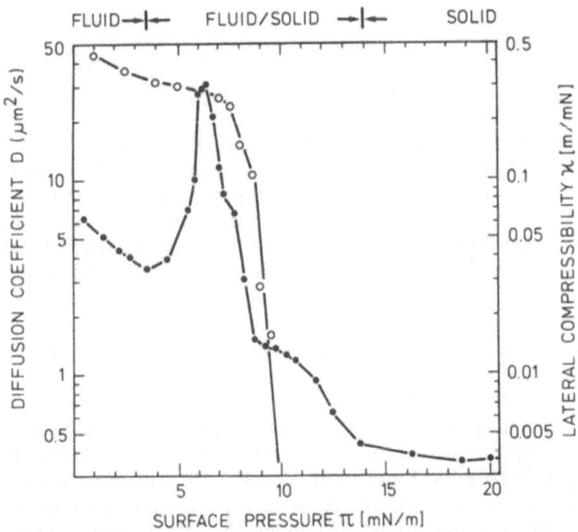

Fig. 6. Correlation of phase separation, translational diffusion, and lateral compressibility. By increasing the molecular packing density of the monolayer discussed in Figs. 4 and 5, in the fluid phase the lateral compressibility $\varkappa$ (*filled dots*) and the diffusion coefficient D (*open dots*) decrease. The creation of small solid islands at a surface pressure Π of about 3–4 mN/m results in a step increase of Π, whereas the course of D is stable. After reaching a maximum at $\Pi \simeq 6$ mN/m, $\varkappa$ decreases rapidly but ends in the nearby exponential decrease ($\Pi \sim 9$ mN/m), which it shows in the fluid phase. At this point the fluid area fraction Φ_F reaches a percolation threshold with $\Phi_c \simeq 0.6$ (see Fig. 5) and D decreases rapidly by a factor of 100 within an interval $\Delta\Pi \leq 3$ mN/m. A further monolayer compression continues the transition from the fluid into the solid phase accompanied by a decrease of $\varkappa$. At $\Pi \simeq 15$ mN/m the homogeneous solid phase is reached and $\varkappa$ as well as D remain constant ($D < 0.03$ μm²/s; not shown). Flow effects induced by convection interfere with an accurate measurement of diffusion coefficients smaller than 0.03 μm²/s in the use of the experimental setup. [Data according to Beck 1984 ($\varkappa$) and Beck and Peters 1985 (D); redrawn]

sibility $\varkappa$ (filled dots) as a function of the surface pressure π, whereas $\varkappa$ decreases with increasing Π in the homogeneous fluid phase, a hyperexponential increase of $\varkappa$ can be found at the beginning of the coexistence region of fluid and solid domains. The average diffusion coefficient D, indicated by open dots, does not significantly change at this event, but shows a rapid decrease when the region of high compressibility is finished. Here, the fluid area fraction Φ_F reaches the percolation threshold Φ_c (see Fig. 4). With a further compression, the solid area fraction increases and finally builds up the whole monolayer. From the beginning of this homogeneous solid phase on, compressibility and diffusion ($D < 0.03\ \mu m^2/s$) remain constant in the time scale of the experiment (not shown).

An analogous relationship between compressibility, diffusion, and phase separation can be found at constant temperature and constant surface pressure, when the fluid area fraction is changed by mixing the fluid host lipid monolayer with different amounts of another lipid which is in the solid phase (cf. Fig. 7 in Beck and Peters 1985 with Fig. 4 in Peters and Beck 1983).

1.3.3 Phase Separation in Biological Membranes

Up to now a simultaneous recording of the static and dynamic behavior during phase separation only was performed in the cited monolayer studies. Indirect evidence for a coexistence region of fluid and solid domains has been already given by interpreting electron spin resonance (e.g. Shimshick and McConnell 1973) and calorimetric data (e.g. End 1986), the observation of a hexagonal packing of lipid chains by electron diffraction (e.g. Hui et al. 1974), the selectivity of fluorescent probes (Klausner and Wolf 1980), and the visualization of domains in the electron microscope (see Fig. 5d–f) (e.g. Neuman et al. 1984). In these studies on liposomes, large domains (diameter larger than 300 nm) as well as aggregates consisting of only a few molecules could be found. Applied to biological membranes, these techniques give evidence that some of them contain a patchwork of small immiscible gel domains in a fluid environment (e.g. Stier and Sackmann 1973; Metcalf et al. 1986).

The present studies on phase separation in biological membranes only allow some hypotheses on its biological relevance. It may be a factor in membrane flow (Zagyansky and Jard 1979), fusion (Cullis and Hope 1978), lipid–protein and protein–protein interaction (Wickner 1977), especially in capping of surface immunoglobulin in lymphocytes (Klausner et al. 1980), and in the activity of enzymes (for review, see Sackmann 1984). The large increase of the lateral compressibility accompanied by the appearance of small gellike domains could facilitate the insertion and extraction of larger membrane areas without changing the thermodynamic equilibrium of the membrane significantly (e.g. oxo- and endocytosis, cytokinesis). Solid domains could function as a physiological trap for toxic molecules and enzymes. Although the wide-range diffusion decreases during phase separation, the microscopic motion of the molecules in the fluid phase could be enhanced: if a_{fl} is increased according to Eq. (14), the free area around these molecules and thus the probability for a diffusional jump (see Fig. 2) is also increased. In this way diffusion becomes canalized by the solid islands.

Saxton (1982) explained the reported growth dependence on temperature of *Escherichia coli* (Jackson and Cronan 1978) in terms of the percolation theory: growth is enhanced when the fluid membrane area fraction is greater than 0.8, whereas at $\Phi_F < 0.45$ growing stops. As a further example he treats the glucosid transport through the membrane of fatty acid auxotroph mutants of *E. coli* (Linden et al. 1973). A "plate model" of membrane structure emphasizing the long-range order in biomembranes was established by Jain and White (1977). They pointed out particularly that a membrane consisting of fluid and gellike areas is more stable to compression, stress, and strain than pure fluid or solid arrays. If the hydrophobic effect is unconsidered, a homogeneous fluid membrane would have its energy minimum in a spherical shape; a solid membrane, however, should tend to be planar. The distinct appearance of cell and organelle shape and their morphological changes would be reached with less energy by a fluid/solid array.

1.4 Concluding Comments

Accurate observations and measurements of membrane-coupled events are the most important prerequisites for understanding the living membrane; but facts are no explanation and often they are even artifacts. The aim of this chapter was to outline some concepts on the mechanisms of membrane mobility. For reaching an understanding of the inherent principles which maintain the integrity of membranous systems, the author's proposal is to investigate, on the one hand, *simple* – possibly biologically "unimportant" – membranes under well-defined conditions, and, on the other hand, to combine these results with those from model membrane studies. The underlying principles should be worked out by specialized theoreticians which above all should reduce the dimensionality of well-established concepts.

Acknowledgments. The author wishes to thank Drs. J. Engel and J. Fessler for discussions and advice. The original work presented in this chapter was achieved during the author's support by a fellowship of the „Max-Planck-Gesellschaft zur Förderung der Wissenschaften", Munich (FRG).

References

Adam G, Delbrück M (1968) Reduction of dimensionality in biological diffusion processes. In: Rich A, Davidson N (eds) Structural chemistry and molecular biology. Freeman, San Francisco, pp 198–215

Agrawal ML, Neuman RD (1987) Surface diffusion in fatty acid monolayers. In: Mittal KD (ed) Surfactants in solution, vol. 5. Plenum, New York

Aizenbud BM, Gershon ND (1985) Diffusion of molecules on biological membranes of nonplanar form. II. Diffusion anisotropy. Biophys J 48:543–546

Albrecht O, Gruler H, Sackmann E (1978) Polymorphism of phospholipid monolayers. J Phys (Paris) 39:301–313

Beck K (1984) Translationale Diffusion und Phasentrennung in Phosphatidylcholin-Monoschichten an der Luft/Wasser-Grenzfläche. Naturwiss Dissertation, Universität Frankfurt

Beck K, Peters R (1983) Translational diffusion in pure and mixed phosphatidylcholine monolayers. Eur J Cell Biol Suppl 2:5

Beck K, Peters R (1985) Translational diffusion and phase separation in phospholipid monolayers: a fluorescence microphotolysis study. In: Bayley PM, Dale RE (eds) Spectroscopy and the dynamics of molecular biological systems. Academic Press, London, pp 177–196

Blank M, Britten JS (1965) Transport properties of condensed monolayers. J Colloid Sci 20:789–800

Bloom JA, Webb WW (1983) Lipid diffusibility in the intact erythrocyte membrane. Biophys J 42:295–305

Bruggeman DAG (1935) Berechnung verschiedener physikalischer Konstanten von heterogenen Substanzen. I. Dielektrizitätskonstanten und Leitfähigkeiten der Mischkörper aus isotropen Substanzen. Ann Phys (Leipzig) 24:636–679

Chang CH, Takeuchi H, Ito T, Machida K, Ohnishi S (1981) Lateral mobility of erythrocyte membrane proteins studied by the fluorescence photobleaching recovery technique. J Biochem (Tokyo) 90:997–1004

Clegg RM, Vaz WLC (1985) Translational diffusion of proteins and lipids in artificial lipid bilayer membranes. A comparison of experiment with theory. In: Watts A, de Pont JJHNM (eds) Progress in protein–lipid interactions, vol I. Elsevier Science, Amsterdam, pp 173–229

Cohen MH, Turnbull D (1959) Molecular transport in liquids and glasses. J Chem Phys 31:1164–1169

Cullis PR, Hope MJ (1978) Effects of fusogenic agent on membrane structure of erythrocyte ghosts and the mechanism of membrane fusion. Nature 271:672–674

De Laat SW, Van der Saag PT, Elson EL, Schlessinger J (1980) Lateral diffusion of membrane lipids and proteins during the cell cycle of neuroblastoma cells. Proc Natl Acad Sci USA 77:1526–1528

DeLisi C, Perelson A (1976) The kinetics of aggregation phenomena. I. Minimal models for patch formation on lymphocyte membranes. J Theor Biol 62:159–210

Dictus WJAG, van Zoelen EJJ, Tetteroo PAT, Tertoolen LGJ, de Laat S, Bluemink JG (1984) Lateral mobility of plasma membrane lipids in Xenopus eggs: regional differences related to animal/vegetal polarity become extreme upon fertilization. Dev Biol 101:201–211

Eldridge CA, Elson EL, Webb WW (1980) Fluorescence photobleaching recovery measurements of surface lateral mobilities on normal and SV40-transformed mouse fibroblasts. Biochemistry 19:2075–2079

End P (1986) Thermodynamische Untersuchungen zum Phasenverhalten von Phosphatidylcholin-Gangliosid-Mischungen. Studien über die Wechselwirkung von Ca^{++} Peanutlectin und Choleratoxin mit gangliosidhaltigen Membranen. Diplomarbeit, Universität Freiburg

Fulford AJC, Peel WE (1980) Lateral pressures in biomembranes estimated from the dynamics of fluorescent probes. Biochim Biophys Acta 598:237–246

Galla HJ, Hartman W, Theilen U, Sackmann E (1979) On two-dimensional passive random walk in lipid bilayers and fluid pathways in biomembranes. J Membr Biol 48:215–236

Golan DE, Veatch W (1980) Lateral mobility of band 3 in the human erythrocyte membrane studied by fluorescence photobleaching recovery: evidence for control by cytoskeletal interactions. Proc Natl Acad Sci USA 77:2537–2541

Hänggi-Mojtabai F (1985) Total internal reflection and fluorescence recovery after photobleaching: instrumentation and determination of molecular motion in biological membranes. PhD Dissertation, University of Basel, Basel

Hughes BD, Pailthorpe BA, White LR (1981) The translational and rotational drug on a cylinder moving in a membrane. J Fluid Mech 110:349–372

Hui SW, Parsons DF, Cowden M (1974) Electron diffraction of wet phospholipid bilayers. Proc Natl Acad Sci USA 71:5068–5072

Jackson MB, Cronan JE Jr (1978) An estimate of the minimum amount of fluid lipid required for the growth of *Escherichia coli.* Biochim Biophys Acta 512:472–479
Jacobson K, Hou Y, Derzko Z, Wojcieszyn J, Organisciak D (1981) Lipid lateral diffusion in the surface membranes of cells and in multibilayers formed from plasma membrane lipids. Biochemistry 20:5268–5275
Jain MK, White HB III (1977) Long-range order in biomembranes. In: Paoletti R, Kritchevsky D (eds) Advances in lipid research, vol 15. Academic Press, London, pp 1–60
Klausner RD, Wolf DE (1980) Selectivity of fluorescent lipid analogues for lipid domains. Biochemistry 19:6199–6203
Klausner RD, Bhalla DK, Dragsten P, Hoover RL, Karnovsky MJ (1980) Model for capping derived from inhibition of surface receptor capping by free fatty acids. Proc Natl Acad Sci USA 77:437–441
Kleinfeld AM, Dragsten P, Klausner RD, Pjura WJ, Matayoshi ED (1981) The lack of relationship between fluorescence polarization and lateral diffusion in biological membranes. Biochim Biophys Acta 648:471–480
Koppel DE, Sheetz MP, Schindler M (1981) Matrix control of protein diffusion in biological membranes. Proc Natl Acad Sci USA 78:3576–3580
Landauer R (1952) The electrical resistance of binary metallic mixtures. J Appl Phys 23:779–784
Lebowitz JL, Helfand E, Praestgaard E (1965) Scaled particle theory of fluid mixtures. J Chem Phys 43:774–779
Linden CD, Wright KL, McConnell HM, Fox CF (1973) Lateral phase separation in membrane lipids and the mechanism of sugar transport in *Escherichia coli.* Proc Natl Acad Sci USA 70:2271–2275
McCloskey M, Poo M (1984) Protein diffusion in cell membranes: some biological implications. Intern Rev Cytol 87:19–81
McCloskey M, Poo M (1986) Rates of membrane associated reactions: reduction of dimensionality revisited. J Cell Biol 102:88–96
McCown JT, Evans E, Diehl S, Wiles HC (1981) Degree of hydration and lateral diffusion in phospholipid multibilayers. Biochemistry 20:3134–3138
Metcalf TN III, Wang JL, Schindler M (1986) Lateral diffusion of phospholipids in the plasma membrane of soybean protoplasts: evidence for membrane lipid domains. Proc Natl Acad Sci USA 83:95–99
Nadler W, Tavan P,Schulten K (1985) A model for the lateral diffusion of "stiff" chains in a lipid bilayer. Eur Biophys J 12:25–31
Neuman RD, Fereshtehkou S, Ovalle R (1984) Electron microscopic observation of LE/LC phase transition in dipalmitoyl phosphatidylcholine monolayers. J Colloid Interface Sci 101:309–313
Nir S, Stein WD (1971) Two modes of diffusion in liquids. J Chem Phys 55:1598–1603
O'Leary TJ (1987) Lateral diffusion of lipids in complex biological membranes. Proc Natl Acad Sci USA 84:429–433
Peters R, Beck K (1983) Translational diffusion in phospholipid monolayers measured by fluorescence microphotolysis. Proc Natl Acad Sci USA 80:7183–7187
Peters R, Cherry RJ (1982) Lateral and rotational diffusion of bacteriorhodopsin in lipid bilayers: experimental test of the Saffman-Delbrück equations. Proc Natl Acad Sci USA 79:4317–4321
Peters R, Richter H-P (1981) Translational diffusion in the plasma membrane of sea urchin eggs. Dev Biol 86:285–293
Pink DA (1985) Protein lateral movement in lipid bilayers. Simulation studies of its dependence upon protein concentration. Biochim Biophys Acta 818:200–204
Polya G (1921) Über eine Aufgabe der Wahrscheinlichkeitsrechnung betreffend die Irrfahrt im Straßennetz. Mathematische Annalen 84:149–160
Sackmann E (1984) Physical basis of trigger processes and membrane structure. In: Chapman D (ed) Biological membranes, vol 5. Academic Press, London, pp 105–143
Saffman PG, Delbrück M (1975) Brownian motion in biological membranes. Proc Natl Acad Sci USA 72:3111–3113

Saxton MJ (1982) Lateral diffusion in an archipelago. Effects of impermeable patches on diffusion in a cell membrane. Biophys J 39:165–173

Schlessinger J, Barak LS, Hammes GG, Yamada KM, Pastan I, Webb WW, Elson EL (1977) Mobility and distribution of a cell surface glycoprotein and its interaction with other membrane components. Proc Natl Acad Sci USA 74:2909–2913

Shimshick EJ, McConnell HM (1973) Lateral phase separation in phospholipid membranes. Biochemistry 12:2351–2360

Shinitzky M (1984) Membrane fluidity in malignancy: adversative and recuperative. Biochim Biophys Acta 738:251–261

Singer SJ, Nicolson GL (1972) The fluid mosaic model of the structure of cell membranes. Science 175:720–731

Smith BA, Clark WR, McConnell HM (1979) Anisotropic molecular motion on cell surfaces. Proc Natl Acad Sci USA 76:5641–5644

Stier A, Sackmann E (1973) Spin labels as enzyme substrates: heterogeneous lipid distribution in liver microsomal membranes. Biochim Biophys Acta 311:400–408

Vaz WLC, Hallmann D (1983) Experimental evidence against the applicability of the Saffman-Delbrück model to the translational diffusion of lipids in phosphatidylcholine bilayer membranes. FEBS Lett 152:287–290

Vaz WLC, Derzko ZI, Jacobson KA (1982) Photobleaching measurements of the lateral diffusion of lipids and proteins in artificial phospholipid bilayer membranes. In: Poste G, Nicolson GL (eds) Membrane reconstitution. Elsevier Biomedical Press, Amsterdam, pp 83–136

Vaz WLC, Goodsaid-Zalduondo F, Jacobson K (1984) Lateral diffusion of lipids and proteins in bilayer membranes. FEBS Lett 174:199–207

Vaz WLC, Clegg RM, Hallmann D (1985a) Translational diffusion of lipids in liquid crystalline phase phosphatidylcholine multibilayers. A comparison of experiment with theory. Biochemistry 24:781–786

Vaz WLC, Hallmann D, Clegg RM, Gambacorta A, De Rosa M (1985b) A comparison of the translational diffusion of a normal and a membrane spanning lipid in L_α phase 1-palmitoyl-2-oleoylphosphatidylcholine bilayers. Eur Biophys J 12:19–24

Wey C, Cone RA (1981) Lateral diffusion of rhodopsin in photoreceptor cells measured by fluorescence photobleaching and recovery. Biophys J 33:225–232

Wickner WT (1977) Role of hydrophobic forces in membrane protein asymmetry. Biochemistry 16:254–258

Zaccai G, Büldt G, Seelig A, Seelig J (1979) Neutron diffraction studies on phosphatidylcholine model membranes. II. Chain conformation and segmental disorder. J Mol Biol 134:693–706

Zagyansky YA, Jard S (1979) Does lectin-receptor complex formation produce zones of restricted mobility within the membrane? Nature 280:591–594

Zocher H, Stiebel F (1930) Dunkelfeldmikroskopie dünnster Filme auf Flüssigkeitsoberflächen. Z Physikal Chem Abt A 147:401–435

II.2 Implications of Microtubules in Cytomechanics: Static and Motile Aspects

DIETER G. WEISS[1], GEORGE M. LANGFORD[2], and ROBERT D. ALLEN[3,4]

2.1 Microtubule Structure: Statics and Elasticity

2.1.1 Substructure of Microtubules

Microtubules (MTs) are hollow core cylinders that measure 25 nm in diameter. The walls are primarily composed of dimers of the proteins α- and β-tubulin (see Dustin 1984 for review). The dimers are arranged in 13 strands or protofilaments that are aligned almost parallel to the longitudinal axis of the MT (Tilney et al. 1973; Crepeau et al. 1978). The individual monomers in adjacent protofilaments are not aligned in register along the perpendicular axis of the tubule, but are staggered at an oblique angle and follow a helical path around the tubule (Amos 1979). This monomer helix has been shown to be a 3-start helical family with a pitch of 12 nm and a pitch angle of 10.5° for 13 protofilament MTs.

The dimer lattice in outer doublet MTs of cilia and flagella axonemes differs in the A- and B-tubules of the doublet. In the A-tubule the dimers in adjacent protofilaments are in a half-staggered arrangement (along a 5-start helix), the dimers in adjacent protofilaments of the B-tubule are lined up obliquely along the 3-start helix (Amos and Klug 1974). The dimer lattice in cytoplasmic MTs has not been determined. In precursor ribbons of protofilaments assembled from cycle-purified brain MT proteins the monomer lattice is a 3-start helix (Erickson 1974).

In electron micrographs of native MTs, the protofilaments forming the wall of the tubule appear globular in cross-section and are surrounded by an electron-dense ring of stain (Fig. 1 a). In samples of reconstituted brain MTs, 95% contain 14 protofilaments which are aligned at a slight angle to the long axis of the tubule. This twist or super-helix has been seen in both 13 and 14 protofilament MTs (Langford 1980; Mandelkow and Mandelkow 1985; Langford et al. 1986).

MTs assembled from cycle-purified brain MT proteins are decorated with two high molecular weight, microtubule-associated proteins (MAPs) (Dentler et al. 1975). Electron microscopy of negatively stained MTs shows that MAPs

[1] Institut für Zoologie, Technische Universität München, D-8046 Garching, FRG
[2] Department of Physiology, School of Medicine, University of North Carolina, Chapel Hill, North Carolina 27514, USA
[3] Department of Biology, Dartmouth College, Hanover, New Hampshire 03755, USA
[4] Posthumously (see Laudatio)

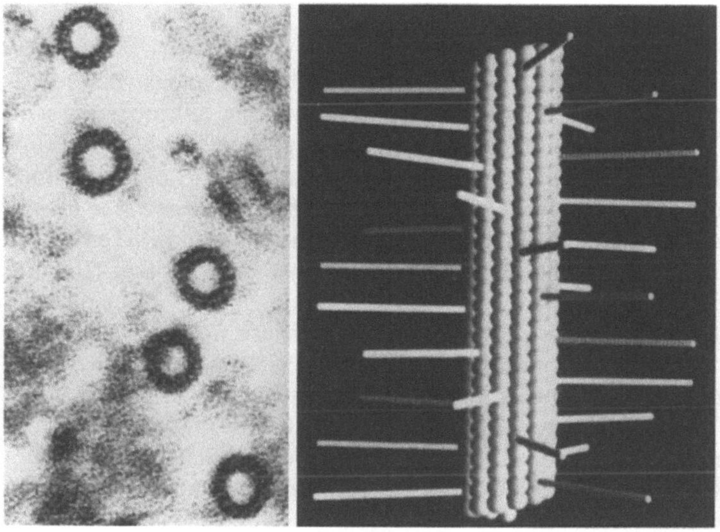

a b

Fig. 1. a Electron-micrograph of a pellet of reassembled shark brain MTs fixed in glutardialdehyde plus 8% tannic acid and thin sectioned according to Tilney et al. (1973). 95% of the population contains 14 protofilaments. The other 5% contains either 13 or 15 protofilaments. The protofilaments do not appear with equal clarity around the entire circumference of the tubule because of the slight inclination of the protofilaments. Therefore, when the plane of section is perpendicular to the protofilament on one side of the tubule, it is at an oblique angle to the protofilament on the opposite side. The diameter of an MT is 25 nm. **b** Computer-generated model of an MT [courtesy of Richard T. Feldman (NIH) and George M. Langford]. MTs are hollow tubes of 25 nm diameter consisting of 13 rows (protofilaments) of tubulin subunits (spheres). A major portion of the microtubule-associated proteins (MAPs) extends out from the wall as filamentous projections. The MAPs are drawn as stiff projections, but should be regarded as flexible filaments. A smaller part of the molecule is assumed to attach along the groove between protofilaments. The high molecular weight MAPs can be seen in electron-micrographs to protrude 20 – 100 nm

are arranged in a helical configuration with an axial repeat distance of 12 tubulin dimers or 96 nm (Amos 1977; Langford 1983) (Fig. 1 b). The average spacing between projections along the side of the tubule is 32 nm (Kim et al. 1979). Factors which control the spacing of MAPs along MTs have not been determined.

It can be expected from the fact that tubulin and tubulin dimers are asymmetrical globular proteins that the stereospecific alignment of dimers within the protofilaments gives intrinsic structural polarity to the MT itself. Indeed, it was shown that this is the case since they have a slowly growing (−) end and a fast growing (+) end (for review, see Dustin 1984, Chap. 2). In cytoplasmic MTs the (−) end is usually situated in the cell center (at the centrosome), while the (+) end points to the periphery. This polarity of MTs together with their stiffness makes them best suited to integrate the cell's interior, to define directions for intracellular movements, and to guarantee proper export of newly synthesised material from the cell center to its periphery (see also Chap. I.1).

2.1.2 Rigidity of Microtubules

Microtubules are found in many instances at cellular localizations where stiffening and skeletal functions are required, especially in long cell processes such as in heliozoan axopods, nerve axons, and chromatophores. But MTs are also located at the periphery of cells where the cell's shape has to be maintained such as in erythrocytes or ciliates (e.g. Cohen et al. 1982; for review, see Dustin 1984, Chaps. 4 and 6). Bereiter-Hahn (1978) has pointed out that MTs seem to be the only cellular constituents that are rigid against lateral forces (flexural rigidity).

Rigidity is absent from the protofilaments which are flexible and have a tendency to form spirals and rings. Bereiter-Hahn (1978) suggested, therefore, that the assembly of protofilaments into an MT forces them to straighten under tension in order to yield the parallel alignment (Fig. 1 a). Such exertion of strain during MT formation could endow the composed structure with its rigidity (Bereiter-Hahn 1978). Furthermore, their tubular form may be another clue to explain the rigidity.

Rigidity of MTs was also implied from electron micrographs which show that MTs even under the tension during drying for negative staining do not exhibit sharp bends, but rather break (Gall 1966; Langford 1983) (Fig. 2a). With the aid of video-enhanced light microscopy MTs could recently be observed directly and it was reported that an MT may break if a force pushes it against an obstacle (see Fig. 8 in Allen et al. 1985). The minimum radius of curvature was found to be of the order of 0.2 μm (Allen et al. 1985).

Direct microscope observation by video techniques (Allen et al. 1985) and in dark-field microscopy (Yamazaki et al. 1982) clearly indicates the rigidity which allows single MTs to resist the enormous forces exerted upon them by Brownian motion. Their slight bending in this situation was also used to measure their flexural rigidity. It was found to be about 10^{-16} dyn cm^2 (10^{-25} Nm^2) which is higher, e.g., than that of actin filaments (Yamazaki et al. 1982).

It can be concluded that the mechanical properties of MTs make it possible for them to provide stiffness to any part of the cell or the cytoplasm in which they are located, while at the same time maintaining their flexural elasticity.

2.1.3 Integration of Microtubules into the Cytoskeleton

From TEM studies MTs have repeatedly been reported to be decorated by wispy, irregularly shaped filaments which sometimes appear to connect MTs with each other, with intermediate filaments, or with adjacent organelles (e.g. Smith et al. 1970, 1975; Tsukita & Ishikawa 1980; Schliwa et al. 1982; Miller and Lasek 1985). These filaments extend from the MT surface and keep MTs from directly contacting other cellular structures (Sasaki et al. 1982). Thus, an exclusion zone is formed around microtubules which is devoid of other structures and often appears as a clear halo, whose radius measures from 10 nm in insect cells (MacGregor and Stebbings 1970) to 50 nm in neurons (Weiss and Gross 1983).

The high molecular weight, MT-associated proteins (MAPs) derived from brain form side arms of 30–90 nm length (Voter and Erickson 1982; Langford 1983). These proteins establish the zones of exclusion around microtubules and maintain the observed spacing between microtubules and cell organelles. Some authors have suggested that the clear zone around MTs could be due to electrostatic repulsion of the anionic MTs (Stebbings and Hunt 1982). A microtubule-associated ATPase which protrudes at least 15–25 nm has been observed by Brady (1985). The ATPase may be involved in MT-related movements (see below).

The electrostatic, hydrostatic, and chemical interaction of MTs and their MAPs with other components of the cytoplasm, therefore, can be considered to be at least partially responsible for the viscoelastic properties of cytoplasm. These can best be demonstrated in squid giant axons. The axoplasm remains for hours as a stable, highly viscous gel cylinder when extruded and kept in a medium mimicking the intracellular milieu (Bear et al. 1937; Morris and Lasek 1982).

In some cases MT-associated proteins have been found to link MTs to intermediate filaments (Leterrier et al. 1982). The fact that multiple crosslinks can be formed explains in part the high viscosity of the cytoplasmic ground substance. The macroviscosity of cytoplasm, i.e., the viscosity particles experience when moved through the cytoplasm, is indeed enormous (see Chap. I.1, Table 1).

2.2 Microtubule-Associated Dynamics: Motion and Tension

For almost all forms of MT-associated motility the force-generating mechanism is not exactly known. In this chapter we will only give an overview of the variety of events without attempting to discuss the molecular mechanisms. However, it is probable that only a few principles and enzymes of force generation will ultimately be found to be at work (Allen et al. 1985; Weiss et al. 1987). As a general rule we can state that in the animal kingdom cell locomotion is an actin-myosin-based process, while intracellular movements are MT-associated (cf. Schliwa 1984; Weiss 1985; Weiss et al. 1986).

2.2.1 Elongation of Microtubules

If assembly of tubulin subunits could occur while one end of the MT is fixed at some structure, this could generate a pushing force. In organelle motion there is little evidence for such movement, as organelles are rarely attached to MT ends. However, in mitosis of at least some primitive organisms, such as yeast and diatoms, the elongation of MTs may result in stemming (King et al. 1982; Pickett-Heaps et al. 1982). Subunit incorporation has been shown at the kinetochore end of spindle MTs (Mitchison et al. 1986) and in the centrosome (see also Chap. I.1.4.3).

2.2.2 Shortening of Microtubules

Dynamic instability of MTs (see Chap. I.1.5.3) does not necessarily result in shape changes or in a pulling force if cytoplasmic MTs are involved. It may change the cell shape dramatically if stiffening MTs are disassembled rapidly. Dynamic instability causes shortening of MTs in the range of 260 dimers per second (Cassimeris et al. 1985) or 0.15 $\mu m\ s^{-1}$ (Mitchison and Kirschner 1984).

The contraction of axopods in Heliozoa caused by disassembly of the axonemal MT bundles is considerably faster and has a velocity of the order of up to 5000 μm^{-1} (see Febvre-Chevalier and Febvre 1986 for review). The axopods of the heliozoan *Actinocoryne* contract even faster (up to 50000 $\mu m\ s^{-1}$), but it seems as if other mechanisms are involved in addition to MT disassembly (Febvre-Chevalier and Febvre 1986).

2.2.3 Treadmilling of Microtubules

Theoretical treatment (Wegner 1976) as well as direct observation of purified MTs in vitro show that there are conditions under which tubulin subunits assemble at one end, while disassembly proceeds at the other end (treadmilling) (Caplow et al. 1982; see Margolis and Wilson 1981 for review). Treadmilling could not be shown in living cells (Scherson et al. 1984; Wadsworth and Salmon 1986) and its efficiency for force generation was estimated to be disappointingly low (Caplow et al. 1982; Hill and Kirschner 1982).

2.2.4 Organelle Movement Along Microtubules

Movements of organelles in axons (Weiss 1986a), chromatophores (Schliwa 1984), and *Foraminifera* (Travis et al. 1983) have long been known to be MT-dependent. It was only since AVEC microscopy (Allen 1985; Weiss 1986b) was introduced that the details of this movement could be observed directly (Allen et al. 1985; Allen and Weiss 1985; Koonce and Schliwa 1985).

In axoplasm of the giant axon of the squid extruded into a dissociation buffer closely mimicking the intracellular fluid, single MTs are observed by AVEC-DIC (Allen video-enhanced contrast differential interference contrast) microscopy with the Hamamatsu Photonic Microscope System C 1966 (Allen et al. 1985; Weiss 1986b) as was verified by electron microscopy. Neurofilaments were also abundant on electron micrographs, but not detectable on the video recordings. Organelles and particles which, according to their size and abundancy, were classified as mitochondria, small vesicles (50–100 nm), and larger round organelles (> 200 nm) were seen to attach to the MTs and to be transported along the MTs to their ends (Fig. 2b). Particles easily switched from one MT to an intersecting one. They were transported unidirectionally in fresh, ATP-containing preparations at velocities of 1–2 $\mu m\ s^{-1}$. In older preparations, or without added ATP, an increase of MTs transporting bidirectionally and a decrease in velocity were observed. Fluorescently labeled polystyrene

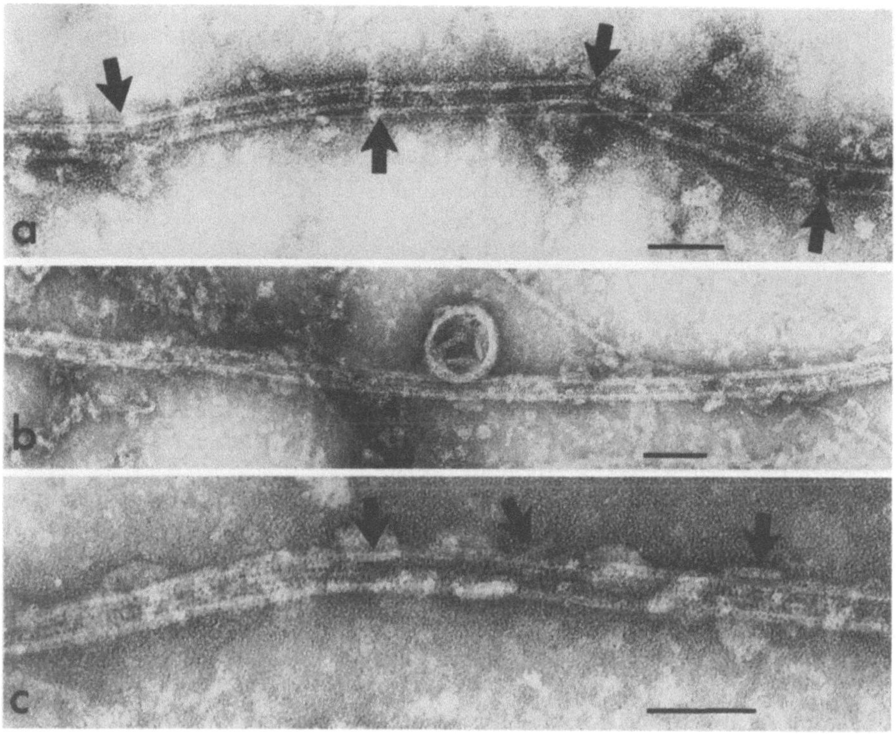

Fig. 2a–c. Electron micrographs of negatively stained, native MTs from dissociated squid axoplasm. **a** MTs show breaks which are almost regularly spaced rather than bends. Breaks are more often observed in native preparations as compared with purified MTs, possibly because both the grid and MT surfaces are less smooth in this more heterogeneous situation. *Bar* = 100 nm. **b** MT with an associated vesicle. Also an amorphous, possibly helical fuzzy coat is present around the MT. Neurofilaments are not associated with the MTs or organelles. *Bar* = 100 nm. **c** MTs in buffer containing AMP-PNP (adenylyl imidodiphosphate). Several rodlike filaments (*arrows*) are attached to the surface of each MT, similar to the side arms observed on the vesicles in the presence of ATP (Langford et al. 1987). One filament is oriented parallel to the long axis. The globular crescent-shaped projections alternate on opposite sides of the MT in an almost helical pattern. Some of these globular projections have a filamentous substructure (Langford et al. 1987)

particles were also transported. Thirty μM taxol did not interfere with this movement. These findings show that the properties of intracellular organelle movement, e.g., axoplasmic transport, can to a great extent be brought about by a single, native, cytoplasmic microtubule with its associated proteins.

Since the behavior of organelles is similar in many cell types (Rebhun 1972; Schliwa 1984), although a direct involvement of MTs has not yet been visualized, we can assume that in most animal cells organelle motion is due to similar mechanisms. The fact that a protein similar to that thought to be involved in force generation in axons is also found in spindles, gives rise to new speculations about possible similarities with the mechanism of mitosis (Scholey et al. 1985).

The mechanism of rapid axonal transport is probably a crossbridge-type mechanism involving a partially soluble and partially organelle-bound ATPase which is often called kinesin. In the presence of nonhydrolyzable ATP-analogs movement stops and kinesin, and possibly other proteins forming the crossbridge, are found attached selectively to the MTs (Brady 1985; Miller and Lasek 1985; Vale et al. 1985 a; Kuznetsov and Gelfand 1986; Weiss 1987; Weiss et al. 1987; Langford et al. 1987) (Fig. 2c).

2.2.5 Gliding of Microtubules

In the summer of 1984 Allen and Weiss observed for the first time active gliding movements of cytoplasmic MTs from extruded and dissociated or gently homogenized squid axoplasm (Allen and Weiss 1985; Allen et al. 1985; Weiss and Allen 1985; Allen 1987; Weiss et al. 1987):

In fresh preparations or when ATP was added, MT gliding near the glass surface was observed irrespective of whether the MTs transported organelles or not. MTs (0.5 – 30 μm in length) which dissociated from the bulk axoplasm showed longitudinal gliding over the glass surface at velocities between 0.2 and 0.6 μm s^{-1}. MT gliding was not inhibited by taxol and, therefore, was considered to be independent of MT assembly, disassembly, or treadmilling. The velocity was not correlated with the MT length so that we concluded that the generation of force must occur all along the length of the MT. Movements of MTs protruding from pieces of cytoplasm appeared as gentle lateral motions resembling waving seaweed and MTs often glided out of the bulk of the filamentous meshwork while one end remained attached, thus forming a growing loop. When the MTs hit an obstacle the rear end started a fishtailing motion (Fig. 3). Entangled MTs formed rings or U-shaped structures which rotated in the presence of ATP. Although during all these movements MTs appeared as stiff elastic rods, they could be bent by the pushing forces exerted by their tail segments to a miminal radius of curvature of about 0.2 μm.

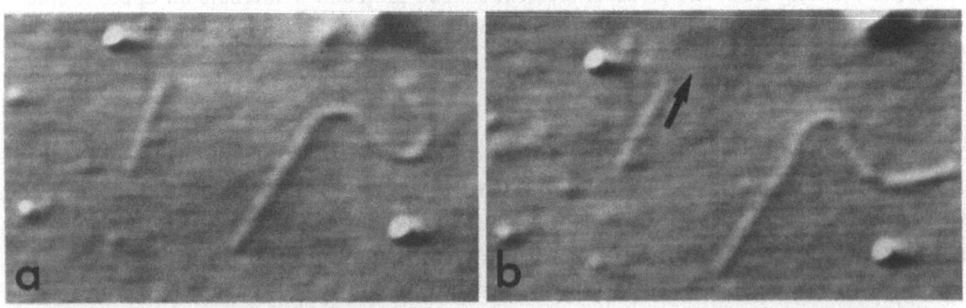

Fig. 3 a, b. AVEC-DIC images showing the behavior of a gliding and a fishtailing native MT as observed for 5 s. The MT on the *right* is attached to the glass surface only at the front end for 3.5 μm, while the rear end, in an attempt to move forward, is forced to move laterally in a serpentine fashion. Note that some of the particles depicted are firmly attached to the glass surface, while others are in Brownian motion. This movement, therefore, is within the first μm above the glass surface. *Bar* = 2 μm

MT gliding has also been found in reassembled MTs to which supernatant fractions or purified kinesins were added (Vale et al. 1985b; Pryer et al. 1986). Although hydrodynamic events have been found to explain most of the behavior of squid axonal MTs (Weiss 1985), it is now generally assumed that the same crossbridge mechanism that moves organelles is also responsible for gliding (Weiss and Allen 1985; Vale et al. 1985b; Weiss 1987).

2.2.6 Sliding of Microtubules

The movement of filaments such as MTs against each other is called sliding. A sliding MT mechanism is known to be responsible for the bending of cilia and flagella (for review, see Dustin 1984, Chap. 4) since it was demonstrated that the MTs of isolated axonemes slide against each other if exposed to mild proteolytic treatment (Warner and Mitchell 1981). The force-generating enzyme is the MT-associated ATPase dynein (Gibbons 1965; Johnson 1985).

In the light of this relatively well-understood type of movement the question arises, whether cytoplasmic MTs would also slide. It was indeed suggested that the wide range of organelle velocities found in the axopods of Foraminifera may be due to the fact that some organelles move along filaments which are themselves riding on another one so that the velocities of two or perhaps more units combine (Allen 1964). This, however, has not yet been proven experimentally.

In the squid axoplasmic preparations also no convincing examples for MT sliding have been observed. If the direction of MT sliding were the same as that of gliding they would move retrogradely, i.e., back to the cell body, which would be contrary to the well-known slow anterograde transport of MTs (for review, see Weiss 1986a). Lasek (1986) speculated that axonal MTs may be pulled, e.g., by actin filaments during slow transport.

While some *Protozoa*, contrary to higher animals, seem to use actin-based systems for organelle movement (Edds 1975), some algae, but not the higher plants, use an MT-based system for generating cytoplasmic streaming. This was suggested to be due to a circular sliding of MTs along themselves close to the cell wall (Kuroda and Manabe 1983). The distinction between streaming (in plants) and saltatory organelle motion (in animals) may no longer be justified, since the impression of a stream can be created by a large number of organelles moving independently, but with similar speed and direction (Weiss et al. 1986).

A sliding mechanism of parallel filaments was also proposed for the movement of chromosomes during mitosis (McIntosh et al. 1969). But since it had turned out that spindle MTs all have the same polarity, this mechanism is considered less probable than others (see Dustin 1984, Chap. 10 for review). However, the very promising recent model of Fuge et al. (1985) proposes that sliding could occur even in MTs of the same polarity if they are arranged at an angle.

2.2.7 Movement of Axostyle Microtubules

Axostyles are bundles of several thousand singlet MTs which form a helical ribbon that extends from the anterior to the posterior end of certain zooflagellates. MTs are organized into rows and within the rows the MTs are connected by regularly spaced linkers (Grimstone and Cleveland 1965; Bloodgood et al. 1974; Langford and Inoue 1979). But, contrary to the heliozoan axopods (see Hausmann, this Vol.), axostyles propagate bending waves along their lengths which provide a primitive form of locomotion to the organism. The bends which form and propagate either as smooth or sawtooth waves are considered to be the result of MT sliding as in cilia and flagella (e.g. Langford and Inoue 1979) since a dyneinlike protein was found (Bloodgood 1975) and since the axostyles are directly connected at one end to the bases of the flagella which often occur in large numbers in these organisms. Recent studies support Langford and Inoue's (1979) earlier hypothesis that the dynein crossbridges must reside within, rather than between, the rows of MTs (Heuser 1986).

2.2.8 Complex Interactions of Microtubules

A wide variety of active movements of MTs seems to be caused by the fact that organelles moving along an MT can have contact to more than one MT (Martz et al. 1984; Allen et al. 1985). By this mechanism native MTs of squid axoplasm can often be seen to be dragged by a particle, exerting considerable tension on MT arrays to which they are attached. If a particle rides along two crossed MTs these are pulled together by a zipperlike action (Fig. 4). Alternatively, a particle crossing a perpendicularly oriented second MT may pull this for some distance until the contact breaks when tension is too high and the MT recoils elastically ("plucking a string," Allen et al. 1985). Such movements caused by interactions between organelles and MTs have mainly been observed in disorganized MT arrays, but they very much resemble many of the MT movements in *Allogromia* axopods (Travis et al. 1983).

This most complex and variable class of interactive movements is far from being well analyzed, but it is possible that it is of key importance for understanding other types of intracellular motility. This behavior shows directly the intrinsic elastic forces and the flexural rigidity of MTs and demonstrates clearly that the forces that keep organelles on the MTs must be strong.

2.2.9 Contraction of Microtubule Arrays

The recently discovered gelation-contraction of MT suspensions could be either due to sliding or MT – organelle interactions. The latter seems to be more probable as is discussed in depth by Weisenberg (Chap. II.3).

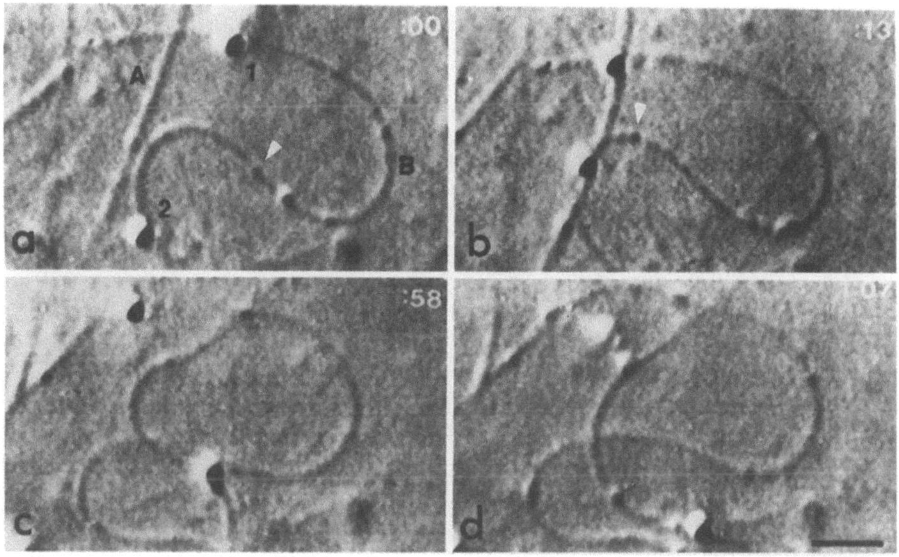

Fig. 4a–d. A sequence of videomicrographs (AVEC-DIC) shows the interactions between two native MTs (*A* and *B*) and two large organelles (*1* and *2*), and bidirectional particle motion on MT *B*. Particles *1* and *2* move in the same direction along *B*; *1* migrates to the end of *B*, but attaches transiently to *A* enough to deform it. A smaller particle moves in the opposite direction of particles *1* and *2* on MT *B* (*arrowhead*). Particle *2* makes dual contact with *A* and *B* in frames **b** and **c**, causing both *A* and *B* to deform due to the force exerted by particle/MT interaction. In frame **d** the attachment of *2* to *B* is suddenly broken, but that to *A* remains. Both MTs recoil elastically. *Bar* = 2 µm (Allen et al. 1985)

2.3 Conclusions

The dual nature of the MTs, namely, that they are stiff and static if stabilized, but also elastic, generating motive force, or quickly growing and shrinking, renders them one of the most mysterious and fascinating cellular components. MTs move organelles through the cell, but form at the same time the cytoskeletal meshwork which to some extent hinders such movement. It may be noteworthy that we have evidence to believe that the as yet unknown forces keeping organelles associated with the MTs are not only high enough to overcome the cytoplasmic viscosity (which is difficult to estimate exactly), but also the friction a large organelle experiences when forced through axons much thinner than its own size (Buchner et al. 1987). These force can also create considerable tension in cytoskeletal scaffolds which may together with the forces produced by the actin-based stress fiber system be involved in the maintenance of cell shape (cf. also Chap. I.1.7).

The other enigma that remains to be solved is the molecular mechanisms by which the MTs move actively and transport organelles. At present, it is unknown whether its location and mode of action are hydrodynamic (Weiss

1985), MT- and surface-associated (Vale et al. 1985b), or exclusively MT-associated (Allen et al. 1985).

Acknowledgments. The original research reported here was supported in part by the Deutsche Forschungsgemeinschaft, a Marine Biological Laboratory Fellowship, and in its initial phase by a Boehringer Ingelheim Fonds fellowship to DGW, by NIH grants GM 28107 to GML, and NS 19962 to RDA.

References

Allen RD (1964) Cytoplasmic streaming and locomotion in marine *Foraminifera.* In: Allen RD, Kamiya N (eds) Primitive motile systems in cell biology. Academic Press, London, pp 407–432

Allen RD (1985) New observations on cell architecture and dynamics by video-enhanced contrast optical microscopy. Ann Rev Biophys Biophys Chem 14:265–290

Allen RD (1987) Microtubules as intracellular engines. Sci Am, 255:42–49

Allen RD, Weiss DG (1985) An experimental analysis of the mechanisms of fast axonal transport in the squid giant axon. In: Ishikawa H, Hatano S, Sato H (eds) Cell motility: mechanism and regulation. 10. Yamada Conference, Sept. 1984. University of Tokyo Press, Tokyo, pp 327–333

Allen RD, Weiss DG, Hayden JH, Brown DT, Fujiwake H, Simpson M (1985) Gliding movement of and bidirectional transport along single native microtubules from squid axoplasm: evidence for an active role of microtubules in cytoplasmic transport. J Cell Biol 100:1736–1752

Amos LA (1977) Arrangement of high molecular weight associated proteins on purified mammalian brain microtubules. J Cell Biol 72:642–654

Amos LA (1979) Structure of microtubules. In: Roberts K, Hyams JS (eds) Microtubules. Academic Press, London, pp 1–64

Amos LA, Klug A (1974) Arrangement of subunits in flagellar microtubules. J Cell Sci 14:523–550

Bear T, Schmitt FO, Young JZ (1937) Investigations on the protein constituents of nerve axoplasm. Proc R Soc Lond B Biol Sci 123:520–529

Bereiter-Hahn J (1978) A model for microtubular rigidity. Cytobiologie 17:298–300

Bloodgood RA (1975) Biochemical analysis of axostyle motility. Cytobios 14:101–120

Bloodgood RA, Miller KR, Fitzharris TB, McIntosh JR (1974) The ultrastructure of *Pyrsonympha* and its associated microorganisms. J Morphol 143:77–106

Brady ST (1985) A novel brain ATPase with properties expected for the fast axonal transport motor. Nature 317:73–75

Buchner K, Seitz-Tutter D, Schönitzer K, Weiss DG (1987) A quantitative study of anterograde and retrograde axonal transport of exogenous proteins in olfactory nerve C-fibers. Neuroscience (in press)

Caplow M, Langford GM, Zeeberg B (1982) Concerning the efficiency of the treadmilling phenomenon with microtubules. J Biol Chem 257:15012–15021

Cassimeris L, Wadsworth P, Salmon ED (1985) Dynamic instability and differential stability of cytoplasmic microtubules in human monocytes. In: De Brabander M, de Mey J (eds) Microtubules and microtubule inhibitors 1985. Elsevier, Amsterdam, pp. 119–125

Cohen WD, Bartelt D, Jaeger R, Langford GM, Nemhauser I (1982) The cytoskeleton system of nucleated erythrocytes. I. Composition and function of mayor elements. J Cell Biol 93:828–838

Crepeau RH, McEwen B, Edelstein SJ (1978) Differences in α and β polypeptide chains of tubulin resolved by electron microscopy with image reconstruction. Proc Natl Acad Sci USA 75:5005–5010

Dentler WL, Granett S, Rosenbaum JL (1975) Ultrastructural localization of the high molecular weight proteins associated with in vitro assembled brain microtubules. J Cell Biol 65:237–241

Dustin P (1984) Microtubules, 2nd edn. Springer, Berlin Heidelberg New York

Edds KT (1975) Motility in *Echinosphaerium nucleofilum*. I. An analysis of particle motions in the axopodia and a direct test for involvement of the axon. J Cell Biol 66:145–155

Erickson HP (1974) Microtubule surface lattice and subunit structure and observations on reassembly. J Cell Biol 60:153–167

Febvre-Chevalier C, Febvre J (1986) Motility mechanisms in the actinopods (*Protozoa*): a review with particular attention to axopodial contraction/extension, and movement of nonactin filament systems. Cell Motil Cytoskel 6:198–208

Fuge H, Bastmeyer M, Steffen W (1985) A model for chromosome movement based on lateral interaction of spindle microtubules. J Theor Biol 115:391–399

Gall JG (1966) Microtubule fine structure. J Cell Biol 31:639–644

Gibbons IR (1965) Chemical dissection of cilia. Arch Biol 76:317–352

Grimstone AV, Cleveland LR (1965) The fine structure and function of the contractile axostyles of certain flagellates. J Cell Biol 24:387–400

Heuser JE (1986) Different structure states of a microtubule cross-linking molecule, captured by quick-freezing motile axostyles in protozoa. J Cell Biol 103:2209–2227

Hill TL, Kirschner MW (1982) Free energy calculations for microtubule treadmilling. Int Rev Cytol 84:185–234

Johnson KA (1985) Pathway of the microtubule-dynein ATPase and the structure of dynein: a comparison with actomyosin. Ann Rev Biophys Biophys Chem 14:161–188

Kim H, Binder LI, Rosenbaum JL (1979) The periodic association of MAPs with brain microtubules in vitro. J Cell Biol 80:266–276

King SM, Hyams JS, Luba A (1982) Absence of microtubule sliding and an analysis of spindle formation and elongation in isolated mitotic spindle from the yeast *Saccharomyces cerevisiae*. J Cell Biol 94:341–349

Koonce MP, Schliwa M (1985) Bidirectional organelle transport can occur in cell processes that contain single microtubules. J Cell Biol 100:322–326

Kuroda K, Manabe E (1983) Microtubule-associated cytoplasmic streaming in *Caulerpa*. Proc Jpn Acad 59 B:131–134

Kuznetsov SA, Gelfand VI (1986) Bovine brain kinesin is a microtubule-activated ATPase. Proc Natl Acad Sci USA 83:8530–8534

Langford GM (1980) Arrangement of subunits in microtubules with 14 protofilaments. J Cell Biol 87:521–526

Langford GM (1983) Length and appearance of projections on neuronal microtubules in vitro after negative staining: evidence against a cross-linking function for MAPs. J Ultrastruct Res 85:1–10

Langford GM, Inoue S (1979) Motility of the microtubular axostyle in *Pyrsonympha*. J Cell Biol 80:521–538

Langford GM, Williams E, Peterkin D (1986) Microtubule-associated proteins (MAPs) of dogfish brain and squid optic ganglia. Ann NY Acad Sci 466:440–443

Langford GM, Allen RD, Weiss DG (1987) Substructure of sidearms on squid axoplasmic vesicles and microtubules visualized by negative contrast electron microscopy. Cell Motil Cytoskel 7:20–30

Lasek RJ (1986) Polymer sliding in axons. J Cell Sci 5:161–179

Leterrier JF, Liem RKH, Shelanski ML (1982) Interactions between neurofilaments and microtubule-associated proteins: a possible mechanism for interorganellar bridging. J Cell Biol 95:982–986

MacGregor HC, Stebbings H (1970) A massive system of microtubules associated with cytoplasmic movement in telotrophic ovarioles. J Cell Sci 6:431–449

McIntosh JR, Hepler PK, van Wie DG (1969) Model for mitosis. Nature 224:659–663

Mandelkow E-M, Mandelkow E (1985) Unstained microtubules studied by cryoelectron microscopy. Substructure, supertwist and disassembly. J Mol Biol 181:123–135

Margolis RL, Wilson L (1981) Microtubule treadmills – possible molecular machinery. Nature 293:705–711

Martz D, Lasek RJ, Brady ST, Allen RD (1984) Mitochondrial motility in axons: membranous organelles may interact with the force generating system through multiple surface binding sites. Cell Motil 4:89–101

Miller RH, Lasek RJ (1985) Crossbridges mediate anterograde and retrograde vesicle transport along microtubules in squid axoplasm. J Cell Biol 101:2181–2193

Mitchison T, Kirschner M (1984) Dynamic instability of microtubule growth. Nature 312:237–242

Mitchison T, Evans L, Schulze E, Kirschner M (1986) Sites of microtubule assembly and disassembly in the mitotic spindle. Cell 45:515–527

Morris JR, Lasek RJ (1982) Stable polymers of the axonal cytoskeleton: the axoplasmic ghost. J Cell Biol 92:192–198

Pickett-Heaps JD, Tippit DH, Porter KR (1982) Rethinking mitosis. Cell 29:729–744

Pryer NK, Wadsworth P, Salmon ED (1986) Polarized microtubule gliding and particle saltations produced by soluble factors from sea urchin eggs and embryos. Cell Motil Cytoskel 6:537–548

Rebhun LI (1972) Polarized intracellular particle transport: saltatory movements and cytoplasmic streaming. Int Rev Cytol 32:93–137

Sasaki S, Stevens JK, Bodick N (1982) Serial reconstruction of microtubule arrays within dendrites of the cat retinal ganglion cell: the cytoskeleton of a vertebrate dendrite. Brain Res 259:193–206

Scherson T, Kreis TE, Schlessinger J, Littauer UZ, Borisy GG, Geiger B (1984) Dynamic interactions of fluorescently labeled microtubule-associated proteins in living cells. J Cell Biol 99:425–434

Schliwa M (1984) Mechanisms of intracellular transport. In: Shay J (ed) Cell and muscle motility, vol 5. Plenum, New York, pp 1–82

Schliwa M, Pryzwansky KB, van Blerkom J (1982) Implications of cytoskeletal interactions for cellular architecture and behaviour. Philos Trans R Soc Lond B Biol Sci 299:199–205

Scholey JM, Porter ME, Grissom PM, McIntosh JR (1985) Identification of kinesin in sea urchin eggs and evidence for its localisation in the mitotic spindle. Nature 318:483–486

Smith DS, Järlfors U, Beranek R (1970) The organization of synaptic axoplasm in the lamprey (*Petromyzon marinus*). J Cell Biol 46:199–219

Smith DS, Järlfors U, Cameron BF (1975) Morphological evidence for the participation of microtubules in axonal transport. Ann NY Acad Sci 253:472–506

Stebbings H, Hunt C (1982) The nature of the clear zone around microtubules. Cell Tissue Res 227:609–617

Tilney LG, Bryan J, Bush DJ, Fujiwara K, Mooseker MS, Murphy DB, Snyder DH (1973) Microtubules: evidence for thirteen protofilaments. J Cell Biol 59:267–275

Travis JL, Kenealy JFX, Allen RD (1983) Studies on the motility of the *Foraminifera.* II. The dynamic microtubular cytoskeleton of the reticulopodial network of *Allogromia laticollaris.* J Cell Biol 97:1668–1676

Tsukita S, Ishikawa H (1980) The movement of membranous organelles in axons. Electron microscopic identification of anterogradely and retrogradely transported organells. J Cell Biol 84:513–530

Vale RD, Reese TS, Sheetz MP (1985a) Identification of a novel, force-generating protein, kinesin, involved in microtubule-based motility. Cell 42:39–50

Vale RD, Schnapp BJ, Reese TS, Sheetz MP (1985b) Organelle, bead and microtubule translocations promoted by soluble factors from the squid giant axon. Cell 40:559–569

Voter WA, Erickson HP (1982) Electron microscopy of MAP 2 (microtubule associated protein 2). J Ultrastruct Res 80:374–382

Wadsworth P, Salmon ED (1986) Analysis of the treadmilling model during the metaphase of mitosis using fluorescence redistrubution after photobleaching. J Cell Biol 102:1032–1038

Warner FD, Mitchell DR (1981) Polarity of dynein-microtubule interactions in vitro: crossbridges between parallel and antiparallel microtubules. J Cell Biol 89:35–44

Wegner A (1976) Head to tail polymerization of actin. J Mol Biol 108:139–150

Weiss DG (1985) Dynamics and cooperativity in the organization of cytoplasmic structures and flows. In: Haken H (ed) Complex systems – operational approaches in neurobiology, physics, and computers 1985. Springer, Berlin Heidelberg New York, pp 179–181

Weiss DG (1986a) The mechanism of axoplasmic transport. In: Iqbal Z (ed) Axoplasmic transport. CRC, Boca Raton FL, pp 275–307
Weiss DG (1986b) Visualization of the living cytoskeleton by video-enhanced microscopy and digital image processing. J Cell Sci 5:1–15
Weiss DG (1987) Visualisation of microtubule gliding and organelle transport along microtubules from squid giant axons. In: Wohlfarth-Bottermann K-E (ed) Nature and function of cytoskeletal proteins in motility and transport. Prog Zool 34: (in press)
Weiss DG, Allen RD (1985) The organization of force generation in microtubule-based motility. In: De Brabander M, De Mey J (eds) Microtubules and microtubule inhibitors 1985. Elsevier, Amsterdam, pp 232–240
Weiss DG, Gross GW (1983) Intracellular transport in axonal microtubular domains. I. Theoretical considerations on the essential properties of a force generating mechanism. Protoplasma 114:179–197
Weiss DG, Keller F, Gulden J, Maile W (1986) Towards a new classification of intracellular particle movements based on quantitative analyses. Cell Motil Cytoskel 6:128–135
Weiss DG, Seitz-Tutter D, Langford GM, Allen RD (1987) The native microtubule as the engine for bidirectional organelle movements. In: Smith RS (ed) Axonal transport. Alan R Liss, NY, pp 91–111
Yamazaki S, Maeda T, Miki-Noumura T (1982) Flexural rigidity of singlet microtubules estimated from statistical analysis of fluctuating images. In: Sakai H, Mohri H, Borisy GG (eds) Biological functions of microtubules and related structures. Academic Press, Tokyo, pp 41–48

II.3 The Nature and Significance of ATP-Induced Contraction of Microtubule Gels

RICHARD C. WEISENBERG[1]

3.1 Introduction

Among the most essential, and dramatic, events in the life of a cell are the processes of cell motility. Cytoplasmic microtubules (we will not consider flagella and cilia microtubules in the present discussion) are a key element in cellular motility (for a review, see Hyams and Stebbings 1979). They appear to be involved in such diverse phenomena as chromosome movement during mitosis, transport of vesicles in axons (Allen et al. 1985) and other cells (Hayden et al. 1983), and elongation of retina cells (Warren and Burnside 1978). Although extensively investigated little is known about either the process controlling microtubule organization in cells, or about the mechanism by which they produce movement.

A major reason for our lack of knowledge about the cellular processes controlling microtubule organization and function is that until recently, there have not been in vitro systems for producing physiological-type behavior of microtubules under controlled conditions. This situation has begun to change, and rapid progress in understanding the physiology of microtubules is certain to occur in the next few years. Apparent progress has already been made in identifying the components involved in the rapid microtubule-based transport of vesicles in axons (Brady 1985; Vale et al. 1985). I will discuss here a new system discovered in my lab (Weisenberg and Cianci 1984) which may provide a model for another aspect of microtubule function in living cells.

3.2 Microtubule Gelation-Contraction

3.2.1 In Vitro Experiments

Our experiments are performed with crude microtubule proteins prepared by three cycles of assembly and disassembly in the presence of glycerol. It is important to use a relatively crude preparation because, as I will discuss, it is not any of the usual microtubule-associated proteins (MAPs) that appear responsible for the observed phenomenon. The process we have discovered we refer to as microtubule gelation-contraction, by analogy with actomyosin gelation-contraction. Like actomyosin gelation-contraction, the microtubule-based process

[1] Department of Biology, Temple University, Philadelphia, Pennsylvania, USA

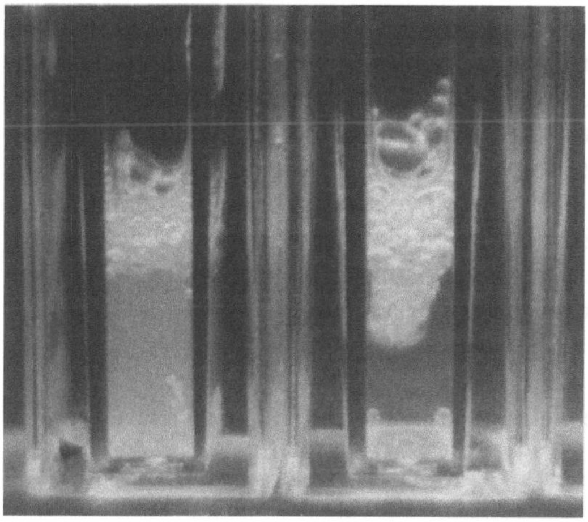

Fig. 1. Contraction of a microtubule gel in the presence of ATP. Both samples were identical preparations of crude microtubule proteins isolated from calf brains, and both were in the presence of GTP to induce microtubule assembly. The sample on the *right* received 2 mM ATP after 30 min incubation. The solutions were photographed 1 h later. The turbid microtubule suspension is visible on the *left*, while the contracted microtubule gel is clearly visible on the *right*. Gels may contract upward, as in this example, or downward, or may split in half with part moving up and part moving down

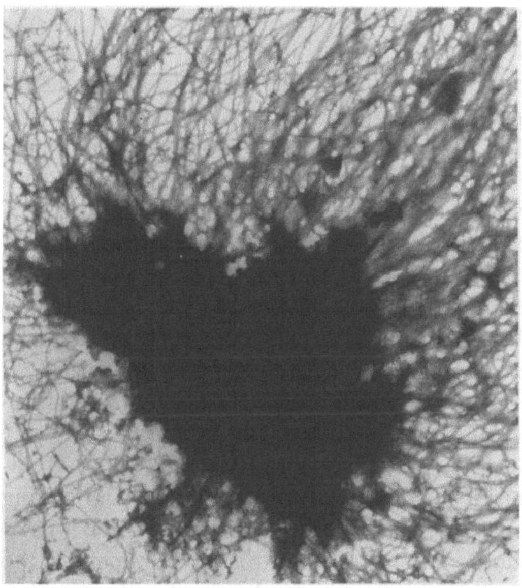

Fig. 2. Electron micrograph of an asterlike structure formed during microtubule gelation-contraction. A contracted gel, as shown in Fig. 1, was dispersed, placed on a carbon-coated grid, and negatively stained with uranyl acetate. The asterlike structures consist of a darkly stained central mass from which the microtubules radiate. Similar structures may be observed by light microscopy (Weisenberg et al. 1986). Magnification is approximately ×10000

requires ATP, and the result of adding ATP is the contraction or syneresis of a gel. An example of a contracted microtubule gel is shown in Fig. 1. Unlike actomyosin contraction, microtubule gelation-contraction does not begin as soon as ATP is added. Instead, it does not become visible until approximately 30 min after ATP addition (although we now know that motility begins as soon as ATP is added). Contraction continues for about another 30 min and at completion the contracted gel may occupy as little as 20% of the initial volume of solution.

We have examined the process of microtubule gelation-contraction by video-enhanced microscopy (Weisenberg et al. 1986). This allows individual microtubules to be visualized (Allen et al. 1981). These observations have revealed that microtubule gelation-contraction is a complex process, and involves particulate structures, which copurify with tubulin as well as microtubules. The identity of the particulates is discussed below. During gelation-contraction the particulates are transported along the microtubule walls, while the microtubules themselves undergo redistribution. Starting from a random state the microtubules and the associated particles move, and gradually become organized into structures which consist of a particulate center from which microtubules radiate (Fig. 2). This structure is very similar to that of a mitotic spindle aster, and we will refer to them as "asterlike structures".

3.2.2 Significance of Microtubule Gelation-Contraction in Living Cells

3.2.2.1 Mitotic Spindle

It is generally conceded that the mitotic spindle forms in vivo by the action of a microtubule organizing center (MTOC). The MTOC is believed to act by initiating, or nucleating, microtubule formation, and subsequent growth of the microtubules from their site of initiation results in the structure known as an aster. In contrast, the asterlike structures formed during ATP-induced microtubule gelation-contraction must form by a different mechanism. Since microtubules are present at the onset of contraction it is unlikely that a nucleation-based mechanism could be involved. Indeed, no evidence for growth of microtubules from the aster centers has been observed. The asterlike structures appear instead to be formed by the unidirectional transport of the particulates along the microtubules, and by the redistribution of the existing microtubules.

The existence of a motility-based mechanism for constructing asterlike structures raises the question: Does this process occur in the cell, and if it does what is its role? The evidence that mitotic spindle asters form by an assembly-based mechanism appears very strong. Certainly asters can be formed this way in vitro (Weisenberg and Rosenfeld 1975; Mitchison and Kirschner 1984). It appears unlikely that the motility-based formation of asters, as it occurs during microtubule gelation-contraction, has a major role during in vivo aster formation. However, we cannot eliminate the possibility that it is a component of the in vivo process, or is a significant mechanism in some kinds of cells. Another possibility is that a similar motile process is involved in the assembly of the microtubule organizing center itself, which has been shown to undergo maturation events during the cell cycle (Weisenberg and Rosenfeld 1975; Spiegelman et al. 1979; Telzer and Rosenbaum 1979). The maturation of the organizing center may involve the accumulation of particulate material around the centriole, and a process such as that observed during microtubule gelation-contraction in vitro could be involved.

Also dramatic is the resemblance of mitotic asters formed in vivo to the structures formed during microtubule gelation-contraction in vitro, it appears likely that this is an artifact, induced by disrupting the normal physiological organization of a microtubule-based system of motility. This leads to the issue of the identification of the motile process involved.

3.2.2.2 Axonal Transport

The clearest demonstration of microtubule-based motility in nerve cells is that of fast axonal transport. During fast axonal transport membrane-limited vesicles (e.g., synaptic vesicles or their precursors) are transported in an anteriograde (from the cell body toward the synapse) direction at a rate of 200–400 mm/day (2–4 μm/s) (Lasek et al. 1984). Evidence for an involvement of microtubules in this process came originally from structural studies and from the use of antimicrotubule drugs. Recently, the role of microtubules in this process has been firmly established by the induction of similar movements along microtubules in vitro (Allen et al. 1985), and a protein, termed kinesin, has been isolated which has been proposed to be the ATP-transducing enzyme involved (Vale et al. 1985), although a demonstration that kinesin has ATPase activity has yet to been presented (Brady 1985).

It seems unlikely that microtubule gelation-contraction is directly related to fast axonal transport. The material transported is not vesicular, and the rate of movement is typically about 1 mm/min (Weisenberg et al. 1986), or over 100 times less than that of fast axonal transport. In addition to fast axonal transport, at least two slower transport processes have been demonstrated in axons, these have been termed slow component a (SCa) and slow component b (SCb) (Black and Lasek 1980). These have also been termed axonal transport components V and IV (Lorenz and Willard 1978). SCa is the slower of the two components, with a typical rate of 1 mm/day (0.7 μm/min), while SCb is two to ten times than SCa. SCa and SCb differ dramatically in the components transported. SCa is composed of a limited number of peptides, the most prominent of which are tubulin and neurofilament peptides, which make up about 75% of the total. Also present in SCa are peptides which have been identified as brain spectrin, or fodrin, and peptides identified as "tau" proteins, a class of microtubule-associated proteins (Lasek et al. 1984). SCb, in comparison, consists of over 200 peptides, including actin, calmodulin, and various enzymes (Lasek et al. 1984).

We have recently isolated the particulates which undergo transport along microtubule walls during gelation-contraction. The isolated particulates added to pure tubulin will undergo ATP-dependent gelation-contraction, while neither the particulates nor pure tubulin will do so alone. The peptide composition of the particulates has been analyzed by one- and two-dimensional gel electrophoresis. They consist primarily of tubulin and neurofilament peptides (which make up about 75% of the total). Additional peptides present have been provisionally identified as brain spectrin and a tau protein. In its major peptides this pattern is almost identical to that which is carried by SCa transport.

Both the rate of transport and the composition of the isolated particulates are very similar to the rate and composition of the SCa system of axonal transport. We therefore believe that we have isolated the components involved in SCa transport. This conclusion raises several significant questions. When the composition of SCa was first demonstrated it was assumed that it reflected the slow transport of the microtubule-neurofilament cytoskeleton (Black and Lasek 1980; Lasek et al. 1984). Since the neurofilaments and the microtubules of the axon are highly interconnected by crossbridges (Ellisman and Porter 1980;

Hirokawa 1982), it seemed reasonable that their transport should be at the same rate. However, the view that slow transport reflected movement of the assembled cytoskeleton was not uniformly accepted, and was criticized by Ochs (1982), who felt that this view was inconsistent with morphological data. Ochs proposed instead that the transport of neurofilament and tubulin peptides occurred in the form of soluble subunits of these polymers.

However, the conclusion of Ochs is itself inconsistent with recent results, which indicate that tubulin transported in SCa differs in its solubility and peptide composition from the bulk of the microtubules in the axon (Brady et al. 1984; Tashiro et al. 1984). SCa tubulin is to a large extent insoluble, or particulate, in buffers that extract normal microtubules and the peptide composition of SCa tubulin differs from the soluble, nontransported tubulin. Likewise, the SCa-transported component of neurofilament protein appears to differ from the neurofilament lattice as such, which appears to be nearly stationary (Nixon and Logvinenko 1986). These results are consistent with ours, and together they provide strong evidence that SCa transport involves the movement of a unique structural complex consisting of tubulin, neurofilament, and spectrin peptides. For convenience we will refer to this structure as the SCAP (slow component a protein) complex.

Our results indicate that SCa transport occurs by movement of the SCAP complex along microtubule walls. Evidence that slow transport is microtubule-dependent has been previously presented, although there have also been contradictory results (Ochs 1982). Much of this confusion may be the result of failure to distinguish between SCa and SCb transport. The mechanism of this movement is not yet known. Our isolated SCAP particles contain significant ATPase activity, but we have not yet been able to disperse the SCAP complex and identify the ATPase, or to demonstrate that it is involved in motility. Although a protein corresponding in mulecular weight to the fast transport component "kinesin" (Vale et al. 1985) is present in our crude microtubule preparations, it is removed during isolation of the SCAP complex, and appears to have no role in their transport.

Many questions are raised by these results: What is the role of SCa-transported peptides? Are the SCa-transported neurofilament peptides and tubulin transferred to the ends, or walls, of the assembled axonal cytoskeleton? If this is true, why does the SCa tubulin apparently contain unique alpha subunits? What is the role of the other SCa peptides, particularly brain spectrin? What is the structure of the SCAP complex? Is there a stoichiometric relationship between the various peptides? Is there a stoichiometric relationship between the various peptides? Does a similar transport system exist in nonneuronal cells? If so, what is its function? What relationship, if any, does the SCAP system have to neuropathic conditions such as Alzheimer's disease, diabetic neuropathy, and other degenerative neurological conditions which may involve malfunctions in the neurofilament and microtubule systems (Elam and Cancalon 1984; Medori et al. 1985; Vitadello et al. 1985)? This is only a partial list of the many questions raised by our results, and while no answers can be given to these now, the availability of an in vitro model system for SCa axonal transport should help us to obtain the answers to many of these questions.

References

Allen RD, Allen NS, Travis JL (1981) Video-enhanced contrast, differential contrast (AVEC-DIC) microscopy: a new method capable of analyzing microtubule-related motility in the reticulopodial network of Allogromina latic collaris. Cell Motil 1:291–302

Allen RD, Weiss DG, Hayden JH, Brown DT, Fugiwake H, Simpson M (1985) Gliding movement of an bidirectional organelle transport along single native microtubules from squid axoplasm: evidence for an active role of microtubules in cytoplasmic transport. J Cell Biol 100:1736–1752

Black M, Lasek RJ (1980) Slow component of axonal transport: two cytoskeletal networks. J Cell Biol 86:616–621

Brady ST (1985) A novel ATPase with properties expected for the fast axonal transport motor. Nature 317:73–75

Brady ST, Tytell M, Lasek RJ (1984) Axonal tubulin and axonal microtubules: biochemical evidence for cold stability. J Cell Biol 99:1716–1724

Elam JS, Cancalon P (1984) Advances in neurochemistry, vol 6. Axonal transport in neuronal growth and regeneration. Plenum, New York

Ellisman MH, Porter KR (1980) Microtrabecular structure of the axoplasmic matrix: visualization of cross-linking structures and their distribution. J Cell Biol 87:464–479

Hayden JH, Allen RD, Goldman RD (1983) Cytoplasmic transport in keratocytes: direct visualization of particle translocation along microtubules. Cell Motil 3:1–19

Hirokawa N (1982) Cross-linker system between neurofilaments, microtubules, and membranous organelles in frog axons revealed by the quick-freeze, deep-etching method. J Cell Biol 94:129–142

Hyams JS, Stebbings H (1979) Microtubule associated cytoplasmic transport. In: Roberts K, Hyams JS (eds) Microtubules. Academic Press, London, pp 487–530

Lasek RJ, Garner JA, Brady ST (1984) Axonal transport of the cytoplasmic matrix. J Cell Biol 99:212–221

Lorenz T, Willard M (1978) Subcellular fractionation of intra-axonally transported polypeptides in the rabbit visual system. Proc Natl Acad Sci USA 75:505–509

Medori R, Autilio-Gambetti L, Monaco S, Gambetti P (1985) Experimental diabetic neuropathy: impairment of slow transport with changes in axon cross-sectional area. Proc Natl Acad Sci USA 82:7716–7720

Mitchison T, Kirschner M (1984) Microtubule assembly nucleated by isolated centrosomes. Nature 312:232–237

Nixon RA, Logvinenko KB (1986) Multiple fates of newly synthesized neurofilament proteins: evidence for a stationary neurofilament network distribution nonuniformly along axons of retinal ganglion cell neurons. J Cell Biol 101:647–659

Ochs S (1982) Axoplasmic transport and relation to other nerve functions. Wiley, New York

Spiegelman BM, Lopata MA, Kirschner M (1979) Aggregation of microtubule initiation sites preceding neurite outgrowth in mouse neuroblastoma cells. Cell 16:253–263

Tashiro T, Kurokawa M, Komiya Y (1984) Two populations of axonally transported tubulin differentiated by their interactions with neurofilaments. J Neurochem 43:1220–1225

Telzer BR, Rosenbaum J (1979) Cell cycle-dependent in vitro assembly of microtubules onto the pericentriolar material of HeLa cells. J Cell Biol 81:484–497

Vale RD, Reese TS, Sheetz MP (1985) Identification of a novel force-generating protein, Kinesin, involved in microtubule motility. Cell 42:39–50

Vitadello M, Filliatreau G, Dupont JL, Hassig R, Gorio A, Di Giamberardino L (1985) Altered axonal transport of cytoskeletal proteins in the mutant diabetic mouse. J Neurochem 45:860–868

Warren RH, Burnside B (1978) Microtubules in cone myiod elongation in teleost retina. J Cell Biol 78:247–258

Weisenberg RC, Cianci C (1984) ATP-dependent gelation-contraction of microtubules assembled in vitro. J Cell Biol 99:1527–1533

Weisenberg RC, Rosenfield AC (1975) In vitro polymerization of microtubules into asters and spindles in homogenates of surf clam eggs. J Cell Biol 64:146–158

Weisenberg RC, Allen RD, Inoue S (1986) ATP-induced formation and motility of aster-like structures in vitro. Proc Natl Acad Sci USA 83:1728–1732

II.4 Generation of Propulsive Forces by Cilia and Flagella

MICHAEL E. J. HOLWILL[1] and PETER SATIR[2]

4.1 Introduction

Eukaryotic cilia and flagella are long (5 – 200 μm), thin (0.2 μm), motile cellular organelles which are used to produce relative motion between a cell and its fluid environment, as observed for free-swimming flagellated protozoa or the mucus currents flowing over the ciliated epithelial surface of the mammalian lung. Since the structure and function of cilia and flagella are identical, it will be convenient in this chapter to use the word "cilium" to include both types of organelle. Cilia execute oscillatory behavior in which bends are propagated along the organelles at frequencies which are generally between 5 and 100 Hz. Bends may be two- or three-dimensional in character, and exhibit various degrees of spatial symmetry. Thus, on some cilia almost symmetric waves (sometimes erroneously described as sinusoidal) are seen, while on others considerable asymmetry is observed. When the waves are asymmetrical, the larger bend is referred to as the "principal bend" and the smaller as the "reverse bend." The so-called typical ciliary beat consists of an effective stroke, in which the principal bend forms at the base of the organelle with the distal portion of the shaft held practically straight, followed by a recovery stroke in which the bend propagates to the cilium tip; usually there is no sharp distinction between the two strokes, and the movement appears as a smooth cycle.

The structure of cilia as revealed by electron microscopy is remarkably consistent among species, as noted in the last paragraph, with a set of nine doublet microtubules surrounding a pair of singlet microtubules – the so-called 9 + 2 structure or axoneme (Fig. 1). Associated with the doublet microtubules are the dynein arms, which project toward the neighboring doublet, spokes, which project toward the center of the axoneme, and interdoublet links. Hooplike structures, known collectively as the central sheath, and to which the spokes may attach, connect the central pair. Fine linkages join each doublet to the ciliary membrane, which is contiguous with that of the cell. Some cilia, such as mammalian sperm tails and trypanosome flagella, contain additional structures, usually in the form of thick, dense (as judged by electron microscopy) fibers which lie along part or all of the organelle outside the axoneme, but within the membrane. Certain cilia bear hairlike processes on the external surfaces of their membranes; these may be apparently stiff mastigonemes, as seen in

[1] Physics Department, King's College, Strand, London WC2R 2LS, England
[2] Department of Anatomy and Structural Biology, Albert Einstein College of Medicine, Bronx, New York 10461, USA

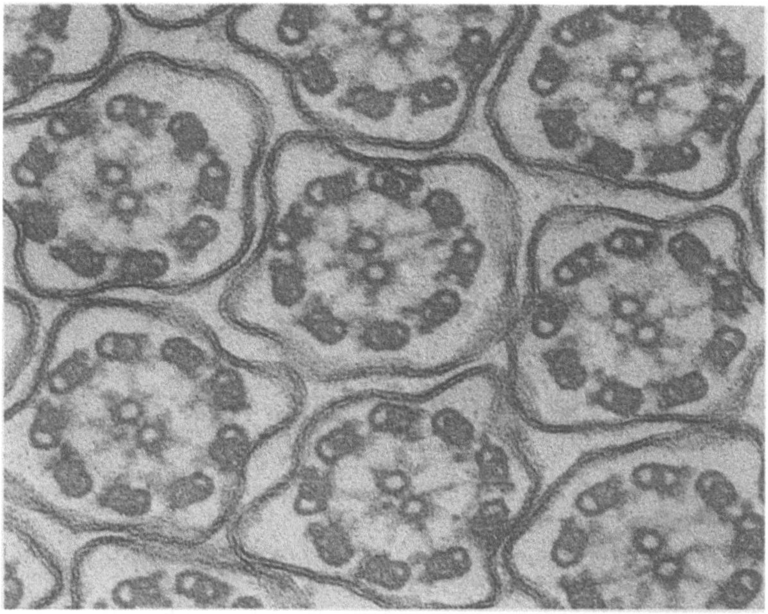

Fig. 1. Transverse section of *Elliptio* gill cilia, showing the 9 + 2 microtubular structure. Also visible are the dynein arms, the radial spokes, the interdoublet links, and the central sheath (× 150500)

Ochromonas, or rather flexible, and wrapped around the ciliary shaft, as in *Chlamydomonas* or *Euglena*.

The sustained amplitudes of bends propagated along many cilia indicate that active bending forces are distributed throughout the ciliary length (e.g. Machin 1963). There is considerable evidence (e.g. Satir 1982) to support the sliding-microtubule model of ciliary activity, in which the dynein arms undergo a cyclic activity involving mechanochemical changes and interact with the neighboring doublet to induce a shear force, and hence a tendency to slide, between neighboring doublets. The restriction of free sliding by ciliary structures, which have not yet been unequivocally identified, converts the shearing force into a bending moment, which, subject to appropriate control mechanisms, leads to the observed coordinated patterns of movement.

A primary aim of research on ciliary activity is to obtain a detailed description of the molecular interactions which lead to the movement of the organelle. Since these interactions provide the forces which bend the cilium, it is useful to investigate the resistance to motion offered by both the fluid and the elasticity of the cilium, as this will provide minimal levels of force and energy to be provided by the molecular process. Hydrodynamic analyses of beating cilia have been used to obtain estimates of the forces generated by these organelles, while mathematical and computer modeling techniques have provided tentative descriptions of the elastic and other mechanical properties which the organelles require to exhibit their observed motility. These analyses are supported by a few studies in which the forces exerted by a cilium are measured directly. In this chapter, it is appropriate to consider first the forces exerted between a cilium and its fluid environment, and then to examine the properties required of the cilium for bend propagation to occur.

4.2 Hydrodynamic Interactions

Interactions between the cilium and the fluid produce the forces responsible for relative motion between the organelle and its environment. Because of their small size and speed of movement, cilia have a low Reynolds number (*Re*) – the parameter which gives an estimate of the relative importance of inertial forces and viscous forces. For cilia, *Re* is of the order 10^{-3}, indicating that inertial forces can be neglected in comparison with the viscous ones (Sleigh and Blake 1977). This provides considerable simplification in respect of the equations which govern fluid motion in general, and has allowed the derivation of equations specific to the movement of cilia (e.g. Gray and Hancock 1955; Holwill and Sleigh 1967; Blake 1972; Lighthill 1976).

In a comprehensive hydrodynamic investigation of the motion of ciliated microorganisms, Lighthill (1976) demonstrated that the fluid flow produced by a motile cilium can be represented by a set of stokeslets distributed along the centerline of the organelle, together with dipoles of appropriate strength to satisfy the boundary conditions. A stokeslet represents the effect of a force acting at a single point of the fluid and for a force **F** gives rise to a velocity distribution

$$\mathbf{u} = \frac{\mathbf{F}}{6\pi\eta r} + \frac{1}{4} r^2 \operatorname{grad}\left[\operatorname{div}\left(\frac{\mathbf{F}}{6\pi\eta r}\right)\right], \qquad (1)$$

where **u** is the velocity at a distance r from the point of application of the force, while η is the coefficient of viscosity of the fluid. For a cilium **F** depends on position (s) along the centerline and is conveniently expressed as the force, $\mathbf{f}(s)$, exerted per unit length on the fluid. If the distribution of stokeslets of strength $\mathbf{f}(s)$ is accompanied by dipoles of strength, $(a^2\,\mathbf{f}_n(s))/(4\,\eta)$, the boundary condition of no-slip at the cilium surface is almost satisfied. $\mathbf{f}_n(s)$ is the vector obtained by resolving $\mathbf{f}(s)$ onto the plane normal to the centerline, while a is the radius of the cilium.

To gain information relating to the mechanical properties of a cilium from a consideration of the forces and couples acting upon it, it is necessary to specify the distribution of stokeslets along the organelle. This can be done in principle, since Eq. (1) can be used to describe the velocities at the ciliary surface, and these can be measured experimentally using photographic or video recording techniques. As Eq. (1) gives the fluid velocity in terms of the stokeslet distribution, and not *vice versa,* the computation of the latter is not a simple matter, and has been performed only for a few special cases, such as a helically beating cilium.

An alternative approach to the derivation of appropriate equations of motion, developed by Gray and Hancock (1955) before Lighthill's more rigorous analysis, uses resistance coefficients to obtain an expression for the force exerted by a cilium. A resistance coefficient is the force exerted by the fluid on unit length of a straight cilium when moving at unit velocity in a particular direction. It is convenient to define two coefficients, one (C_T) for motion parallel to the axis of the organelle, and the other (C_N) for motion normal to this axis, so that the force exerted normal to its axis by an element of a cilium of length ds, moving with a velocity V_N in the direction of the normal is

$$dF_N = C_N V_N ds \tag{2}$$

with a similar expression for dF_T. Taking components of dF_N and dF_T in the direction required, and integrating over the length of the organelle and a complete beat cycle, an expression for the ciliary force in that direction is obtained; the form of the expression depends on the motion of the individual cilium considered, and can be determined for any geometry. The only parameters not accessible to direct measurement are the resistance coefficients C_N and C_T, and their values have been the subject of considerable discussion, as it is not clear how these parameters should be represented for a cylinder which is being continually deformed.

Lighthill (1976) showed that the original values for the resistance coefficients derived by Gray and Hancock (1955) were in error, and, realizing the usefulness of this approach, made use of his work to obtain improved values for these parameters. It should be emphasized that the rigorous approach based on Eq. (1) is to be preferred, as it provides better information about the forces exerted on the cilium and also yields the fluid flow pattern about the cilium. However, the ease with which resistance coefficients can be applied to general ciliary motions has led to their widespread use by many authors whose work is relevant to the present discussion. Comparison of the rigorous approach with the resistance coefficient method indicates that the refined coefficients give satisfactory estimates of force generation for helical beating of arbitrary amplitude and for planar beating where the amplitude is small compared with the wavelength; Lighthill also considered that they can be applied with reasonable confidence to planar undulations of arbitrary amplitude, but there is at present no direct analytical support for the supposition.

Of particular significance are the absolute values of the two resistance coefficients, together with their ratio, as these appear in any equation derived in the manner outlined above. Lighthill gave the following expressions for a cilium of radius a moving in a medium of viscosity η:

$$C_N = \frac{4\pi\eta}{\ln(2q/a) + \frac{1}{2}} \quad \text{and} \quad C_T = \frac{2\pi\eta}{\ln(2q/a)}. \tag{3}$$

For an undulation of wavelength λ measured *along* the cilium, q takes the value $0.09\,\lambda$, while for a straight cilium of length l, its value is $0.22\,l$.

In earlier studies using the values for the resistance coefficients given by Gray and Hancock (1955), good agreement was obtained between the predict-

Table 1. Velocities and power dissipation for some ciliated micro organisms

Organism	$(C_T/C_N)_L$	V_{GH}	V_L	V_M	W_{GH}	W_L	R_W	Reference
Lytechinus sperm	0.58	178	149	158	44	61	1.40	Brokaw 1965
Ciona sperm	0.58	168	145	165	40	56	1.41	Brokaw 1965
Chaetopterus sperm	0.58	118	100	105	36	51	1.42	Brokaw 1965
Psammechinus sperm	0.58	192	158	191	31	43	1.41	Gray and Hancock 1955
Crithidia oncopelti	0.59	19.7	19.2	17.0	22	31	1.44	Holwill 1966
Trypanosoma vivax	0.61	12.1	11.5	18.0	25	38	1.52	Holwill 1966

Key: C_T Resistance coefficient for parallel motion; C_N resistance coefficient for normal motion; V velocity in μm s^{-1}; W relative power dissipation; *GH* Gray and Hancock; *L* Lighthill; *M* measured; R_W power ratio = W_L/W_{GH}; $(C_T/C_N)_{GH}$ is always 0.5.

ed swimming speeds of cells and those observed (e.g. Gray and Hancock 1955; Brokaw 1965; Holwill 1966) and this agreement led to misplaced confidence in the values for C_N and C_T. It is therefore of interest to compare the velocities predicted in the earlier work with those using Lighthill's improved coefficients. This is done for several organisms in Table 1. Despite the more rigorous approach by Lighthill, it can be seen that the agreement with the experimental values is not greatly improved, and that the two predictions are similar. The agreement occurs because the expression for the velocity depends essentially on the ratio of the coefficients, which has a value of 0.5 in the Gray and Hancock treatment and roughly the same value in Lighthill's analysis. The different values for the coefficients become important in assessing the power dissipated by the cilium against viscous forces, because in this case the appropriate expression depends on the coefficients separately, rather than on their ratio. This is clear from a consideration of the expression for the power dissipation P, given by

$$P = \int (C_N V_N^2 + C_T V_T^2)\, ds\,, \tag{4}$$

which can be integrated if the geometry of the ciliary motion is known. Table 1 contains the results of power calculations for undulating cilia using the two sets of resistance coefficients discussed; inspection of the values indicates that in earlier studies the power expenditure was underestimated by between 40% and 50%. The energy needed to overcome viscous resistance is provided ultimately by the breakdown of ATP molecules; values in Table 1 represent the minimum energy to be made available, assuming that the energy needed to overcome elastic resistance is negligible. These rates of energy supply are available from the dephosphorylation of ATP molecules, which can provide up to 40 kJ mol^{-1} under physiological conditions. Brokaw and Gibbons (1975) noted that the efficiency of conversion of chemical energy to mechanical energy by flagella, estimated to be about 30%, is lower than that for muscle, between 45% and 70% (Kushmeric and Davies 1969; Pybus and Tregear 1972). Brokaw and Gibbons (1975) based their estimate on Gray and Hancock's (1955) expressions for the resistance coefficients, and a revised value using Lighthill's (1976) approach suggests an efficiency of 40% or 45%. A comparative argument (Brokaw 1975) indicates that the elastic and viscous bending moments are of the same order of magnitude at normal beat frequencies; the energy required to overcome elastic resistances is therefore likely to be of the same order as that dissipated against viscous forces. If allowance is made for this, a mechanochemical conversion efficiency as high as 80 or 90% is indicated.

4.3 Passive Elastic Properties

It was natural that early theoretical models of a bending cilium should consider the motion of the organelle in terms of an elastic rod immersed in a viscous fluid; it was this approach that enabled Machin (1963) to demonstrate that the observed motile behavior required an energy source distributed along the length of the cilium. To incorporate the elastic properties of a cilium, a model is created in which an actively generated bending moment is balanced by elastic and viscous bending moments. In principle, these moments can be expressed in terms of the physical and chemical characteristics of the system, but its structural complexity requires that many assumptions be made in the formulation. As noted by Brokaw (1975) and others, an homogeneous elastic rod is neither a good nor a helpful model for a cilium, and some effort has been expended in achieving a more realistic representation. A difficulty associated with this endeavor is that the number of parameters needed to describe the system is so large that the set of values required to constrain the model to behave realistically is not unique; of importance, however, is the fact that realistically beating models have been produced using reasonable assumptions (e.g. Brokaw 1985; Lubliner and Blum 1972) and some deductions about the mechanism can be made.

The importance of the elastic properties of an organelle can, in principle, be assessed by a critical investigation of its waveshape. One way in which this can be achieved has been described by Silvester and Holwill (1972), who used a Fourier analysis method to differentiate between likely ciliary waveforms. Applying this method to *Crithidia oncopelti*, Johnston et al. (1979) showed definitively that the flagellum of this organism propagated bends that were more closely matched by a waveform consisting of circular arcs joined by straight lines than by a sinusoid or a meander. It is not an easy matter to differentiate between these three waveshapes by eye, and each may have different implications for the flagellar mechanism. Some authors, using visual curve-fitting procedures, have described flagellar waveshapes as meanderlike (e.g. Rikmenspoel 1971) and sine-generated – a shape close to the meander (e.g. Hiramoto and Baba 1978). However, earlier, other workers (Brokaw 1965; Satir 1967) described the arc-line form. In cilia in particular, measurements of the internal alignment of spokes of different doublets demonstrated clearly the existence of circular arcs and long, entirely straight regions (Warner and Satir 1974). The meander is the shape which might be expected if elastic properties dominate bend formation, since it is the shape which minimizes the stored elastic potential energy in a beam with freely hinged ends. The arc-line shape actually found reinforces the suggestion that the organelle is not behaving as a bending beam.

Changes in the shape of an organelle with time are also affected by its elastic properties. Consider a small element of a cilium, assumed to be elastic and bending in a viscous environment under the influence of an active bending moment. If, on reaching the maximum curvature, the active moment is suddenly removed, the element would straighten passively under the influence of its own elasticity, with the curvature decaying exponentially with time (t), i.e., as e^{-kt}, where k is a constant determined by the elasticity and viscosity. In their study, Jonston et al. (1979) showed that the bending of the *Crithidia* flagellum had the form predicted, but that the curvature during straghtening was not an exponential function of time, suggesting that straightening is not determined solely by the passive elastic properties of the axonemal element. Elastic effects would give graduated, rather than abrupt, transitions between circular arcs and straight regions on a cilium, and this result shows that the transitions are more abrupt than anticipated by passive elasticity alone.

4.4 Active Mechanical Properties

A recent approach to the interpretation of ciliary mechanisms relies on current understanding of how sliding microtubules operate. In the presence of 1 mM ATP, the dynein arms of a doublet microtubule (N) will push an adjacent doublet (N + 1) tipward with a velocity of about 14 μm s^{-1} (Yano and Miki-Noumura 1980). In the absence of constraints on sliding, this is accomplished as a function of the dynein arm crossbridge cycle (Gibbons and Gibbons 1974; Satir et al. 1981). The maximal force generated by a dynein arm has been measured by Kamimura and Takahashi (1981) to be about 1 pN. In the crossbridge cycle, a dynein arm undergoes successive conformational changes which allow attachment to and detachment from doublet N + 1. The state where the arm is virtually orthogonal to the microtubule and attached at both ends is called *rigor*, which seems uniformly achieved in the absence of ATP. Since all arms in an axoneme operate with a fixed cycle polarity during ciliary beating (e.g. Sale and Satir 1977), they must function asynchronously in order that a continuous shearing force can be exerted between two adjacent microtubules. A simple model suggests that in the presence of ATP, one-half of the axoneme contains doublets with active arms while the other half has inactive arms during most of the beat, with the halves switching back and forth at the ends of the effective and recovery strokes (Satir 1985). The two halves are separated by the plane of bending. A slightly more complex version of this model can be matched to an actual beat form in a *Paramecium* ciliary cycle (Sugino and Naitoh 1982). Thus, in the presence of ATP, the dynein arm may be in at least two states: an active cycling state and a relaxed or inactive state. In an intact axoneme, the radial spokes and interdoublet links may constrain active sliding, although the mechanisms by which such constraint is achieved are not entirely known.

The formation of transient crossbridges by the dynein arms and, possibly, the radial spokes and interdoublet links will cause the gross elastic properties of a cilium section to vary with time. If no crossbridges are made, the elastic behavior of the system will depend on the rigidity of the doublet and singlet microtubules. Experimental studies of the elasticity of bull and echinoderm spermatozoa have shown that the stiffness of the sperm tail in a medium without ATP is some ten to fifteen times greater than the value obtained with ATP present (Lindemann et al. 1973; Okuno and Hiramoto 1979; Okuno 1980). The stiffness of the bull sperm tail, at 30×10^{-21} N m^2 in the rigor state, and 2×10^{-21} N m^2 in the presence of ATP, is, as might be expected from its additional structures, greater than that of the echinoderm sperm flagellum (15×10^{-21} N m^2 in rigor, 0.9×10^{-21} N m^2 relaxed). The magnitudes in the presence of ATP have been used to estimate the Young modulus of the microtubules, assuming them to be the only contributors to the stiffness. Values in the region of 5×10^6 N m^{-2} are obtained, which are not unreasonable for protein structures. However, in the presence of ATP or ATP and vanadate, the state of the axoneme according to the switching models just discussed may be nonuniform, and some crossbridges between microtubules may still be present, so that the stiffness measured may not be for the microtubules alone, and their Young modulus may accordingly be an overestimate.

We have indicated that the moment balancing method for analyzing ciliary motion is inaccurate insofar as the active bending moment is difficult to assess and the elastic bending moment of the passive organelle treated as a bending beam is not a good representation of the true elastic bending moment of the active axoneme. To apply the moment balancing method accurately to a beating cilium, with a view to gaining information about internal mechanical processes, it is necessary to know where bends are generated in relation to dynein arm activity. One implication of the switch-point hypothesis based on asynchronous active sliding microtubules is that dynein arm activity in one location of the axoneme can be converted into bending at another, different location (Spungin and Satir 1986, unpublished). An example where this occurs is given by Shingyoji et al. (1977) who pipetted ATP locally onto a sperm axoneme in rigor. The concentration gradient of ATP is such that active sliding is evidently induced near the point where the pipette is closest to the axoneme, while the ends of the axoneme, far from the point of application, remain in rigor. Oppositely directed bends form at either side of the active zone, between the region of active sliding and the region of rigor. An interpretation of this is that straight regions with dynein arm activity are relatively stiff. The active bending moment generated by the shearing due to sliding is transferred via the microtubules to a region remote from dynein arm activity. In this region between the active and rigor zones, the stiffness must be sufficiently small to allow bending to occur. The longitudinal elastic resistance of the microtubules must be great enough to withstand the applied longitudinal forces without collapsing and sufficiently incompressible to transmit the force generated at one crossbridge to a remote part of the system.

Measurements made on the drawings and photographs published by Shingyoji et al. (1977) for a typical experiment reveal that the two oppositely curved regions each have a radius of about 14 μm and are separated by a straight section about 10 μm long. Under the conditions of their experiments the viscous resistance to bending is negligible compared with the elastic resistance, and the bending moment required to produce the observed curvature can be estimated at 7×10^{-17} N m using our earlier value for the Young modulus of a "relaxed" axoneme. Assuming the bending to be induced by active sliding of one-half of the axoneme (i.e., five microtubules) in the straight region, and that one-half of the dynein arms are attached at any one time, the force exerted by an individual dynein arm is about 1 pN. This is consistent with the direct measurement made by Kamimura and Takahashi (1981) and supports the assumption that half the dynein arms are exerting force at any one time. One implication of this is that the attachment phase occupies about one-half of the mechanochemical cycle of the arm.

From calculations of this type, we should be able to construct a more accurate three-dimensional pattern than hitherto available of arm activity during the beat of cilia with a variety of wave patterns. It may also be possible to specify the physical location and nature of the structures offering elastic resistance to shear. Given the growing reliability of the information, there is a realistic expectation that we shall be able to incorporate the chemical events into the model, thereby providing a complete mechanochemical specification of the system.

4.5 Conclusions

1. Considerations of the hydrodynamic aspects of ciliary motion should use Lighthill's (1976) rigorous treatment; this leads to improved estimates of the power expenditure by the cilium and suggests an efficiency of energy conversion comparable to that of muscle.
2. The waveshape of a bending cilium consists of circular arcs joined by straight lines, rather than a meander. The arc-line shape and the suggestion that straightening is not determined solely by passive elastic properties reinforce the conclusion that the organelles are not behaving as a bending beam.
3. Switching models of dynein arm activity suggest that in the presence of ATP the elastic resistance along the axoneme may be nonuniform. The Young modulus of axonemal microtubules (5×10^6 N m^{-2}) is somewhat overestimated.
4. Straight regions with dynein arm activity along a beating cilium are relatively stiff. Dynein arm activity at one location of the axoneme can be converted into bending at another location remote from the site of activity.
5. Calculations of the active bending moment yield a value of about 7×10^{-17} N m, with a corresponding value of about 1 pN for the force exerted by an individual dynein arm.
6. There is a realistic expectation that we will be able to incorporate both chemical events and mechanical parameters into a comprehensive description of the system.

Abbreviations and Symbols

a	Radius of cilium
ATP	Adenosine triphosphate
C_T	Resistance coefficient for parallel motion
C_N	Resistance coefficient for normal motion
$\left(\frac{C_T}{C_N}\right)_L$	Ratio of Lighthill's force coefficients
$\left(\frac{C_T}{C_N}\right)_{GH}$	Ratio of Gray and Hancock's force coefficients
$\mathbf{f}(s)$	Force exerted on fluid by unit length of cilium
$\mathbf{f}_n(s)$	Normal component of $\mathbf{f}(s)$
$\mathbf{F}$	Point force in fluid
F_N	Force exerted normal to axis of cilium
F_T	Force exerted parallel to axis of cilium
Hz	Hertz
k	Rate constant dependent on elasticity and viscosity
kJ	Kilojoule
l	Length of cilium
m	Meter
mM	Millimolar
mol	Mole
N	Newton
P	Power dissipation
pN	Piconewton (= 10^{-12} newton)
q	Length parameter dependent on ciliary shape
r	Distance from point force
Re	Reynolds number
R_W	Power ratio $-\left(\frac{W_L}{W_{GH}}\right)$
s	second
s	Distance along cilium centerline
t	Time
$\mathbf{u}$	Fluid velocity
V_{GH}	Velocity of organism calculated using Gray and Hancock's coefficients
V_L	Velocity of organism calculated using Lighthill's coefficients
V_M	Measured velocity of organism
V_N	Velocity of element of cilium normal to its axis
V_T	Velocity of element of cilium parallel to its axis
W_{GH}	Relative power dissipation calculated using Gray and Hancock's coefficients
W_L	Relative power dissipation calculated using Lighthill's coefficients
η	Coefficient of viscosity
Λ	Wavelength measured along cilium
μm	Micrometer

Acknowledgments. This work was supported by grants from the United States Public Health Service (HL 22560) to PS and from the Science and Engineering Research Council to MEJH.

References

Blake J (1972) A model for the micro-structure in ciliated organisms. J Fluid Mech 55:1–23

Brokaw CJ (1965) Non-sinusoidal bending waves of sperm flagella. J Exp Biol 43:155–169

Brokaw CJ (1975) Spermatozoan motility: a biophysical survey. Biol J Linn Soc (Suppl 1) 7:423–439

Brokaw CJ (1985) Computer simulation of flagellar movement, VI. Biophys J 48:633–642

Brokaw CJ, Gibbons IR (1975) Mechanisms of movement in flagella and cilia. In: Wu TY-T, Brokaw CJ, Brennan C (eds) Swimming and flying in nature, vol 1. Plenum, New York, p 89

Gibbons BH, Gibbons IR (1974) Properties of flagellar "rigor waves" formed by abrupt removal of adenosine triphosphate from actively-swimming sea-urchin sperm. J Cell Biol 63:970–985

Gray J, Hancock GJ (1955) The propulsion of sea-urchin spermatozoa. J Exp Biol 32:802–814

Hiramoto Y, Baba S (1978) A quantitative analysis of flagella movement in echinoderm spermatozoa. J Exp Biol 76:85–104

Holwill MEJ (1966) Physical aspects of flagellar movement. Physiol Rev 46:696–785
Holwill MEJ, Sleigh MA (1967) Propulsion by hispid flagella. J Exp Biol 47:267–276
Johnston DN, Silvester NR, Holwill MEJ (1979) An analysis of the shape and propagation of waves on the flagellum of *Crithidia oncopelti*. J Exp Biol 80:299–315
Kamimura S, Takahashi K (1981) Direct measurement of the force of microtubule sliding in flagella. Nature 293:566–568
Kushmeric MJ, Davies RE (1969) The chemical energetics of muscular contraction. II. Proc R Soc Lond B Biol Sci 174:315–353
Lighthill J (1976) Flagellar hydrodynamics. SIAM Rev 18:161–230
Lindemann CB, Rudd WG, Rikmenspoel R (1973) The stiffness of the flagella of impaled bull sperm. Biophys J 13:437–448
Lubliner J, Blum JJ (1972) Model of flagellar waves. J Theor Biol 34:515–534
Machin KE (1963) The control and synchronisation of flagellar movement. Proc R Soc Lond B Biol Sci 158:88–104
Okuno K (1980) Inhibition and relaxation of sea urchin sperm flagella by vanadate. J Cell Biol 85:712–725
Okuno K, Hiramoto Y (1979) Direct measurements of the stiffness of echinoderm sperm flagella. J Exp Biol 79:235–243
Pybus J, Tregear R (1972) Estimates of force and time of actomyosin interaction in an active muscle and of the number interacting at any one time. Cold Spring Harbor Symp Quant Biol 37:655–660
Rikmenspoel R (1971) Contractile mechanisms in flagella. Biophys J 11:446–463
Sale WS, Satir P (1977) Direction of active sliding of microtubules in *Tetrahymena* cilia. Proc Natl Acad Sci USA 74:2045–2049
Satir P (1967) Morphological aspects of ciliary motility. J Gen Physiol 50:241–248
Satir P (1982) Mechanisms and controls of microtubule sliding in cilia. Symp Soc Exp Biol 35:172–201
Satir P (1985) Switching mechanisms in the control of ciliary motility. Mod Cell Biol 4:1–46
Satir P, Wais-Steider J, Lebduska S, Nasr A, Avolio J (1981) The mechanochemical cycle of the dynein arm. Cell Motil 1:303–327
Shingyoji C, Murakami A, Takahashi K (1977) Local reactivation of Triton-extracted flagella by iontophoretic application of ATP. Nature 265:269–270
Silvester NR, Holwill MEJ (1972) An analysis of hypothetical flagellar waveforms. J Theor Biol 35:505–523
Sleigh MA, Blake JR (1977) Methods of ciliary propulsion and their size limitations. In: Pedley TJ (ed) Scale effects in animal locomotion. Academic Press, London, p 243
Sugino K, Naitoh Y (1982) Simulated cross-bridge patterns corresponding to ciliary beating in *Paramecium*. Nature 295:609–611
Warner FD, Satir P (1974) The structural basis of ciliary bend formation. J Cell Biol 63:35–63
Yano Y, Miki-Noumura T (1980) Sliding velocity between outer doublet microtubules of sea-urchin sperm axonemes. J Cell Sci 44:169–186

II.5 The Cortical Cytoplasmic Actin Gel

THOMAS P. STOSSEL, PAUL A. JANMEY, and KEN S. ZANER[1]

5.1 Historical Background

For 2 centuries, cell biologists have inferred that some kind of reversibly forming "gel" in the cell periphery is important for cell shape, cell movements, and other cell functions. Johan August Roesel von Rosenhof of Nuremberg, Germany (1705–1759) was evidently the first to recognize the special character of the periphery of an amoeboid cell. His illustrated description of what he termed "a delicate outer skin", captures the organelle exclusion characteristic of this region (1755) (Fig. 1). Historical accounts have credited Felix Dujardin (1801–1867) as the originator of the concept of "gel" in cell structure (DeBruyn 1947; Taylor and Condeelis 1979; Porter 1984). Dujardin's concept of "sarcode," the "diaphenous and gelatinous" substance of the cell as the essence of life (1835) is perhaps the most mystical and metaphysical expression of what later came to be called protoplasm. However, with respect to descriptions of cell consistency, Dujardin was anticipated by nearly a century. Abraham Trembley (1700–1784) ascribed the elasticity, contractility, and propulsive activity of the tentacles of the green hydra to a mucoid substance ("matière glaireuse") (1744). Martin Ledermüller (1719–1769) alluded to the "elasticity" of protozoa (1761), and Johann Eichhorn (1718–1790) described an amoeboid cell as "mere slime" (*ein bloßer Schleim*) (1781). An early use of the term gel to describe a cell is in Linnaeus' 1760 classification of animals which states that the protist *Volvox* has a "free gelatinous body" (corpus liberum gelatinosum). This usage is echoed by the Danish naturalist, Otto Friedrich Müller (1730–1784), who produced an early classification of protozoa. In a commentary concerning an amoeboid cell named "proteus tenax," Müller pictured a "clear, gelatinous, transparent body (corpus gelatinosum pellucidum) filled with dark granules ... from which extends a sharp spicule, the content of which is glassy (hyalini) ... and which is retracted to the cell body (apex spiculi rursus in globuum retrahitur & c)" (1786) (Fig. 1).

It is unclear what the word gel meant to eighteenth century naturalists. The Oxford Dictionary identifies one derivation from the latin verb "gelare," "to freeze," implying that contemporary observers believed the cell to have both liquid and solid properties. Another derivation is from a French word meaning a sticky fish soup. In either case, the idea of the cell as gel must have become widely known, because Jean Baptiste Lamarck (1744–1829) and others alluded

---

[1] Hematology/Oncology Unit, Massachusetts General Hospital, Department of Medicine, Harvard Medical School, Boston, Massachusetts 02114, USA

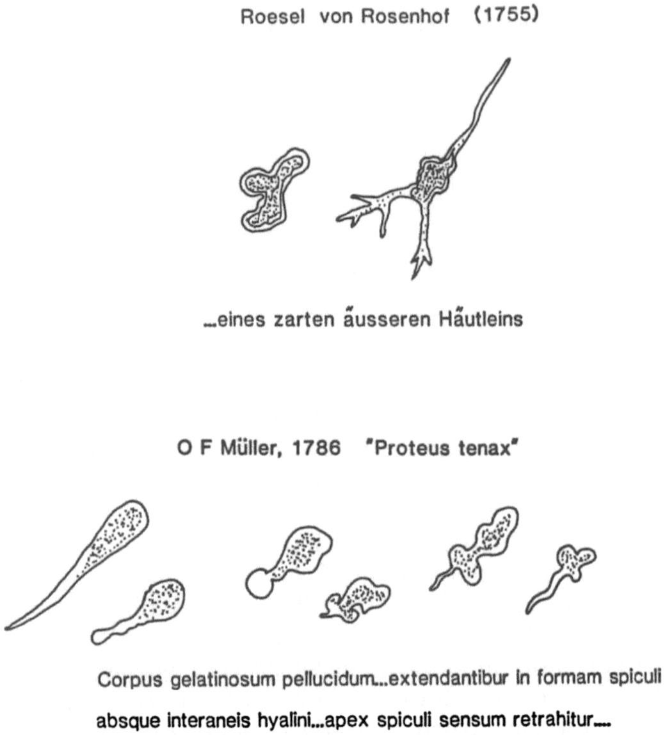

Fig. 1. Early appreciation of the hyaline cortex of eukaryotic cells. Amoeboid cells redrawn from the work of von Rosenhof (*top*) and Müller (*bottom*)

to the gelatinous body of protozoa nearly a quarter of a century before Dujardin (1801).

Until recently, any notions about the physical and chemical nature of cytoplasmic gels were speculative and primarily based on observations with the light microscope. As recounted by Porter (1984), efforts to identify subcellular structures responsible for the physical properties of protoplasm were subjected to fatal sabotage by the accusation that the findings represented artifacts of specimen preparation for microscopy. Attempts to define such structures chemically (e.g. Reinke 1881) lacked adequate methodology and suffered from vitalist criticisms to the effect that "a plasmodium rubbed up in a mortar is no more a plasmodium than a ground up watch is a timepiece" (Heilbrunn 1954).

Although the elasticity and contractility of muscle was repeatedly related to similar properties in nonmuscle cells (e.g. Ecker 1848), the identification of proteins with viscous and contractile properties from muscle (Straub 1942; Szent-Gyorgyi 1947) and nonmuscle cells (Lowey, 1954) represented the beginning of a physical and chemical definition of the cell's mechanical properties. For example, Lowey (1954) concluded that an interaction between actin and myosinlike proteins might be responsible for consistency changes ("morphoplastic transformations") in slime mold extracts. A decade later when the fibrillar nature of muscle proteins was better characterized (e.g. Huxley 1963;

Hanson and Lowy 1963) and when Hatano and Oosawa (1966) and others had purified actin and myosin from nonmuscle cells, it was possible to conclude with greater confidence that "increases in cytoplasmic consistency are caused by the formation of 50–70 angstrom filaments from precursors in the groundplasm" (Pollard and Ito 1970). Furthermore, concomitant with the initial biochemical studies of nonmuscle actin, it became evident that actin filaments are principal structures of the peripheral cytoplasm historically associated with motility and consistency changes. The initial information was that transmission electron micrographs of thin sections of cells revealed 4 to 6 nm diameter filaments as a dominant element in peripheral cytoplasm, especially in microvilli and pseudopodia (reviewed by Wessels et al. 1970). Reaction of permeabilized cells with the myosin fragments, heavy meromyosin, or myosin subfragment I, caused the filaments to appear thickened in thin section electron micrographs, and favorable images revealed an arrowhead configuration on the filaments, characteristic for actomyosin complexes (e.g. Ishikawa et al. 1969). Nevertheless, the power of the sliding filament model as an explanation for muscle contraction caused researchers to focus more on the interaction between actin and myosin as the basis of phase changes in cells than on actin itself (e.g. Taylor et al. 1973).

It was an accidental finding, however, which galvanized investigators to think about a primary role for actin in cytoplasmic structure. In attempting to isolate microtubule protein from Hawaiian sea urchin eggs, Kane discovered that heated egg extracts seemed to solidify and that bundles of actin comprised the major structural component of the egg extract gels (1975). Although the experimental protocols differed, several groups soon prepared actin gels from cytoplasmic extracts of widely different cell types (Stossel and Hartwig 1976; Pollard 1976; Taylor and Condeelis 1976; Boxer and Stossel 1976; Merriam and Clark 1978; Weihing 1977). The graphic impact of these extract gels which trapped air bubbles and withstood inversion of their containers brought home more than earlier viscosity measurements that actin gelation might represent the long-sought basis of peripheral cytoplasmic transformations between sol and gel states. Also of importance in this conceptual shift was the work of Tilney which showed a compelling correlation between actin assembly and the extension of the acrosomal process in marine sperm (Tilney et al. 1973); suggesting that actin assembly per se might be adequate for certain mechanical events in the cell. Moreover, suitably prepared gelled extracts when fortified with ATP underwent a "contraction" in which the gelled material expressed fluid and compressed itself into an aggregate, further suggesting that cellular contractility might somehow be related to actin gel formation.

Another intriguing feature of the cytoplasmic extract gels was the inhibition of their formation and dissolution by the cytochalasins (Hartwig and Stossel 1976; Weihing 1976; Condeelis and Taylor 1977). These fungal metabolites powerfully inhibit cell locomotion and phagocytosis, enhance exocytosis, and cause retraction and blebbing of the cortical cytoplasm (reviewed in Tannenbaum et al. 1979). Even more interesting, in light of attention to calcium as a second messenger in stimulus/response coupling, in general, and in excitation/contraction coupling, in particular, was the observation that the cytoplasmic extracts gelled only when the free calcium concentration was reduced to nanomolar concentrations by chelation.

5.2 The Assembly of Actin and Actin-Binding Proteins Regulating Actin Assembly

In the ensuing decade, much work has gone into defining what was being observed in these extracts and to relate the findings back to the cell. Of major importance in this effort has been a deepening of understanding concerning the mechanism of actin assembly and of the role of actin-binding proteins in modulating this assembly.

Actin is a highly conserved molecule. Although eukaryotic cells possess several genes encoding actin, the proteins produced differ only slightly in primary structure and in functional properties, at least as currently understood (e.g. Vanderkerckhove and Weber 1978). Pure monomeric actin is a slightly asymmetrical protein of Mr 42,500 which, in the presence of neutral salts, spontaneously assembles to form the double helical actin filament. The functional aspects of actin assembly and its regulation by actin-binding proteins has been reviewed in detail (Oosawa and Asakura 1974; Frieden 1985; Stossel et al. 1985; Pollard and Cooper 1986). Briefly, a rate-limiting step in actin assembly is the formation of a nucleus composed of two or three monomers. Therefore, the presence of preexisting actin nuclei, filaments, or other agents which substitute as nuclei can have an important effect on the kinetics of actin assembly. At the completion of polymerization from monomers, monomers continue to exchange with the ends of filaments which have a very large length distribution varying from a few that are greater than 10 μm in length to many that are a few monomers in composition. As in any bimolecular reaction, the rate of assembly is influenced by the concentration of the reactants and by the association and dissociation rate constants. Therefore, the concentration of actin monomers and the number of actin filament ends are important determinants of the assembly process. In the case of actin, the number of filament ends can be increased by a significant degree of spontaneous fracturing of filaments, and such fragmentation can be increased by mechanical disruption. The rate constants for addition and loss of actin monomers are about 25-fold different for the opposite ends of actin nuclei or filaments. In addition, the concentration of G-actin in dynamic equilibrium with actin filaments is different at the opposite filament ends. This concentration, known as the critical monomer concentration, is also about 15 times lower for the barbed as contrasted with the pointed filament end. Since exchange is so much faster at the barbed than at the pointed end, the critical concentration of pure actin is essentially that of the barbed end. Under ionic conditions believed to be reasonably physiological, this concentration is about 0.1 μM.

In many cells about half of the total actin is in the form of filaments that remain in a cellular "cytoskeleton" of cells subjected to detergent solubilization of their membranes. The rest is in a form which is extracted along with other "soluble" proteins by detergent treatment. When platelets are reacted with stimulatory agonists such as thrombin or leukocytes by the chemotaxis-inducing peptide F-met-leu-phe, by tumor-promoting phorbol esters, by a calcium ionophore in the presence of extracellular calcium, and by a number of other agonists, the fraction of detergent-insoluble actin transiently rises (Reviewed by Phillips and Fox 1984; Stossel 1986). Since the total actin concentration in the peripheral cytoplasm may be on the order of 400 μM, all but a miniscule fraction (0.1 μM) ought to be polymerized, based on the critical concentration of pure actin under ionic, temperature, and metabolite conditions thought to exist in the cytoplasm of a cell. Therefore, the assembly of actin in the cell must be regulated to keep so much of the actin unpolymerized in resting cells and to allow for rapid polymerization and depolymerization of actin in response to stimulation. Theoretically, there are two approaches to this regulation. In one, a molecule sequesters actin monomers with sufficient affinity to prevent their nucleation or exchange onto filaments. In a second, molecules bind the ends of filaments to prevent monomer addition or loss. Both mechanisms seem to be active in eukaryotic cells (see Tables 1 and 2).

Table 1. Actin monomer binding proteins

Given Name	Source
Profilin	Spleen, Acanthamoeba, Dictyostelium, marine eggs, blood platelets, lung macrophages, Physarum, bovine brain
Vitamin d binding protein	Human plasma
Depactin [a]	Marine eggs
19 KD protein	Bovine brain
Actophorin [a]	Acanthamoeba
Cofilin [b]	Bovine brain

[a] Severs actin filaments and removes an actin monomer to which it binds.
[b] Also binds to the sides of actin filaments.

Table 2. Some actin filament end-blocking proteins

Given name	Source
Gelsolin [a,d]	Mammalian eukaryotic cell cytoplasm, blood plasma
Villin [a]	Mammalian and avian brush border epithelial cells, amphibian eggs
Fragmin/severin [a]	Physarum, Dictyostelium, marine eggs
Capping protein	Acanthamoeba, bovine brain
Beta-actinin [b]	Striated muscle cells, kidney cells
Cap 42 a,b [c]	Physarum

[a] Calcium-sensitive severing activity present.
[b] Binds to the "pointed" end of myosin fragment-decorated F-actin.
[c] Regulated by phosphorylation.
[d] Inhibited by polyphosphoinositides (Janmey and Stossel, 1987)

The ability of the monomer-sequestering proteins such as profilin described in Table 1 to keep actin unpolymerized has several important implications, especially if its actin-sequestering function can be regulated. First, the unpolymerized actin can diffuse throughout the peripheral cytoplasm to serve as a source of monomers for assembly on demand. Second, the actin can be kept so far above the critical concentration that upon dissociation of the regulatory protein, assembly could be very rapid. However, this characteristic has the possible disadvantage that nucleation of actin would occur randomly in space. One way to gain spatial control is to couple changes in the polymerizability of actin with regulation at the rapidly growing end of actin filaments.

Table 2 summarizes cytoplasmic molecules recognized to confer this kind of regulation. Many of these proteins bind the fast-growing ends of actin filaments and prevent exchange of monomers with that end. In addition to blocking the ends of actin filaments, some of these proteins sever actin/actin bonds in filaments to rupture the filaments, and, in many cases, this severing action is dependent upon calcium. It is obvious how such an action could dissolve an actin network gel and be responsible for the calcium sensitivity noted for cytoplasmic extract actin gels. Since the cytochalasins also bind the fast-growing ends of actin filaments, their effects on gelation of cytyoplasmic extracts can also be ascribed to their ability to shorten the length of actin filaments.

5.3 The Rheology of Actin and its Modulation by Actin-binding Proteins and Other Factors

Although the high viscosity of actin in the presence of salts was ascribed early to its being in a filamentous state (Straub 1942), relatively little effort was made at the time to understand the basis of this mechanical property. A few investigators concluded on the basis of electron microscope, turbidometry, and sedimentation experiments that actin filaments exhibited a tendency to aggregate side-to-side (Mommaerts 1952; Johnson et al. 1963). Although the addition of myosin to F-actin results in the formation of a very solid gel as the actomyosin mixture consumed ATP (Weber and Portzehl 1954), again, the riveting power of the sliding filament model of muscle contraction hardly encouraged attention to the rheologic properties of actin filaments, with or without myosin, removed from their anisotropic state in the sarcomere.

Nevertheless, Kasai, Kawashima, and Oosawa undertook a formal study of the rheology of F-actin which was published in 1960. Using a Couette-type apparatus to measure the responses of solutions to oscillations, they documented a pronounced shear dependence on the deformability of an F-actin solution and reported that the rigidity of such solutions was directly proportional to the solution temperature. They also noted that stationary actin solutions exhibited significant birefringence which diminished under gentle shear. They interpreted the results by modeling a system of overlapping semiflexible rods. A good agreement between the absolute rigidity value and a calculation based on the assumptions of this model, rubber elasticity theory, and the temperature dependence findings led them to conclude that F-actin might be a system of crosslinked rubbery chains and that the links were reversibly broken by shear. Over a decade later, Maruyama, Kaibara, and Fukada confirmed the reversible shear dependence of F-actin's viscosity and dynamic rigidity, obtained a value, 0.1 Nm^{-2}, similar to Kasai et al. for the apparent rigidity of F-actin at a concentration of 1 mg ml^{-1} and also showed that the rigidity was proportional to the actin concentration raised to the 2.5 power (1974). However, in contrast to Kasai et al., they observed an inverse dependence of the rigidity on the solution temperature. Abe and Maruyama (1974) recorded that myosin and proteolytically-derived actin-binding myosin fragments markedly increased the dynamic rigidity of F-actin, that this effect was inhibited by ATP (which decreases the binding affinity of myosin for actin), and that, like F-actin alone, the consistency of the actomyosin "gel" is reversibly diminished by shear. The results were ascribed to "thixotropy," the production and destruction of unspecified crosslinks between filaments, notions rooted in the colloid science of the 1930s and 1940s (e.g. Goodeve 1939; Kruyt 1947).

Thus, when the cytoplasmic extract gelation phenomenon was discovered in the mid-1970s, there was little theoretical or experimental background on which to interpret the results, and it is no surprise that confusion and controversy arose. One common feature of cytoplasmic extract gels in addition to the inhibition of their formation by calcium was their relatively low protein concentration, in general (usually <10 mg ml^{-1}), and their low actin content (usually <2 mg ml^{-1}), in particular. Based on the assumption that some kind of nonideal molecular interaction is required for colloidal behavior, it seemed almost mandatory that the actin be in a highly extended conformation, most likely as long filaments, to explain the gellike consistency of the extracts. Moreover, highly purified muscle actin at comparable concentrations just did not seem qualitatively to have the same degree of consistency as the extract gels. This meant either that the cellular actin was different than muscle actin or else that other factors were stabilizing the actin in the extract gels. Consequently, some attributed the gelation of the extracts to a special property of the extract actin alone (Pollard 1976). Others blamed different actin-crosslinking proteins (Stossel and Hartwig 1976; Bryan and Kane 1978; Maruta and Korn 1977), and still others emphasized the complexity of the phenomenology involved in the formation of these extract gels (Taylor et al. 1976). Contemporary reviews encapsulate the irritation of knowledgable onlookers contemplating the early developments in this field (Hitchcock 1977; Clarke and Spudich 1977).

Impediments to clarification were serious. The actin-binding proteins purported to influence the gel structure of actin were initially purifiable only in small amounts. Also, in contrast to synthetic polymer systems, the actin gels under investigation were relatively weak and fragile, and researchers initially lacked instrumentation for surmounting this problem.

Viscometry and flow birefringence had been used extensively to monitor the assembly of actin. The viscosity and flow birefringence of actin solutions showed a direct proportionality with the F-actin concentration (e.g. Maruyama 1964), although the relative contributions of polymer mass and polymer length could not be estimated with confidence. Both of these methods should theoretically monitor an average filament length, but the usual shear rates applied in the experiments influenced the results by reducing the solution consistency. The extents to which filament realignment, destruction of "crosslinks," and rupture of the filaments contributed to this effect were debated. For example, Maruyama initially concluded that a reduction in F-actin's sedimentation velocity, dynamic rigidity, flow birefringence, and viscosity by beta-actinin, an actin-binding protein of skeletal muscle, was caused by beta-actinin's shortening of the average actin filament length (Maruyama 1971; Maruyama et al. 1975), but later interpreted this result as an inhibition of binding between the "pointed" ends and the sides of actin filaments (Maruyama et al. 1977).

This notion of interfilament crosslinks as end-to-side interactions was strongly promoted by some workers. They interpreted the striking inhibitory effects of a variety of agents known or surmised to bind the "barbed" ends of actin filaments on the viscosity of actin measured in a small volume falling ball device as evidence for important crosslinks between the ends and sides of F-actin which the agents were presumed to destroy (MacLean-Fletcher and Pollard 1980a; 1980b; 1980c; Isenburg et al. 1980; Griffith and Pollard 1982). Others favored the idea that changes in the average filament length caused by these compounds were adequate to explain the rheological effects observed (Hartwig and Stossel 1979; Yin and Stossel 1979; Yin et al. 1980; Nunnally et al. 1981; Zaner and Stossel 1982; Schliwa 1982; Tellam and Frieden 1983), an interpretation based on the main stream of thinking regarding the rheology of rodlike synthetic and biological materials (e.g. Ferry 1980).

A key point in resolving these controversies was whether or not the actin-modulating factors such as the cytochalasins or the proteins listed in Table 2 that affect actin's consistency really change the actin filament length. The mechanism of actin assembly suggests theoretically that the final polymer length distribution at steady state after polymerization ought to be exponential (Miyaki and Stockmayer 1974; Hill 1983), and length measurements of negatively stained actin filaments photographed in the electron microscope have verified such a distribution for purified F-actin (Kawamura and Maruyama 1971; Takebayashi et al. 1975; Hartwig and Stossel 1979; Yin et al. 1980; Nunnally et al. 1981; Janmey et al. 1986). However, there has been less consensus as to how seriously to take absolute values of the length measurements, which have yielded weight-average end-to-end lengths ranging between 2 and 4 μm, and changes in which were taken by some as evidence that the cytochalasins and certain actin-binding proteins decreased the consistency of F-actin by shortening the average filament length. There are reasons to suspect that artifacts related to the purification of the actin, its application to the electron microscope grids, possible selection biases in counting, and probably others might reduce the validity of measurements. One might expect the technique to underestimate the average length, because of fragmentation of and difficulty in measuring very long filaments. In spite of these theoretical problems, there has been a striking congruence in reported average lengths measured in the same and different laboratories, and the results have been in reasonable agreement with estimates made by physical techniques. Furthermore, the criticism of electron microscopy as an approach is probably less serious when the experimental intent emphasizes comparisons of effects of different treatments on the filament lengths rather than the absolute length values per se. As more and more agents have been discovered which have relatively obvious effects on actin filament lengths that are highly correlated with changes in actin's consistency, it seems increasingly reasonable to accept that agents such as the cytochalasins which produce more subtle changes in these parameters really do affect the actin filament length. However, the initial difficulty encountered in accepting this idea must be kept in mind when reviewing the development of concepts of action of the regulatory actin-binding proteins described above.

Another important parameter to be defined is the configuration of actin filaments. Most electron microscope images of F-actin give the impression that the filaments are relatively straight rods, although they occasionally exhibit bends of as much as 45°, and a series of such bends can make some filaments appear to curl. Nevertheless, the overall configuration of F-actin in such specimens tends more to the rod than to the coil, in that the end-to-end lengths of actin filaments are identical to their contour lengths below 4 μm and deviate only slightly downward as the contour length rises above 4 μm (Takebayashi et al. 1975). Again the apparent advantage of being able to visualize F-actin by a variety of electron microscope techniques is offset by the criticism of possible artifacts. However, experiments measuring the flexibility of myosin fragment/F-actin complexes in the light microscope using dark field or fluorescence optics and polarized fluorescence or quasi-elastic light scattering studies of F-actin yield similar estimates of the flexural rigidity and torsional motions of F-actin and an estimate of the elastic modulus for bending of approximately 7.5×10^{-17} dyn cm^{-2}, a value consistent with the actin filament being highly extended (Nagashima and Asakura 1980; Yanagida et al. 1980; Yoshimura et al. 1984; Fujime et al. 1984).

Therefore, it now seems reasonable to conceptualize F-actin as a relatively stiff rod composed of monomeric subunits and to formulate a simple model against which to fit the experimental data (Fig. 2). Dilute monomeric actin should exhibit ideal behavior, that is, its diffusion is what is expected for a

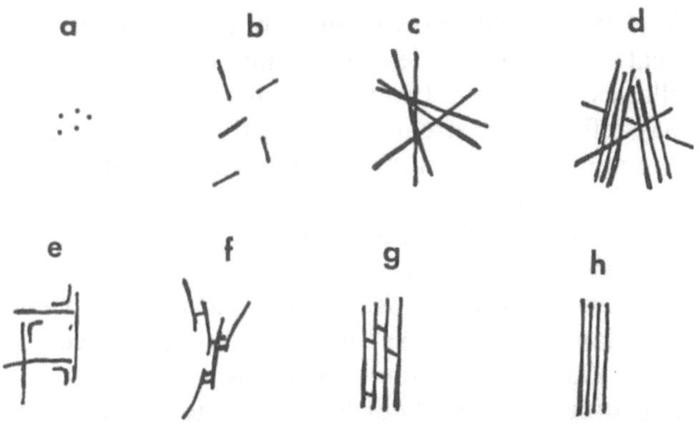

Fig. 2. Proposed conformations of actin: **a** dilute monomeric; **b** dilute polymer rods; **c** semidilute polymers; **d** concentrated and/or very long polymers in a glassy state; **e–h** polymers held by crosslinking factors in decreasingly isotropic configurations

rigid solid interacting only with its solvent and then only as a result of random thermal collisions. Gel filtration, sedimentation velocity, quasi-elastic light and neutron scattering experiments, and fluorescence photobleaching recovery studies support this notion (Rees and Young 1968; Lanni and Ware 1984; Montague et al. 1983; Mozo-Villarias and Ware 1984; Godette et al. 1986), inasmuch as the diffusion constants determined are in agreement with what is expected for the slightly elongated globular actin molecule (Egelman 1985). Provocative exception is the claim of Sato et al. (1985, 1986) that dilute solutions of globular proteins including G-actin have elastic properties after incubations in a rheometer which they interpreted as a result of unexplained interactions between the molecules.

Once rod-shaped actin particles become sufficiently long or concentrated to overlap, significant impediment to rotational diffusion should occur, whether or not any particular "crosslinks," exist between the rods at points of overlap. In this case the ideal behavior changes to one termed semidilute, and the conditions under which this transition occurs have been deduced from geometric considerations as when the concentration (c^*) is greater than the inverse of the cube of the rod length (L) or $c^* \gg 1/L^3$ (Doi and Edwards 1976a, b). Because of the high average length of F-actin, nearly all studies of its rheology have involved filaments in the semidilute range. Using the actin filament severing protein gelsolin to shorten actin filaments stoichiometrically (Yin and Stossel 1979; Yin et al. 1980), Doi and Frieden (1984) and Janmey et al. (1985, 1986) have experimentally documented the transition from dilute to semidilute behavior for F-actin as a function of filament length. In addition, they showed by quasi-elastic light scattering and fluorescence photobleaching recovery experiments that the diffusion of dilute F-actin particles was that expected for noninteracting dilute rods, thereby ruling out the existence of "crosslinks" between actin monomers in unperturbed filaments (Doi and Frieden 1984; Janmey et al. 1986).

If no specific interactions complicate the behavior of pure F-actin, then the high viscosity and viscoelastic properties of F-actin can be explained on the basis of steric constraints, and the results of some fluorescence photobleaching recovery or dynamic rheology experiments showing immobilization or near immobilization of actin filaments (Tait and Frieden 1982; Lanni et al. 1981; Lanni and Ware 1984; Mozo-Villarias and Ware 1984; Doi and Frieden 1984; Jen et al. 1982; Sato et al. 1985) reflect the extremely slow diffusion of sterically impeded filaments which are microns in length. Consonant with this interpretation is the observation that the compliance profile of F-actin, its pattern of strain over time in response to a constant stress, brings out that F-actin exhibits flow, especially at long times after application of a stress. The flow pattern is identical over a range of stresses, including stresses so low that it is inconceivable as to what kind of interfilament "crosslinks" might be broken to account for it (Zaner and Stossel 1983; Zaner 1986). Moreover, it has been pointed out that the slope of the curve relating the viscosity of F-actin to the applied shear is the same for F-actin as for many entangled, but not chemically crosslinked synthetic polymers and that the concentration and length dependence of F-actin's dynamic viscoelasticity and of its compliance fit Jain and Cohen's (1975)

Table 3. Actin filament crosslinking proteins

Given name	Source
Some proteins promoting isotropic actin network formation	
Actin-binding protein (also called filamin)	Vertebrate eukaryotic nonmuscle cells
Spectrin (also called fodrin)	Red blood cells and eukaryotic cells
HMW	Physarum
Some proteins promoting anisotropic F-actin bundle formation	
Alpha-actinin	Eukaryotic cells
Fascin	Marine eggs
Fimbrin	Avian brush border epithelial cells
Villin [a]	Mammalian and avian brush border epithelial cells, amphibian eggs
30 kD Protein	Dictyostelium
55 kD Protein	Brain, epithelial cells

[a] Crosslinking function only expressed when the calcium concentration is $<\mu M$.
For a description of the properties of these proteins, see Stossel et al. (1985).

theoretical approximation of this behavior based simply on topologically constrained rods (Zaner and Stossel 1982, 1983; Zaner 1986). However, the fact remains that very long or very concentrated actin filaments have an extremely high consistency and that their diffusion is so constrained that over reasonable time intervals they could in a practical sense be considered to be in a phenomenologically "gelled" state. In fact, since concentrated long rods are thermodynamically driven to form side-to-side aggregates (Flory 1956), a system of concentrated long actin filaments might have domains of anisotropic bundles coexisting with isotropically interlocked single filaments in which the bundles contribute considerable rigidity. Rheological measurements of concentrated F-actin comprised of very long filaments prepared by subjecting the actin to a degree of purification that removes substoichiometric contaminants that shorten actin filaments (e.g. MacLean Fletcher and Pollard 1980c; Zaner and Stossel 1982) reveal extremely high rigidities (KS Zaner, unpublished results).

Against this "gel"-like property of pure F-actin must be reconciled that cells contain a number of proteins which can crosslink actin filaments. From the foregoing it is evident that the activity of such proteins is most obvious when the F-actin is relatively dilute or when the filaments are relatively short, or both. The study of actin crosslinking proteins has resulted in their being classified into two categories: proteins that promote the formation of anisotropic bundles and others that stabilize networks of varying isotropy (reviewed by Stossel et al. 1985; Table 3). Since the accumulated evidence suggests that actin filaments are essentially straight rods, it has been argued that the degree of isotropy of an actin network depends largely on the angle at which actin filaments become crosslinked. Branching at right angles, a configuration produced by a high molecular weight actin-binding protein of vertebrate cells, provides the greatest degree of isotropy (Stossel et al. 1978; Hartwig et al. 1980; Niederman et al. 1983; Hartwig and Shevlin 1986). This generalization applies to the setting in which the actin filaments are in the semidilute range, such that it is the actin filaments themselves which provide the extension for interaction and crosslinking and in which the crosslinking molecules do not contribute materially to the volume fraction of the network. In another setting, for example, the proposed structure of the spectrin-actin network which laminates the cytoplasmic surface of red blood cell membranes, in which very short actin oligomers are joined by large flexible spectrin tetramers (Bennett 1985; Schanus et al. 1984), the angle of crosslinking is less relevant; indeed, the actin oligomers can be viewed as the "crosslinkers" in this kind of gel.

The mechanism of crosslinking of actin filaments by cell proteins has primarily been analyzed by a combination of electron microscopy and by gel point measurements, since these techniques could be applied to small amounts of the components available. An abrupt increase in consistency at a critical degree of crosslinking is characteristic of polymer gels (Fig. 3), and this gel point was

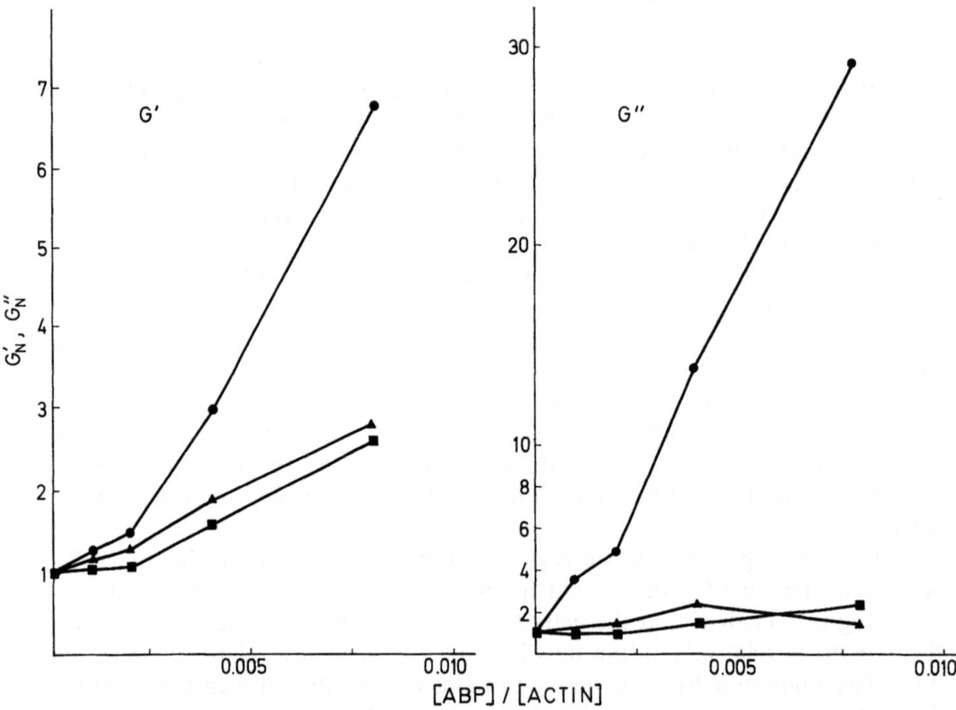

Fig. 3. Gel point determinations of F-actin polymerized in the presence of the indicated molar proportions of a macrophage crosslinking protein (the 540-kD actin-binding protein) (Zaner 1986). The consistency of the solutions was determined in a vibrating reed device (Zaner et al. 1981) by measuring the response of the solutions in phase (*G'*) and out of phase (*G''*) with sinusoidally applied stress at a frequency of 0.03 Hz. *G'* is an elastic component, the dynamic storage modulus, and *G''* the viscous component, the dynamic loss modulus, of the response to the applied stress. The values reported are for solutions containing 1 (■), 2 (▲), and 4 (●) mg ml^{-1} actin polymerized with the indicated molar ratios of actin-binding protein

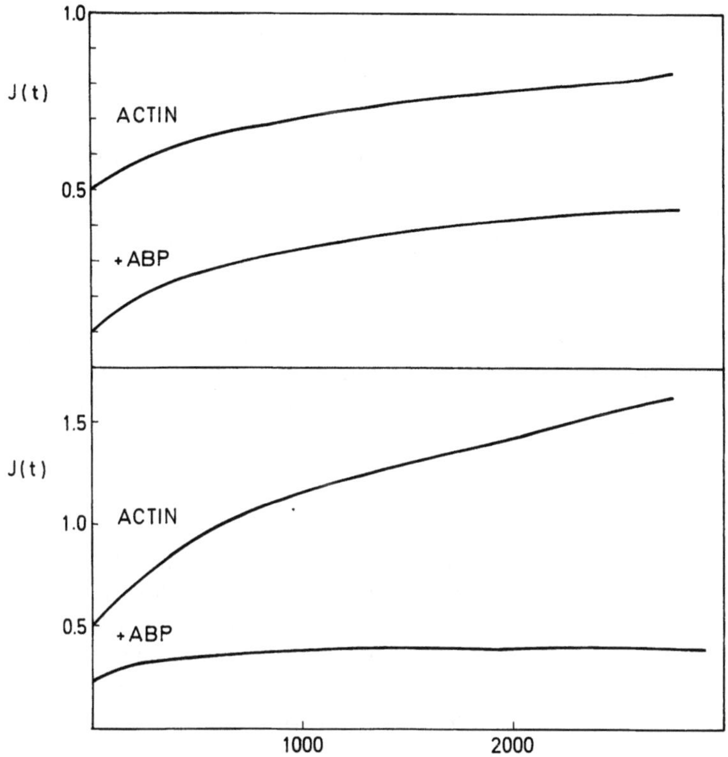

Fig. 4. Compliance J(t) of F-actin in the presence or absence of macrophage actin-binding protein (0.01 mol/mol). In these experiments a constant stress (0.05 dyn cm^{-2}) was applied to the solutions in the rheometer, and the time-dependent strain, the compliance, was determined. As expected for semidilute rods, the compliance of F-actin alone increases with time as the polymers align in parallel permitting an approach to Newtonian flow. In contrast, the compliance of F-actin crosslinked by actin-binding protein reaches an equilibrium value expected for a true gel in which the polymers are immobilized and in the absence of rupture cannot flow beyond the limits of deformability of the gel (Zaner 1986)

shown by Flory to be inversely proportional to the weight-average polymer length (1942). In nearly all cases, the molecules with the lowest minimum gelling concentrations for actin promote the formation of isotropic networks as evidenced by electron microscopy, whereas the agents which must be present in relatively high concentrations to produce gel points make actin filaments line up in bundles. Furthermore, the proteins producing isotropic actin networks tend to be large and relatively flexible, whereas the others are short rods or globular proteins. It may be, therefore, that only molecules with sufficient extension can engage adequate areas of overlapping actin filament helices to bind them together and more specifically to stabilize them in a configuration other than a parallel alignment (Hartwig and Stossel 1981). The conditions for forming an actin filament bundle are much less stringent. All that is required is for filaments to line up in register, after which, multiple potential sites for crossbridging are available (DeRosier et al. 1982). As expected from Flory's theory, the gelation of F-actin by crosslinking agents is inhibited by agents which shorten the actin filament length distribution (Hartwig and Stossel 1979; Yin and Stossel 1979; Yin et al. 1980; Nunnally et al. 1981; Isenburg et al. 1980; Zaner and Stossel 1983).

Since actin filaments are relatively stiff rods, F-actin crosslinked by an actin-gelation protein could achieve very high rigidities, since the deformability of such a network would be limited to short-range bending motions, in contrast to the situation in a rubberlike gel comprised of flexible chains, such as the spectrin-actin network (Schanus et al. 1985). F-actin crosslinked by the high Mr macrophage actin-binding protein achieves a high rigidity and demonstrates an equilibrium compliance value after an initial deformation in response to stress, a finding characteristic of true gels. The high rigidity and equilibrium compliance depend upon the actin concentration being high enough for actin-binding protein to connect a sufficient number of filaments into a continuous network (Figs. 3 and 4) (Zaner 1986).

Hitherto, most experimental work on actin gelation has focused on the mechanical properties of the assembled protein networks. However, other important aspects of gel's are their swelling properties and their effects on the diffusion of entrapped solvents and other molecules. Some interesting ideas have been proposed considering how the swelling of crosslinked cytoplasmic gels might contribute to the movement of the cell (Oster 1984), and it will be interesting to see whether experimental support for these ideas can be generated. Ito et al. (1987) have recently shown that actin filaments can impede water flow in response to osmotic stress and even immobilize water.

5.4 Actin Gelation in the Cell

When all is said and done, experiments with purified constituents in vitro can only suggest what might happen in the cell, and it is necessary to look to the cell as best one can to test the suggestions. Today's biochemical and biophysical information about cytoplasmic proteins add a new dimension to the traditional utility of microscopy in this quest.

The easiest target of examination are actin filament bundles that can be identified in some parts of some cells. Histochemical techniques reveal a concentration of the actin bundle-promoting proteins such as villin and fimbrin in parallel actin filament arrays within microvilli at the apical surfaces of epithelial brush border cells. The fact that these proteins are uniquely located within the microvillus further supports their role in stabilizing the actin bundles within these structures (reviewed by Mooseker 1985). A different actin filament bundle-promoting protein, alpha-actinin, is prominent in somewhat less ordered actin filament bundles called stress fibers that are present on the ventral surface of many cells in tissue cultures. Aggregation of actin filaments by myosin may also be important in the formation and stabilization of this type of bundle as well as of actin fibers oriented in parallel at the cleavage furrow of dividing cells. The specificity of these findings, however, is less clear than for the brush border microvillus actin bundles, since alpha-actinin and myosin are distributed throughout the peripheral cytoplasm in association with an isotropic actin network as well. Furthermore, the high Mr actin-binding protein of vertebrate cells mentioned above can promote perpendicular branching of F-actin at low molar ratios with respect to actin or bundles at higher ratios, and it is present in stress fibers, cleavage furrows, and throughout the peripheral cytoplasm of certain vertebrate cells.

New ultrastructural techniques have permitted investigators to obtain a clearer picture of the three-dimensional structure of cortical cytoplasm than was possible with earlier methods (Fig. 5). Porter and co-workers have utilized high voltage electron microscopy of whole cells to visualize in stereo what they call a "microtrabecular network" within the peripheral cytoplasm. The images give the impression of the existence of a delicate meshwork of short filaments within the bulk of the peripheral cytoplasm, although the chemical composition or detailed morphometry of this network are unknown (Porter 1984). Schliwa and co-workers have utilized critical point drying to preserve the three-dimensionality of fixed and detergent-permeabilized leukocytes and examined the structure of cortical filaments. These studies documented that the peripheral filaments were actin by their ability to find heavy meromyosin molecules and showed that some of the actin filaments tended to branch at right angles (e.g. Schliwa and Van Blerkom 1981; Pyrzwanski et al. 1983). Heuser and Kirschner (1980) presented spectacular electron micrographs of peripheral cytoplasmic networks of cells rapidly frozen after extraction with detergent. In particular, the cytoskeleton within the leading edge of a fibroblast appeared to contain a network of short, presumably actin, filaments branching at high angles.

Hartwig and Shevlin have recently examined the architecture of the cortical cytoplasm of macrophages fixed and detergent extracted while spreading on a surface (1986). These investigators employed both rapid freezing and critical point drying to prepare the specimens for electron microscopy, in order to provide better reassurance that the three-dimensional structure observed was not affected by shrinkage artifacts. They confirmed with immunohistochemical techniques that the majority of filaments in the cell cortex are actin. Using a computer-assisted morphometric technique to analyze stereo-pair electron micrographs, they determined that the actin filament concentration in the macrophage periphery was about 15 mg ml^{-1} ($\sim$200 μM), that the average spacing between filament overlaps was about 120 nm with a narrow distribution, and that the branch angle at filament overlaps was strongly biased toward the perpendicular. Relatively few free filament ends could be seen in these images; most were observed either attaching to the adherent surface of the lamellipod or else extending upward from the free surface. With immunogold labeling, molecules of the high Mr actin-binding protein were localized at many of the points at which filaments branched at close to right angles.

These ultrastructural examinations of the peripheral cytoplasm suggest for the case examined that the bulk of the actin is in the form of short branching segments and that actin-crosslinking proteins such as actin-binding protein which promote high angle branching actin gels in vitro function in this way in the cell as well. The recognition of such a network provides plausible explanations for the apparent rigidity and the organelle exclusion of that region. An

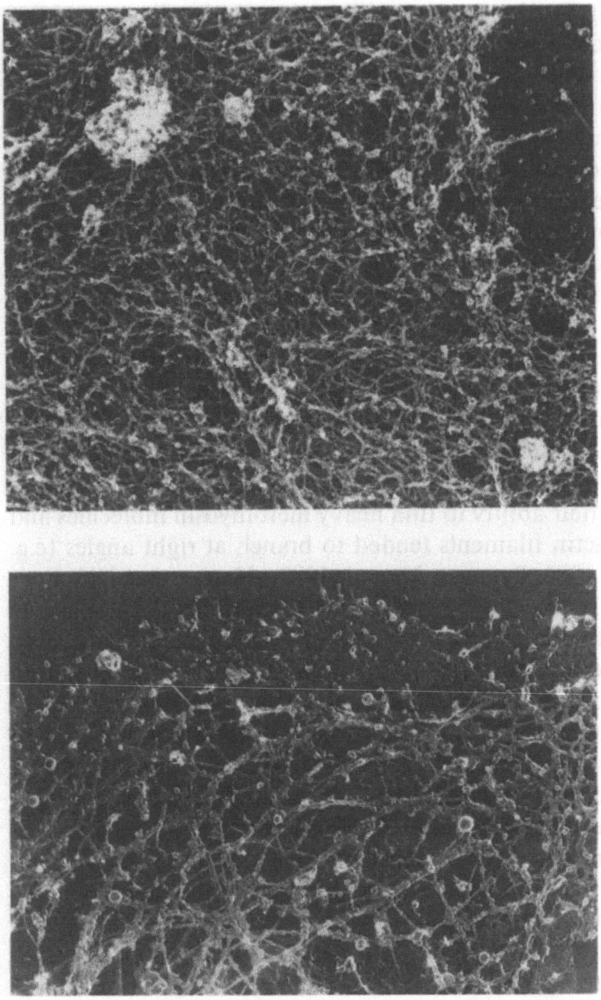

a

b

Fig. 5. Electron micrographs of **a** an orthogonal actin network in the peripheral cytoplasm of a macrophage; **b** in the peripheral cytoplasm of a platelet; **c** purified skeletal muscle actin filaments; **d** an orthogonal network comprised of purified actin and macrophage actin-binding protein; **e** actin filament bundles crosslinked by alpha-actinin. The techniques for preparing the electron micrographs are described in the publication of Hartwig and Shevlin (1986) who provided the photomicrographs

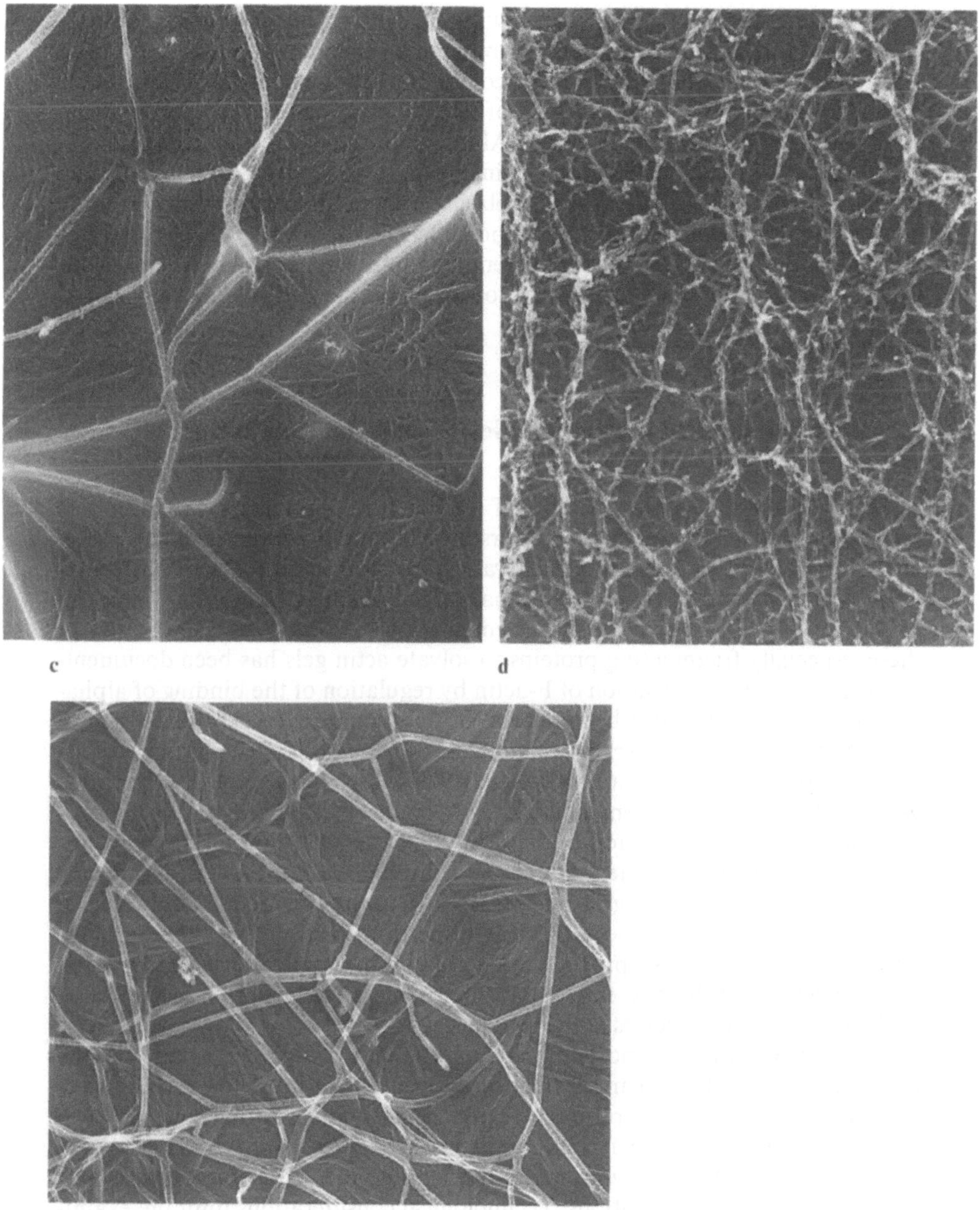

c d e

Fig. 5c–e

orthogonal network of actin filaments with the dimensions observed could provide maximal extension of the lamellipod with an economy of mass, permitting water, solutes and small vesicles to percolate through the network. On the other hand, the pore size of the network is sufficiently small to produce the organelle exclusion observed for so many years. Although the study of Hartwig and Shevlin indicated that an orthogonal actin network comprises the bulk of the pseudopodial network, they have also observed some filaments which align side by side and then diverge, indicating a possible role for other types of crosslinking. In addition, it is possible that different architectural arrangements of actin filaments will be observed as investigators examine the peripheral cytoplasm of other cells.

5.5 Regulation of the Actin Sol/Gel Transformation in the Cell

In principle, the reversible assembly of F-actin from monomers, the reversible fragmentation of F-actin, and the reversible binding of actin-crosslinking molecules to F-actin can regulate peripheral cytoplasmic actin sol/gel transformations. As discussed above, molecules have been purified from cells which can cause the reversible assembly and fragmentation of F-actin, and the ability of these, especially fragmenting proteins to solvate actin gels has been documented in vitro. Reversible gelation of F-actin by regulation of the binding of alpha-actinin to F-actin has also been shown, in that nonmuscle cell alpha-actinins' binding to F-actin is markedly diminished when the alpha-actinin binds micromolar calcium (Mimura and Asano 1979; Rosenberg et al. 1981; Burridge et al. 1982; Pollard 1982; Bennett et al. 1984; Condeelis et al. 1984). It is interesting that the isotropic gelation of actin by high Mr actin-binding proteins appears to be regulated at the level of the actin, whereas the more anisotropic bundling characteristic of gelation by alpha-actinin is evidently controlled at the crosslinker. It has not been possible hitherto to prove other than by inference that these regulatory proteins are acting on the cortical cytoplasm. It can be anticipated that the next few years will provide documentation of the action of these regulatory processes at the cellular level.

Another potentially important approach to the study of actin gelation in the cell is the examination of movement of molecules of different size in the peripheral cytoplasm. In general, much evidence has indicated that some cell water is quite mobile, but that some aqueous molecules may be relatively immobilized. In addition, proteins microinjected into the cell diffuse more slowly than expected for ideal solutions. Technological considerations limit the examination of particular cell domains, and results obtained tend to be controversial (Clegg 1983). Nevertheless, it is interesting that a recent study of the diffusion of microinjected dextrans suggested an equivalent intracellular "pore size" on the order of a few hundred Ångstroms (Luby-Phelps et al. 1986). Considering that some water may be immobilized and that other charge effects could play a role, the value is not too much smaller than the interstices of the cortical actin filament network recently measured by Hartwig and Shevlin.

Evidence that depolymerization and/or fragmentation of F-actin can produce a gel to sol transition in the cell is provided by the effect of cytochalasins on cells. On the basis of many studies it is reasonable to conclude that the principal site of action of cytochalasins in the cell is the fast-growing end of actin filaments and that the most pronounced effect of cytochalasins on actin is shortening of the filaments with a concomitant decrease in F-actin's consistency, especially if the filaments are crosslinked isotropically (Hartwig and Stossel 1979; MacLean-Fletcher and Pollard 1980c). That cytochalasins reduce the stiffness of different eukaryotic cells has been documented by a variety of rheological techniques (Mazur and Williamson 1970; Miller 1979; Pasternak and Elson 1985; Smith et al. 1982; Valberg and Albertini 1985), in addition to fragmenting actin filaments in the cell (Schliwa 1982). Unfortunately, the rheological studies have not specifically documented that the effects are localized to the peripheral cytoplasm. On the other hand, it has been shown that the activation of blood platelets, which is associated with actin polymerization, results in a rise in cytoplasmic viscosity as assessed by a spin probe (Rotman and Heldman 1981). Since blood platelets are essentially bits of megakaryocyte peripheral cytoplasm richly endowed with actin and actin-associated proteins (Fox and Phillips 1983), they constitute a nice model for correlating cortical cytoplasmic mechanical changes with alterations in the state of actin.

Acknowledgments. Supported by USPHS grants HL19429, HL01063, HL23591 GM05923, by grants from the Council for Tobacco Research and the Edwin S. Webster Foundation and by the Richard L. Saltonstall Endowment.

References

Abe SI, Maruyama K (1974) Dynamic viscoelastic study of actoheavy meromyosin in solution. Biochim Biophys Acta 160:160–174

Asakura S, Taniguchi M, Oosawa F (1963) Mechano-chemical behavior of F-actin. J Mol Biol 7:55–69

Bennett JP, Zaner KS, Stossel TP (1984) Isolation and some properties of macrophage a-actinin: evidence that it is not an actin gelling protein. Biochem 23:5081–5086

Bennett V (1985) The membrane skeleton of human erythrocytes and its implications for more complex cells. Ann Rev Biochem 54:273–304

Boxer LA, Stossel TP (1976) Isolation and properties of actin, myosin, and a new actin-binding protein of chronic myelogenous leukemia leukocytes. J Clin Invest 57:5696–5705

Brotschi E, Hartwig JH, Stossel TP (1978) The gelation of actin by actin-binding protein. J Biol Chem 253:8988–8993

Bryan J, Kane RE (1978) Separation and interaction of the major components of sea urchin actin gel. J Mol Biol 125:207–224

Burridge K, Feramisco JR (1981) Non-muscle a-actinins are calcium-sensitive actin-binding proteins. Nature 294:565–567

Clarke M, Spudich JA (1977) Nonmuscle contractile proteins: the role of actin and myosin in cell motility and shape determination. Ann Rev Biochem 46:797–822

Clegg JS (1985) Properties and metabolism of the aqueous cytoplasm and its boundaries. Am J Physiol 246:133–151

Condeelis J, Vahey M, Carboni JM, DeMey J, Ogihara S (1984) Properties of the 120,000- and 95,000-dalton actin-binding proteins from dictyostelium discoideum and their possible functions in assembling the cytoplasmic matrix. J Cell Biol 99:119–126

Condeelis JS, Taylor DL (1977) The contractile basis of amoeboid movement. V. The control of gelation, solation, and contraction in extracts from Dictyostelium discoideum. J Cell Biol 74:901–927
Davies WA, Stossel TP (1977) Peripheral hyaline blebs (podosomes) of macrophages. J Cell Biol 75:941–955
DeBruyn PPH (1947) Theories of ameboid movement. Q Rev Biol 22:1–24
DeRosier D, Mandelkow E, Silliman A, Tilney L, Kane R (1977) Structure of actin-containing filaments from two types of non-muscle cell. J Mol Biol 113:679–695
Doi M, Edwards SF (1978) Dynamics of rod-like macromolecules in concentrated solution, part 2. J Chem Soc Faraday Trans 74:918–932
Doi Y, Frieden C (1984) Actin polymerization: the effect of brevin on filament size and rate of polymerization. J Biol Chem 259:11868–11875
Dujardin MF (1938) Mémoire sur l'organisation des infusoires. Ann Sci Nat Zool Biol Anim 10:230–315
Ecker A (1848) Zur Lehre vom Bau und Leben der contractilen Substanz der niedersten Tiere. Z Wiss Zool 1:218–245
Egelman EH (1985) The structure of F-actin. J Muscle Res Cell Motil 6:129–151
Eichhorn JC (1781) Beiträge zur Naturgeschichte der kleinsten Wassertiere. Walwerth, Berlin
Fechheimer M, Brier J, Rockwell M, Luna EJ, Taylor DL (1982) A calcium- and pH-regulated actin binding protein from D. discoideum. Cell Motil 2:287–308
Ferry JD (1980) Viscoelastic Properties of polymers, 3rd ed. Wiley, New York
Flory PJ (1942) Constitution of three-dimensional polymers and the theory of gelation. J Phys Chem 46:132–140
Flory PJ (1956) Statistical thermodynamics of semi-flexible chain molecules. Proc R Soc Lond Ser A 234:60–72
Fox JEB, Phillips DR (1983) Polymerization and organization of actin filaments within platelets. Semin Hematol 20:243–260
Frieden C (1985) Actin and tubulin polymerization: the use of kinetic methods to determine mechanism. Ann Rev Biophys Chem 14:189–210
Fujime S, Ishiwata S (1971) Dynamic study of F-actin by quasielastic scattering of laser light. J Mol Biol 62:251–265
Fujime S, Ishiwata S, Maeda T (1984) Dynamic light scattering study of muscle F-actin. Biophys Chem 20:1–21
Goddette DW, Überbacher EC, Bunick GJ, Frieden C (1986) Formation of actin dimers as studied by small angle neutron scattering. J Biol Chem 261:2605–2609
Goodeve CF (1939) A general theory of thixotropy and viscosity. Trans Farad Soc 35:342–358
Griffith LM, Pollard TD (1982) Cross-linking of actin filament networks by self-association and actin-binding macromolecules. J Biol Chem 257:9135–9142
Hanson J (1973) Evidence from electron microscope studies on actin paracrystals concerning the origin of the cross-striation in the thin filaments of vertebrate skeletal muscle. Proc R Soc Lond B Biol Sci 183:39–58
Hanson J, Lowy J (1963) The structure of F-actin and of actin filaments isolated from muscle. J Mol Biol 6:46–60
Hartwig JH, Shevlin P (1986) The architecture of actin filaments and the ultrastructure of actin-binding protein in the periphery of lung macrophages. J Cell Biol 103:1007–1020
Hartwig JH, Stossel TP (1979) Cytochalasin B and the structure of actin gels. J Mol Biol 134:539–554
Hartwig JH, Stossel TP (1981) The structure of actin-binding protein molecules in solution and interacting with actin filaments. J Mol Biol 145:563–581
Hartwig JH, Tyler J, Stossel TP (1980) Actin-binding protein promotes the bipolar and perpendicular branching of actin filaments. J Cell Biol 87:841–848
Hatano S, Oosawa F (1966) Isolation and characterization of plasmodium actin. Biochim Biophys Acta 127:488–498
Heilbrunn LV (1956) The dynamics of living protoplasm. Academic Press, London
Hermans PH (1949) Gels. In: Kruyt HR (ed) Colloid science. Elsevier, New York, pp 483–651
Heuser JE, Kirschner MW (1980) Filament organization revealed in platinum replicas of freeze-dried cytoskeletons. J Cell Biol 86:212–234
Hill TL (1984) Introductory analysis of the GTP-cap phase-change kinetics at the end of a microtubule. Proc Natl Acad Sci USA 81:6728–6732

Huxley HE (1963) Electron microscope studies on the structure of natural and synthetic protein filaments from striated muscle. J Mol Biol 3:281–308
Isenburg G, Aebi U, Pollard TD (1980) An actin-binding protein from Acanthamoeba regulates actin filament polymerization and interactions. Nature 288:455–459
Ishikawa H, Bischoff R, Holtzer H (1969) Formation of arrowhead complexes with heavy meromyosin in a variety of cell types. J Cell Biol 43:312–328
Ito T, Zaner KS, Stossel TP (1987) Nonideality of volume flows and phase transitions of F-actin solutions in response to osmotic stress. Biophys J 51:745–753
Jain S, Cohen C (1981) Rheology of rod-like macromolecules in concentrated solution. Macromolecules 14:759–765
Janmey PA, Lind SE, Yin HL, Stossel TP (1985) Effects of semi-dilute actin solutions on the mobility of fibrin protofibrils during clot formation. Biochim Biophys Acta 841:151–158
Janmey PA, Peetermans J, Zaner KS, Stossel TP, Tanaka T (1986) Structure and mobility of actin as measured by quasielastic light scattering, viscometry and electron microscopy. J Biol Chem 261:8357–8362
Janmey PA, Stossel TP (1987) Modulation of gelsolin function by Phosphatidylinositol-4,5-bis phosphate. Nature 325:362–364
Jen CJ, McIntire LV, Bryan J (1982) The viscoelastic properties of actin solutions. Arch Biochem Biophys 216:126–132
Johnson P, Napper DH, Rowe AJ (1963) Sedimentation studies on polymerised actin solutions. Biochim Biophys Acta 74:365–373
Kane RE (1975) Preparation and purification of polymerized actin from urchin egg extracts. J Cell Biol 66:305–315
Kasai M, Kawashima H, Oosawa F (1960) Structure of f-actin solutions. Polymer Sci 44:51–69
Kawamura M, Maruyama K (1970) Electron microscopic particle length of F-actin polymerized in vitro. J Biochem 67:437–457
Korn ED (1982) Actin polymerization and its regulation by proteins from nonmuscle cells. Physiol Rev 62:672–737
Lamarck JB (1801) Histoire naturelle des animaux sans vertèbres. Bailliere, Paris
Lanni F, Ware BR (1984) Detection and characterization of actin monomers, oligomers, and filaments in solution by measurement of fluorescence photobleaching recovery. Biophys J 46:97–110
Ledermüller MF (1761) Mikroskopische Gemueths- und Augen-Ergoetzung. de Launoy, Berlin
Linnaeus C (1760) Systema Naturae. II Curt, Holmiae, Magdeburg
Luby-Phelbs K, Taylor DL, Lanni F (1986) Probing the structure of cytoplasm. J Cell Biol 102:2015–2022
MacLean-Fletcher S, Pollard TD (1980a) Mechanism of actin of cytochalasin b on actin. Cell 20:329–341
MacLean-Fletcher S, Pollard TD (1980b) Identification of a factor in conventional muscle actin preparations which inhibits actin filament self-association. Biochem Biophys Res Commun 96:18–27
McLean-Fletcher SD, Pollard TD (1980c) Viscometric analysis of the gelation of Acanthamoeba extracts and purification of two gelation factors. J Cell Biol 85:414–428
Maruta H, Korn ED (1977) Purification from Acanthamoeba castellanii of proteins that induce gelation and syneresis of f-actin. J Biol Chem 252:399–402
Maruyama K (1964) A flow birefringence study of F-actin. J Biochem 55:277–286
Maruyama K (1971) A study of beta-actinin, myofibrillar protein from rabbit skeletal muscle. J Biochem 69:369–386
Maruyama K, Kaibara M, Fukada E (1974) Rheology of F-actin. 1. Network of F-actin in solution. Biochim Biophys Acta 371:20–29
Maruyama K, Abe S-I, Ishii T (1975) Dynamic viscoelastic study of the effect of beta-actinin on the interaction between F-actin and heavy meromyosin. J Biochem 77:131–136
Maruyama K, Kimura S, Ishii T, Kuroda M, Ohashi K, Muramatsu S (1977) Beta actinin, a regulatory protein of muscle. J Biochem 81:215–232
Mazur MT, Williamson JR (1977) Macrophage deformability and phagocytosis. J Cell Biol 75:185–198
Merriam RW, Clark TG (1978) Actin in Xenopus oocytes II. Intracellular distribution and polymerizability. J Cell Biol 77:439–447

Miller ME (1979) Cell elastimetry in the study of normal and abnormal movement of human heutrophils. Clin Immunol Immunopathol 14:302–310
Mimura N, Asano A (1979) Ca^{2+}-sensitive gelation of actin filaments by a new protein factor. Nature 282:44–47
Miyake A, Stockmayer WH (1965) Theoretical reaction kinetics of revisible living polymerization. Makromol Chem 88:90–116
Mommaerts WFHM (1952) The molecular transformations of actin II. The polymerization process. J Biol Chem 198:459–467
Montague C, Rhee KW, Carlson FD (1983) Measurement of the translational diffusion constant of G-actin by photon correlation spectroscopy. J Muscle Res Cell Motil 4:95–101
Mooseker MS (1985) Organization, chemistry, and assembly of the cytoskeletal apparatus of the intestinal brush border. Ann Rev Cell Biol 1:209–241
Mozo-Villarias A, Ware BR (1985) Actin oligomers below the critical concentration detected by fluorescence photobleaching recovery. Biochem 24:1544–1548
Müller OF (1786) Animalcula infusoria fluviatilia et marina, quae detexit systematice describset et ad vivum delineari curavit. Molleri, Copenhagen
Nagashima H, Asakura S (1980) Dark-field light microscopic study of the flexibility of F-actin complexes. J Mol Biol 136:169–182
Niederman R, Amrein P, Hartwig JH (1983) The three-dimensional structure of actin filaments in solution and an actin gel made with actin-binding protein. J Cell Biol 96:1400–1413
Oosawa F, Asakura S (1975) Thermodynamics of the polymerization of protein. Academic Press, New York
Oster GF (1984) On the crawling of cells. J Embryol Exp Morphol 83:329–364
Pasternak C, Elson E (1985 a) Lymphocyte mechanical response triggered by cross-linking surface receptors. J Cell Biol 100:860–872
Pasternak C, Elson E (1985 b) Lymphocyte mechanical response triggered by cross-linking surface receptors. J Cell Biol 100:860–872
Pollard TD (1976) The role of actin in the temperature-dependent gelation and contraction of Acanthamoeba extracts. J Cell Biol 68:579–601
Pollard TD (1981) Purification of a calcium-sensitive actin gelation protein from acanthamoeba. J Biol Chem 256:7666–7670
Pollard TD, Ito S (1970) Protoplasmic filaments of Amoeba proteus. The role of filaments in consistency changes and movement. J Cell Biol 46:267–289
Porter KR (1984) Historical development of concepts related to the cytomatrix. J Cell Biol 99:1–20
Pryzwansky KB, Schliwa M, Porter KR (1983) Comparison of the three-dimensional organization of unextracted and triton-extracted human neutrophilic polymorphonuclear leukocytes. Eur J Cell Biol 30:112–125
Reinke J (1881) Protoplasma. Untersuchungen Bot Lab Göttingen 2:79–99
Rosenberg S, Stracher A, Burridge K (1981) Isolation and characterization of a calcium-sensitive a-actinin-like protein from human platelet cytoskeletons. J Biol Chem 256:12986–12991
Rotman A, Heldman J (1981) Intracellular viscosity changes during activation of blood platelets: studies by fluorescence polarization. Biochemistry 20:5995–5999
Sato M, Leimbach G, Schwarz WH, Pollard TD (1985) Mechanical properties of actin. J Biol Chem 260:8585–8592
Sato M, Schwarz WH, Pollard TD (1986) Acanthamoeba profilin affects the mechanical properties of nonfilamentous actin. J Biol Chem 261:10701–10706
Schliwa M (1982) Action of cytochalasin D on cytoskeletal networks. J Cell Biol 92:79–91
Schliwa M, Van Blerkom J (1981) Structural interactions of cytoskeletal components. J Cell Biol 90:225–235
Schanus E, Booth S, Hallaway B, Rosenberg A (1985) The elasticity of spectrin-actin gels at high protein concentration. J Biol Chem 260:3724–3730
Stossel TP (1987) The mechanical responses of white blood cells. In: Gallin J, Goldstein I, Snyderman R (eds) Inflammation: Basic principles and clinical correlates. Raven Press, New York, in press
Stossel TP, Hartwig JH (1976) Interaction of actin, myosin, and a new actin-binding protein of rabbit pulmonary macrophages. II. Role in cytoplasmic movement and phagocytosis. J Cell Biol 68:602–614

Stossel TP, Hartwig JH, Yin HL, Davies WA (1979) Actin-binding protein. In: Hatano S, Ishikawa H, Sato H (eds) Motility: molecules and organization. University of Tokyo Press, Tokyo, pp 189–209

Stossel TP, Chaponnier C, Ezzell RM, Hartwig JH, Janmey PA, Kwiatkowski DJ, Lind SE (1985) Nonmuscle actin-binding proteins. Ann Rev Cell Biol 1:353–402

Straub FB (1942) Actin. Stud Szeged 2:3–15

Szent-Gyorgyi A (1947) Chemistry of muscular contraction. Academic Press, London

Tait JF, Frieden C (1982) Polymerization and gelation of actin studies by fluorescence photobleaching recovery. Biochemistry 21:3666–3674

Takebayashi T, Morita Y, Oosawa F (1977) Electron microscopic investigation of the flexibility of F-actin. Biochim Biophys Acta 492:357–363

Tannenbaum SW (1978) Cytochalasins – biochemical and cell biological aspects. Elsevier, Amsterdam, North-Holland

Taylor DL, Condeelis JS (1979) Cytoplasmic structure and contractility in ameboid cells. Int Rev Cytol 56:57–144

Taylor DL, Fechheimer M (1982) Cytoplasmic structure and contractility: the solation–contraction coupling hypothesis. Philos Trans R Soc Lond B Biol Sci 299:185–197

Taylor DL, Condeelis JS, Moore PL, Allen RD (1973) The contractile basis of amoeboid movement. I. The chemical control of motility in isolated cytoplasm. J Cell Biol 59:378–394

Taylor DL, Rhodes JA, Hammond SA (1976) The contractile basis of ameboid movement. II. Structure and contractility of motile extracts and of plasmalemma-ectoplasm ghosts. J Cell Biol 70:112–143

Tellam R, Frieden C (1982) Cytochalasin D and platelet gelsolin accelerate actin polymer formation. A model for regulation of the extent of actin polymer formation in vivo. Biochem 21:3207–3214

Tilney LG, Hatano S, Ishikawa H, Mooseker MS (1973) The polymerization of actin: its role in the generation of the acrosomal process of certain echinoderm sperm. J Cell Biol 59:109–126

Trembley A (1744) Memoires pour servir à l'histoire d'un genre de polypes d'eau douce à bras en forme de cornes. Verbeek, Leiden

Valberg PA, Albertini DF (1985) Cytoplasmic motions, rheology, and structure probed by a novel magnetic particle method. J Cell Biol 101:130–140

Vanderkerckhove J, Weber K (1978) At least six different actins are expressed in a higher mammal: an analysis based on the amino acid sequence of the amino-terminal peptide. J Mol Biol 126:783–802

Von Rosenhof A, Roessel J (1792) Insecten Belustigungen. Kleeman, Nürnberg (Neue Edition)

Weber HH, Portzehl H (1954) The transference of the muscle energy in the contraction cycle. Progr Biophys Chem 4:60–111

Weihing RR (1976) Cytochalasin B inhibits actin-related gelation of HeLa cell extracts. J Cell Biol 71:303–307

Weihing RR (1977) Effects of myosin and heavy meromyosin on actin-related gelation of HeLa cell extracts. J Cell Biol 75:95–103

Yanagida T, Nakase M, Nishiyama K, Oosawa F (1984) Direct observation of motion of single F-actin filaments in the presence of myosin. Nature 307:58–60

Yin HL, Stossel TP (1979) Control of cytoplasmic actin gel–sol transformation by gelsolin, a calcium-dependent regulatory protein. Nature 281:583–586

Yin HL, Zaner KS, Stossel TP (1980) Ca^{+} control of actin gelation. J Biol Chem 255:9494–9500

Yoshimura H, Nishio T, Mihashi K, Kinosita K, Ikegami A (1984) Torsional motion of eosin-labelled F-actin as detected in the time-resolved anisotropy decay of the probe in the sub-millisecond time range. J Mol Biol 179:453–467

Zaner KS (1986) The effect of the 540 kilodalton actin crosslinking protein, actin-binding protein (ABP) on the mechanical properties of F-actin. J Biol Chem 261:2615–2620

Zaner KS, Stossel TP (1982) Some perspectives on the viscosity of actin filaments. J Cell Biol 93:987–991

Zaner KS, Stossel TP (1983) Physical basis of the rheologic properties of F-actin. J Biol Chem 258:11004–11009

Zaner KS, Fotlund R, Stossel TP (1981) A low-shear, small volume viscoelastometer. Rev Sci Instr 52:85–87

II.6 Dynamic Organization and Force Production in Cytoplasmic Strands

KARL-ERNST WOHLFARTH-BOTTERMANN[1]

6.1 Nature and Locomotory Phenomena of Physarum Plasmodia

The multinuclear plasmodia of acellular slime molds can reach a size of several m^2 and are able to migrate with a velocity of up to 1 cm h^{-1}. The basis of this rapid locomotion is a permanent translocation of cytoplasm plus nuclei in the form of a shuttle streaming: the endoplasm flows back and forth with a periodicity of ½ to 2 min, especially visible in plasmodial strands which characterize the rear part of a plasmodium. The frontal part of a plasmodium consists of a more or less continuous protoplasmic sheet which also encloses the pathways of an endoplasmic shuttle streaming. This migration polarity of a plasmodium is subject to changes due to, e.g., chemical stimuli.

Figure 1 demonstrates the light microscopic structure of plasmodial strands and the arrangement of their cytoplasmic actomyosin fibrils in the ectoplasmic tube (Wohlfarth-Bottermann 1975a). Note that the central endoplasmic channel in plasmodial strands in situ is free of cytoplasmic actomyosin fibrils and does not contain plasmalemma invaginations. In this channel, the endoplasm is transported back and forth. The locomotion of the total plasmodium results from the fact that the volume of protoplasm transported into the direction of migration is larger than the volume of the shuttle streaming in the opposite direction (regulated variables: diameter of the channel and duration of the streaming in that direction in which the plasmodium intends to move).

6.2 The Generation of Hydrostatic Pressure Flow

The motive force generation for shuttle streaming is achieved by an intrinsic and rhythmic contractility of the cytoplasmic actomyosin in the ectoplasmic tube. Note in Fig. 1 that the actomyosin fibrils have an arrangement as in a muscle tube. An increasing contraction of the ectoplasmic tube as a structural entity generates an increasing pressure within the endoplasmic channel. This leads to a translocation of the more fluid endoplasm into regions of the plasmodium with a lower hydrostatic pressure (pressure flow mechanism, Kamiya 1959). The regular back and forth streaming of the endoplasm is a consequence of slight, but rather regular and precisely controlled pressure differences in

[1] Institute of Cytology, University of Bonn, Ulrich-Haberland-Straße 61 a, D-5300 Bonn 1, FRG

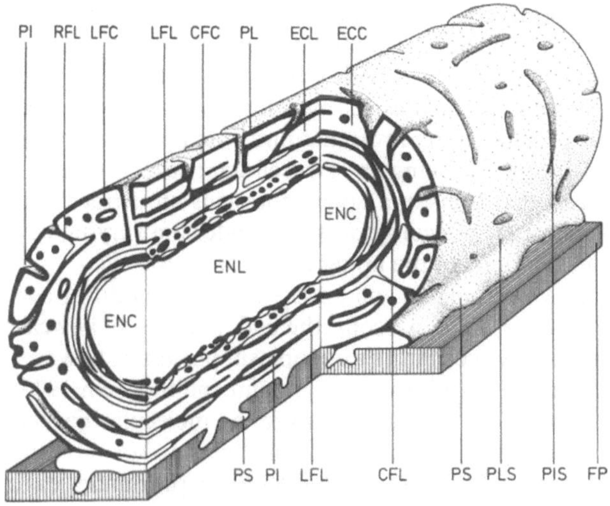

Fig. 1. Arrangement of cytoplasmic actomyosin fibrils in a plasmodial strand of Physarum polycephalum. *PI* Plasmalemma invagination; *RFL* radial actomyosin fibril in longitudinal section; *LFC* longitudinal fibril in cross-section; *LFL* longitudinal fibril in longitudinal section; *CFC* circular fibril in cross-section; *PL* plasmalemma; *CFL* circular fibril in longitudinal section; *ECL* ectoplasmic tube in longitudinal section; *ECC* in cross-section; *ENC* endoplasmic tube in cross-section; *ENL* endoplasmic tube in longitudinal section; *PS* pseudopodia; *CFC* circular fibril in cross-section; *PLS* surface view onto the plasmalemma; *PIS* surface view onto plasmalemma invaginations; *FP* filter paper (substrate under laboratory conditions) (Fleischer and Wohlfarth-Bottermann 1975)

various regions of the plasmodium. Precondition for an efficient functioning of such a mechanism is a highly regulated, i.e., more or less synchronous, contractile activity of the cytoplasmic actomyosin in different regions. In fact, the direction of endoplasmic streaming after a standstill is determined by slight phase differences of the onset of a contraction period in opposite directions of a plasmodium (Achenbach and Wohlfarth-Bottermann 1981).

The ectoplasmic tube is not completely separated by plasmalemma invaginations from the endoplasmic channel (Wohlfarth-Bottermann and Achenbach 1982). There are apertures through which streamlets of endoplasm enter the ectoplasmic tube (during dilatation, i.e., radial relaxation of the strand); these streamlets, running transverse to the main direction of the shuttling endoplasm, flow back into the endoplasmic channel (during radial contraction of the strand). Thus, there is a permanent exchange of ectoplasm and endoplasm by partial intermixing of ectoplasm and endoplasm, and this is also the physical basis, namely, the information pathway, for the synchronization of contraction phenomena (Achenbach and Wohlfarth-Bottermann 1981). It has been shown that the endoplasmic stream transmits the synchronizing factor and also, e.g., the factor for the transmission of a blue-light stimulus to nonilluminated sites of the plasmodium (Wohlfarth-Bottermann and Block 1981). From this fact it follows that the contractile apparatus shows a high degree of adaptability to external stimuli; e.g., to blue light and chemicals, indicating the surface of its habitat and food sources, respectively. The contact of the plasmalemma with the substratum sensu strictu is mediated by a slime layer excreted at all external surfaces of a plasmodium (a giant glycocalix with a thickness in the millimeter range); in this way, the plasmodium creates its own microenvironment. The slime also fills the numerous plasmalemma invaginations which protrude deeply into the cell; their content, in spite of being an extracellular space, is certainly under strict cellular control, so that the slime layer represents an extracellular matrix of the plasmodium, assuredly of ecological value under the varying conditions of the natural habitat, the clefts of the upper layers of forest soils.

The mechanical properties of ectoplasm and endoplasm (Fig. 1) differ considerably: whereas the ectoplasm is highly viscous, the endoplasm – in any case when flowing – has a relative low viscosity. The endoplasm possesses, however, the ability to gelate within seconds (Wohlfarth-Bottermann and Stockem 1970). For cytomechanical problems it is of interest to analyze the molecular basis of the transition mechanism of endoplasm into ectoplasm ("plasma sol → plasma gel transformation") and vice versa. This can be done in Physarum under quite natural conditions, during the spontaneous transition of a freshly generated endoplasmic drop into a small, locomoting plasmodium (Wohlfarth-Bottermann 1983). Endoplasmic drops are at the beginning only a bulk of endoplasm protruded from the endoplasmic channel, sporadically this also happens spontaneously. Immediately after the experimental generation of an endoplasmic drop (after puncturing the ectoplasmic tube with a glass needle), the endoplasm gelates within a few seconds. It was shown that this strong and rapid increase in viscosity is achieved by a polymerization of low polymeric actin in the previous endoplasm (Isenberg and Wohlfarth-Bottermann 1976). This polymerization occurs within 0.1 – 2 min after the generation of the drop. Even within a shorter time span (6 s) a new plasmalemma is generated (Wohlfarth-Bottermann and Stockem 1970). The capability of the cytoplasm to increase its viscosity so rapidly is of major importance for considerations in the area of cytomechanics. What could be the nature of the signal for the actin polymerization? The time range of seconds makes a chemical factor improbable. Plausibly, the sudden decrease in hydrostatic pressure, i.e., when the endoplasm protrudes from its channel, could be the physical signal, but this remains unknown to date.

Inner pressure differences in different regions of a plasmodium as well as the contractile activity of cytoplasmic actomyosin of the ectoplasmic tube can be directly registered and demonstrated: pressure differences by the "double chamber method" (Kamiya 1959) and the contractile activity by tensiometry (Wohlfarth-Bottermann 1975b). For proof of the ability of cytoplasmic actomyosin fibrils to contract, see Isenberg et al. (1976), for contraction of "synthetic models," see Beck et al. (1970).

The importance of intracellular hydrostatic pressure for the locomotory mechanism and the shape of other eukaryotes, e.g., of tissue cells, has been stressed by Bereiter-Hahn (1985, Chap. V.3 this Vol.).

6.3 Contractile Activities as Measured by Tensiometry

The regularity of the intrinsic contractile activity of a plasmodial strand can be shown after cutting a strand of approximately 1-cm-length out of a plasmodium and mounting it in a highly sensitive tensiometer (Fig. 2).

Strands of 1-cm-length and 0.1-mm-diameter produce an active tension of approximately 10 millipond (mp) (1 kp = 9.80665 Newton) during each contraction phase (Wohlfarth-Bottermann 1979). This force output is delivered every 2 min for many hours under conditions of starvation of the specimens in a flow-through chamber or in humid atmosphere (consumption of glycogen reserves). The ATP supply of the contractile machinery can be derived alternatively from glycolysis or respiration (Korohoda et al. 1983). For a deeper insight into the nature of the physiological oscillator driving this periodic activity in the minute range, see Korohoda et al. (1983) and Baranowski (1985).

The ability of individual cytoplasmic actomyosin fibrils to contract isotonically was shown after their isolation by laser microsurgery (Isenberg et al. 1976). For a coordination of force production, on the one hand, and the behavior of the force-producing structures, on the other hand, it was of interest to fix the specimens during their tensiometric measurement at different stages of their normal contraction-relaxation cycle (Fig. 3). A comparison of the fibrillar pattern analyzed under isometric conditions of measurement revealed that at the beginning of a contraction phase many actomyosin fibrils appear (Fig. 3 B), which, however, disappear again at the start of a relaxation phase (Fig. 3 A). This means that the fibrillar pattern (at the light microscopic level) changes strongly in the minute rhythm; in other words, the cytoplasmic actomyosin

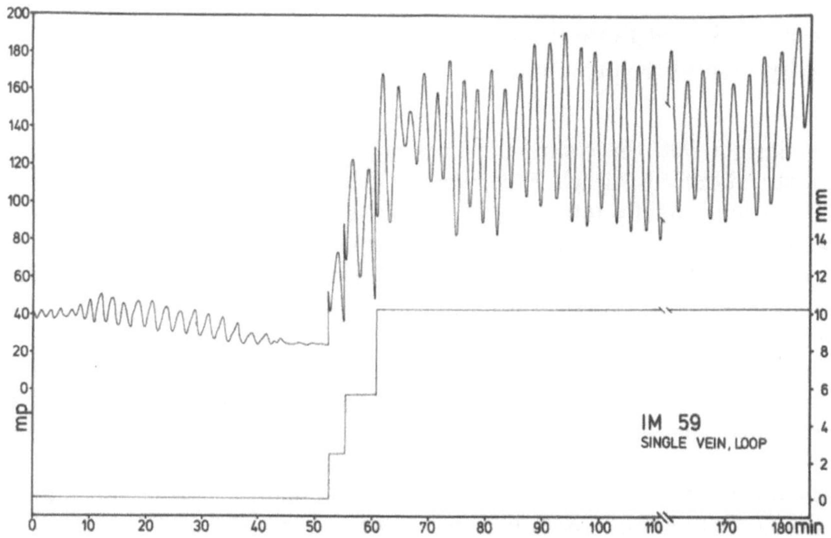

Fig. 2. Tensiometrically measured isometric force production (longitudinal activity) of a plasmodial strand, before (0 to 52 min) and after the application of stretch stimuli (Wohlfarth-Bottermann 1975 b)

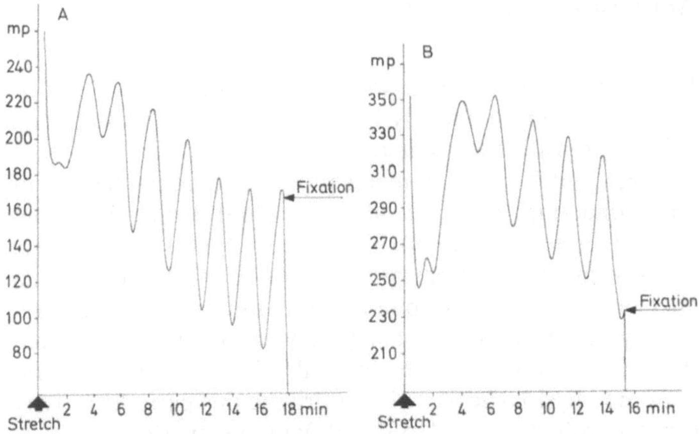

Fig. 3. Tension curves of two plasmodial strands (longitudinal contraction, isometric conditions), which remained in the tensiometer 12–18 min after a stretch stimulus and were then fixed: **A** at the beginning of a relaxation phase; **B** at the beginning of a contraction phase. For the structural comparison of both strands, see text and Fig. 4 (Wohlfarth-Bottermann and Fleischer 1976)

fibrils are highly dynamic structures during their rhythmic force generation (Fleischer and Wohlfarth-Bottermann 1975; Wohlfarth-Bottermann and Fleischer 1976). This is of importance when discussing their additional static functions, which follow below.

6.4 Analysis of Morphological Alterations Induced by Stretch Experiments

A deeper insight into the contraction mechanism of cytoplasmic actomyosin was brought about by stretch experiments, using the aforementioned tensiometers, but additionally introducing isotonic conditions at certain stages. It can already be seen in Fig. 2 that the force output during the longitudinal contraction of a strand is strongly enforced after a mechanical stretch. Thus, a systematic morphological analysis was performed during the time period immediately following a stretch (Fig. 4). The first tensiometrically registrable reaction of the strand upon the stretch stimulus is a "stress relaxation phase" (duration approximately 30 s). This is followed by a contraction phase (30–80 s after a stretch of 50% of the original length of the strand). The diagrammatic curve in Fig. 4 demonstrates further the following relaxation phase, the time points 1–6 of morphological stages of analysis, but not the fact that the normal oscillatory contraction activity is then continued (as shown in Fig. 2). The diagrams 1–6 symbolize the pattern of actomyosin fibrils in longitudinal sections of strands when fixed in the tensiometer at defined time points (see curve) after stretching (Wohlfarth-Bottermann and Fleischer 1976). The first observation is that

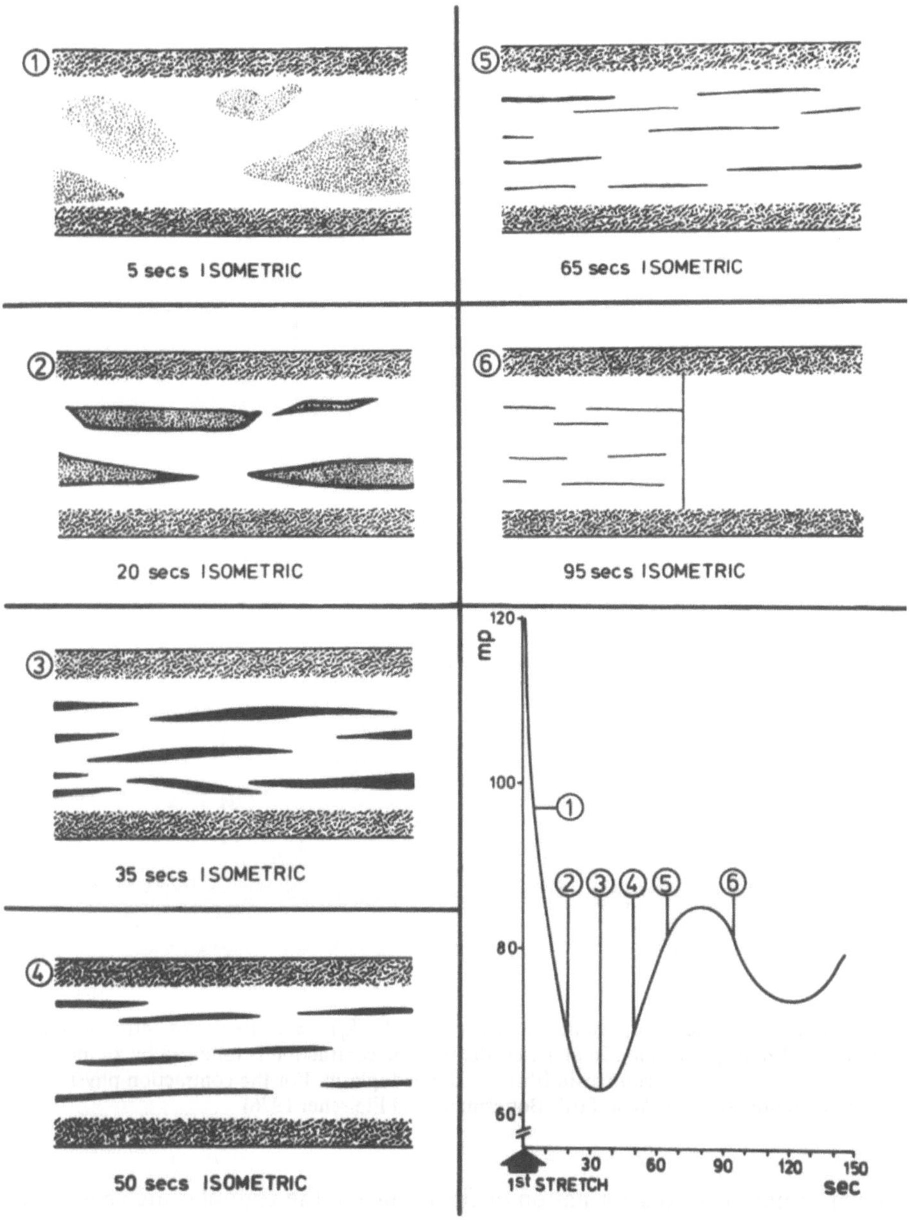

Fig. 4. Assembly and disassembly behavior of cytoplasmic actomyosin fibrils in dependence of a preceding stretch-induced contraction phase (isometric conditions of measurement). The tensiometric curve indicates the tension reaction of a strand to a stretch stimulus given at 0 s: 0–35 s = stress relaxation phase; 35–80 s = first contraction phase following the stretch; beginning at 80 s: first relaxation phase after the stretch; thereafter, the strand continues to perform regular oscillatory contraction activity as shown in Fig. 2. The *encircled numbers 1–6* indicate stages at which the tensiometrically measured force production was morphologically analyzed and thus correlated with ultrastructural alterations in the strands; characteristic alterations are symbolized as longitudinal sections (5 to 95 s after stretching). The peripheral *dotted borders* of the diagrams symbolize the ectoplasmic tube, the *central white area*, the endoplasm and its (upon the stretch stimulus) de novo generated population of actomyosin fibrils. For further explanation, see text (Wohlfarth-Bottermann and Fleischer 1976)

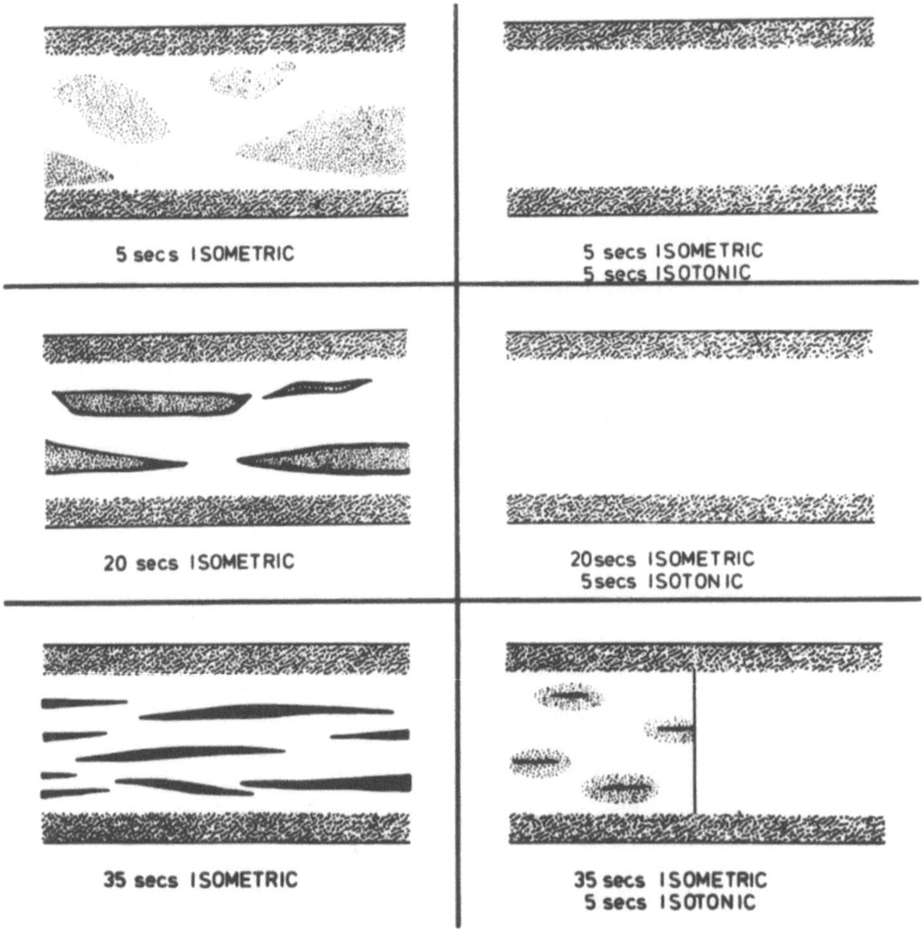

Fig. 5. Reaction of the endoplasmic actomyosin fibrils (*left side*) after a 5-s introduction of isotonic conditions (*right side*). In all cases, the isotonic contraction is followed by an immediate disaggregation of the actomyosin fibrils in the endoplasm. For the contraction physiological interpretation, see text (Wohlfarth-Bottermann and Fleischer 1976)

stretch stimuli lead to a formation of fibrils also in the endoplasmic channel, a phenomenon explaining the conspicuous higher force output of stretched strands.

Of more interest are the dynamics of this de novo generated fibrillar pattern, in temporal dependence of the preceding stretch. During stage 1 de novo generation of fibrils in the endoplasm starts already 5 s after the stretch (as visualized by light microscopy, in reality certainly earlier). These light microscopically viewed "precursors" (stages 1–2 in Fig. 4) transform to a dense fibrillar pattern (stage 3) within 30 s (the onset of the first contraction phase) and fade out at the end of this contraction phase, to disappear during the following relaxation phase. The electron microscopic control reveals parallel-organized F-actin filaments during stages 3 and 4; the precursor stages (1, 2) and the remnants of former fibrils during the relaxation phase (6) do not reveal parallel arrangement of F-actin, but entangled filaments, as analyzed in thin sections.

Figure 5 symbolizes the behavior of the fibrillar pattern as demonstrated in Fig. 4 upon the sudden introduction of *isotonic* conditions at different stages (5–35 s) of the former isometric regime. The introduction of isotonic conditions, i.e., a 5-s isotonic contraction of the strands, leads immediately to the disappearance of most of the fibrils.

6.5 Nature and Implications of the Contraction Cycle

An interpretation of the contraction cycle of cytoplasmic actomyosin is of interest for many different cell biological problems, independent of the main topic of this book. In order to understand the dynamic and thus the mechanical functions that are of primary interest in cytomechanics, a short discussion of the contraction cycle is necessary.

We have to consider that cytoplasmic actomyosin resembles muscle actomyosin, but that there are differences in the myosin and to a smaller extent in the actin. The differences cannot be discussed here, but play an important role in the dynamic behavior of cytoplasmic actomyosin. This dynamic behavior was elucidated in the experiments demonstrated in Figs. 3–5.

From these experiments, the following contraction-physiological interpretations can be derived:

1. The isometric contracted state of cytoplasmic actomyosin is fibrillar (parallel texture of F-actin).
2. The relaxed state under isometric conditions is *not* fibrillar.
3. Most interesting: the isotonic contracted state is also *not* fibrillar.
4. We have, to date, no information about the isotonic relaxed state.

From these results follows that *only the isometric contracted* state shows a parallel texture of F-actin and that this stage implies a function of cytoplasmic actomyosin fibrils to transmit a more or less constant tension between their points of insertion. This knowledge is primarily of practical interest for the dynamic-functional interpretation of electron micrographs, but also for the elucidation of the molecular contraction cycle: cytoplasmic actomyosin shows much higher dynamics in comparison to muscle actomyosin, where a complete structural transformation of the contractile units, the sarcomeres, is absent.

The light and electron microscope observations described above do not justify the postulation of a "complete" transformation of the actin chain, but other experiments using actin-specific phallotoxins (Götz v. Olenhusen and Wohlfarth-Bottermann 1979) demonstrate that a transformation of the actin chains to lower polymeric forms is necessary for a complete, i.e., a repetition of, the contraction cycle. This means that we must consider not only a deparallelization of the actin chains in the fibrils during relaxation, but also an actin depolymerization. This dynamic actin behavior seems to be a characteristic difference between cytoplasmic and muscle actomyosin.

The problem of the molecular mechanics of the contraction (sliding mechanism?) will not be dealt with here; it should only be mentioned that a sliding mechanism seems possible (as derived from indirect evidence; see Beck et al. 1970), but it has not yet been established with certainty.

Of interest in this context is the dynamic assembly and disassembly behavior of actin in the fibrils in the course of a complete contraction cycle. Correspondingly, it was found that the ground plasm contains a considerable pool of low polymeric actin (Isenberg and Wohlfarth-Bottermann 1976), being in permanent equilibrium with the filamentous form. This dynamic and very sensitive equilibrium makes it easier to understand the transport of the contractile proteins in migrating cells, in which not only motive force has to be generated, but in which the motive force-generating structures, at the same time, have to be translocated in the course of cell migration (Naib-Majani et al. 1984).

6.6 The Widely Unknown Regulation

The regulation of the contraction cycle seems to be predominantly Ca^{2+}-mediated, but rather different from cross-striated muscle. Ca^{2+} ions play a role in the control of the cytoplasmic actin equilibrium which is governed by Ca^{2+}-sensitive "actin-regulating" proteins (e.g., fragmin; Hasegawa et al. 1980; Hinssen 1980, 1981) *and* seem to be decisive for the regulation of myosin (Kessler et al. 1980). In contrast to muscle actomyosin, however, Ca^{2+} does not induce the contraction phase, but rather the relaxation phase! It has already been mentioned that the interpretation of the molecular events in the contraction cycle of cytoplasmic actomyosin are rather complicated, and that the precise, possibly sequentially different roles of Ca^{2+} are not yet clearly established. This information may be sufficient to indicate the present state of knowledge, inasmuch as the mechanical interpretations to be discussed in this book can now be derived from clearly established phenomena, i.e., from the morphological behavior of the fibrils as seen under continuous tensiometric force measurements. The precisely timed, sequential interaction of cytoplasmic actomyosin and Ca^{2+} is in any case of great importance for many cytomechanical problems.

6.7 Cytomechanical Implications

The foregoing results demonstrate that cytoplasmic strands isolated from plasmodia of Physarum polycephalum offer a unique possibility to measure changing tension forces in living cells, to coordinate the varying force production closely with contraction-relaxation cycles of cytoplasmic actomyosin and to analyze concomitant morphological changes (Wohlfarth-Bottermann 1983). In

Physarum plasmodia, microtubules do not play a role with the exception in mitotic movement. The viscoelastic properties of the cytoplasm of Physarum, however, are not only influenced by the state of actin equilibrium, but also by a network of a Titin-like protein, which was recently detected by Gassner et al. (1985). The possible dynamic capacities of this protein network consisting of "superfine filaments" 2–3 nm in diameter have yet to be elucidated. The present findings are in favor of the thesis that the Titin network, in addition to actin, is of major importance for the viscoelastic properties of the cytoplasm.

The mechanical functions of the cytoplasmic actomyosin fibrils ("stress fibrils" in animal tissue cells) must be interpreted as isometrically contracting „*Verspannungselemente*", i.e., structures delivering static rigidity for the preservation of the form of the cell. Simultaneously, however, these actomyosin fibrils are able to perform isotonic contractions, i.e., they demonstrate a high dynamic capability. Their isometric as well as their isotonic contraction behavior serves simultaneously for the motive force generation needed for motility: isometric contraction for the preservation of the form of the cell against an increasing interior hydrostatic pressure at the time of influx of endoplasm; isotonic contraction results in a diminution of the diameter of the strands (volume decrease) at the moment of outflow of endoplasm.

Isometrically contracting fibrils counterbalance an increasing hydrostatic pressure in the endoplasmic channel, in cooperation with cortical and internal actomyosin sheets located at the plasmalemma and its invaginations. This shows that both isometric as well as isotonic contraction behavior are simultaneously responsible for static *and* motile functions and that both functions cannot be considered separately. This seems to be characteristic for functions of cytoplasmic actomyosin, and is not only valid for cytoplasmic strands of Physarum, but also for other eukaryotic cells.

Greenspan and Folkman (1977) considered theoretically stress fibrils to be "visible manifestations of tension field lines in the basal membrane." Cytoplasmic actomyosin fibrils, however, are not manifestations of forces produced elsewhere or "in the past," but force-generating structures themselves, which disappear immediately as soon as their temporally limited function is completed (see Fig. 5).

The generation of the fibrils, i.e., the molecular events for their „self-organization," are far from being clear, but surely this process is closely coordinated with the self-assembling properties of actin and its control.

The consideration of the importance of cytoplasmic actomyosin *fibrils* for the mechanical properties of cells, however, should not lead to the opinion that only the *fibrillar* form of cytoplasmic actomyosin is of functional importance. In fact, the *nonfibrillar* form, i.e., an irregular network of actin filaments associated with myosin and regulatory proteins, seems to be much more important because this is the normal, most widely spread form in which cytoplasmic actomyosin is found in eukaryotic plant and animal cells. It was established rather early that this reticular form is also able to perform volume contractions (Beck et al. 1970), in biochemical investigations previously but erroneously designated as "superprecipitation." The morphology of this contractile network in the ground plasm is far less conspicuous compared with the fibrillar form, but much more important. Because of its relative low degree of structural order, however, a functional analysis in situ is even more difficult. Concluding from its ubiquitous occurrence, however, the cytomechanical importance of the irregular cytoplasmic actomyosin network and its dynamic properties cannot be overestimated.

The macroscopic as well as the microscopic architecture of a plasmodial strand (Fig. 1) is determined by an interactive cooperation of the ectoplasmic tube, on the one hand, and the inner more fluid endoplasm, compressed and thus translocated by this tube, on the other hand. Gravitational force leads to a conspicuous flattening of the ectoplasmic tube in the area of its substrate contact. The form of the ectoplasmic tube at sites without substrate contact is more or less uniformly round and results from an interplay of the compressed endoplasmic stream and the complex structural continuum of the ectoplasmic tube: its plasmalemma, its plasmalemma invaginations, the adhering and continuous cortical and inner actomyosin sheets (irregular actomyosin network) and actomyosin fibrils which are finally also closely interconnected with each other (Fig. 1). The actin filaments of the cortical actomyosin sheets as well of the fibrils terminate finally at the plasma membrane to ensure the connection of the contractile elements of the ectoplasmic tube with its peripheral plasmalemma, the outer cell boundary. The plasmodial strands are a tubular, tensile-element construction, the properties of which are at least determined by two proteins: (1) cytoplasmic actomyosin, i.e., the interaction of actin and myosin and possibly other actin-binding proteins, and (2) the network of the Titin-like 2 – 3 nm filaments, of which the possible dynamic changes and/or interactions with other structural components are yet unknown.

The longitudinally, radially, and spirally arranged actomyosin fibrils in the ectoplasmic tube of strands in situ ramify, form extended sheets, and closely contact the peripheral plasmalemma, thus transmitting their tension forces finally to plasmalemmal structures.

In isolated strands, the fibrils appear also in the endoplasmic channel, after stretching the strand predominantly in the longitudinal direction (the direction of the stretch). There is no doubt that the actomyosin fibrils are tensile elements, simultaneously significant for the preservation of cell architecture and the generation of the motive force for the movement of the cell.

As mentioned above, microtubules do not play a role in the tension force generation of cytoplasmic strands of Physarum because cytoplasmic microtubules are lacking. This underlines the importance of cytoplasmic actomyosin (probably in cooperation with the Titin-like system of 2 – 3 nm filaments) for the biomechanical properties of the cytoplasm and moreover many phenomena in cellular mechanics. The cytoplasmic actomyosin is surely one of the most prominent "self-organizing" proteins that produce force and form at the same time. Appropriate sensors for the control of its activity have to be anticipated in the plasmalemma, but may also be present at certain sites of the energy-producing mechanism (see Korohoda et al. 1983) to ensure an appropriate reaction of the locomotory machinery to external, changing conditions upon which the cell has to react in an appropriate manner in order to survive.

References

Achenbach U, Wohlfarth-Bottermann KE (1981) Synchronization and signal transmission in protoplasmic strands of Physarum. The endoplasmic streaming as a pacemaker and the importance of phase deviations for the control of streaming reversal. Planta (Berl) 151:584–594

Baranowski Z (1985) Consequences of impeding in mitochondrial functions in Physarum polycephalum. III. Reversible cessation of the contraction-relaxation cycle and the temperature sensitivity of the alternate respiratory pathway. Eur J Cell Biol 39:283–289

Beck R, Hinssen H, Komnick H, Stockem W, Wohlfarth-Bottermann KE (1970) Weitreichende, fibrilläre Protoplasmadifferenzierungen und ihre Bedeutung für die Protoplasmaströmung. V. Kontraktion, ATPase-Aktivität und Feinstruktur isolierter Actomyosin-Fäden von Physarum polycephalum. Cytobiologie 2:259–274

Bereiter-Hahn J (1985) Architecture of tissue cells. The structural basis which determines shape and locomotion of cells. Acta Biotheor 34:139–148

Fleischer M, Wohlfarth-Bottermann KE (1975) Correlation between tension force generation, fibrillogenesis and ultrastructure of cytoplasmic actomyosin during isometric and isotonic contractions of protoplasmic strands. Cytobiologie 10:339–365

Gassner D, Shraideh Z, Wohlfarth-Bottermann KE (1985) A giant titin-like protein in Physarum polycephalum: evidence for its candidacy as a major component of an elastic cytoskeletal superthin filament lattice. Eur J Cell Biol 37:44–62

Götz von Olenhusen K, Wohlfarth-Bottermann KE (1979) Evidence for actin transformation during the contraction-relaxation cycle of cytoplasmic actomyosin: cycle blockage by Phalloidin injection. Cell Tissue Res 196:455–470

Greenspan HP, Folkman J (1977) Hypotheses on Cell Adhesion and Actin Cables. J Theor Biol 65:397–398

Hasegawa T, Takahashi S, Hatano S (1980) Fragmin: a calcium ion-sensitive regulatory factor on the formation of actin filaments. Biochemistry 19:2677–2683

Hinssen H (1980) The regulation of actin transformation by a calcium-dependent actin-modulating protein from the slime mould Physarum polycephalum. Eur J Cell Biol 22:327

Hinssen H (1981) An actin-modulating protein from Physarum polycephalum. II. Ca^{++}-dependence and other properties. Eur J Cell Biol 23:234–240

Isenberg G, Wohlfarth-Bottermann KE (1976) Transformation of Cytoplasmic Actin. Importance of the organization of the contractile gel reticulum and the contraction-relaxation cycle of cytoplasmic actomyosin. Cell Tissue Res 173:495–528

Isenberg G, Rathke PC, Hülsmann N, Franke WW, Wohlfarth-Bottermann KE (1976) Cytoplasmic actomyosin fibrils in tissue culture cells. Direct proof of contractility by visualization of ATP-induced contraction in fibrils isolated by laser microbeam dessection. Cell Tissue Res 166:427–443

Kamiya N (1959) Protoplasmic streaming. In: Heilbrunn LV, Weber F (eds) Protoplasmatologie VII. 3a. Springer, Wien

Kessler D, Eisenlohr LC, Lathwell MJ, Huang G, Taylor HC, Godfrey SD, Spady ML (1980) Physarum myosin light chain binds calcium. Cell Motil 1:63–71

Korohoda W, Shraideh Z, Baranowski Z, Wohlfarth-Bottermann KE (1983) Energy metabolic regulation of oscillatory contraction activity in Physarum polycephalum. Cell Tissue Res 231:675–691

Naib-Majani W, Achenbach F, Weber K, Wohlfarth-Bottermann KE, Stockem W (1984) Immunocytochemistry of the acellular slime mold Physarum polycephalum IV. Differentiation and dynamics of the polygonal actomyosin system. Differentiation 26:11–22

Wohlfarth-Bottermann KE (1975a) Weitreichende fibrilläre Protoplamadifferenzierungen und ihre Bedeutung für die Protoplasmaströmung X. Die Anordnung der Actomyosin-Fibrillen in experimentell unbeeinflußten Protoplasmaadern. Protistologica 11:19–30

Wohlfarth-Bottermann KE (1975b) Tensiometric demonstration of endogenous, oscillating contractions in plasmodia of Physarum polycephalum. Z Pflanzenphysiol 76:14–27

Wohlfarth-Bottermann KE (1979) Oscillatory contraction activity in Physarum. J Exp Biol 81:15–32

Wohlfarth-Bottermann KE (1983) Dynamic cellular phenomena in Physarum possibly accessible to laser techniques. In: Earnshaw JC, Steer MW (eds) The application of laser light scattering to the study of biological motion. NATO ASI Ser A: Life Sciences, vol 59. Plenum, New York, pp 501–517
Wohlfarth-Bottermann KE, Achenbach F (1982) Lateral apertures as passage-ways between ectoplasm and endoplasm in plasmodial strands of Physarum. Cell Biol Int Rpts 6:57–61
Wohlfarth-Bottermann KE, Block I (1981) The pathway of photo-sensory transduction in Physarum polycephalum. Cell Biol Int Rpts 5:365–373
Wohlfarth-Bottermann KE, Fleischer M (1976) Cycling Aggregation Patterns of cytoplasmic F-Actin coordinated with oscillating tension force generation. Cell Tissue Res 165:327–344
Wohlfarth-Bottermann KE, Stockem W (1970) Die Regeneration des Plasmalemms von Physarum polycephalum. Wilhelm Roux' Arch 164:321–340

Note Added in Proof. The functional interpretation of stress fibrils presented in this chapter is in accordance with the function of stress fibrils in vertebrate tissue cells in situ and in culture:

In splenic sinus endothelium cells in situ, stress fibrils "can create tension that might allow the endothelium to resist the fluid shear forces of blood flow" (Drenckhahn D, Wagner J, 1986) Stress Fibers in the Splenic Sinus Endothelium In Situ: Molecular Structure, Relationship to the Extracellular Matrix, and Contractility. J Cell Biol 102: 1738–1747).

In epidermal epithelia cells moving in culture, F-actin bundles are "aligned parallel to tension" and this "can be induced within seconds" (Kolega J, 1986, Effects of Mechanical Tension on Protrusive Activity and Microfilament and Intermediate Filament Organization in an Epidermal Epithelium Moving in Culture. J Cell Biol 102: 1400–1411). This author gives further references concerning the cytomechanics of vertebrate stress fibrils.

III. Mechanical Factors Determining Morphogenesis of Protists

III.1 Determination of Body Shape in Protists by Cortical Structures

KLAUS HAUSMANN[1,2] and MARIA MULISCH[1]

1.1 Introduction

Identification of protists can be made by their light microscopic anatomy. This statement seems to be trivial. However, protists need skeletal elements to maintain their shape. This shape may be rather complex or even bizarre. The reasons why they have such shapes is not discussed, rather this chapter deals with how the intra- and extracellular cortical organization is responsible for the appearance of protists. For. convenience, in this chapter protists will be divided into the (algal and protozoan) Flagellata, Sarcodina, Sporozoa, and Ciliata.

1.2 Intracellular Cortex Structures

A wide variety of shapes and sizes in protists is known (Fig. 1). Many of them are pleomorphic. The most striking in this respect are the Sarcodina most of which can change their body shape: *Amoeba proteus* (Fig. 2) alters its outline within some seconds. Nevertheless, many (but not all) species of Sarcodina can be distinguished by their light microscopic appearance, i.e., the form and number of the pseudopodia, motile behavior, life cycle patterns, the relative position of some bigger organelles (e.g., nucleus, contractile vacuole), and by the form of their uroid. There are no obvious cytoskeletal elements in the cortex of naked amoebae. But most probably the spatial distribution of the actin or actomyosin network is – besides its important role in locomotion (Stockem et al. 1981) – responsible for maintaining the cell's shape. This is not inconsistent with the changeability of shape as this network in certain amoeboid protists at least may depolymerize and polymerize again within a few seconds (Sheterline 1983).

More is known about a special type of pseudopod, the axopods. These cell extensions usually radiate from the mostly free-floating Actinopoda, i.e., the radiolaria and heliozoa (Fig. 3a). Axopods are characterized by geometrically packed microtubules (Fig. 3b). They are stiff but may shorten rapidly by depolymerization of the microtubules, which can be induced by external and internal factors (Dustin 1984). Within the Actinopoda, there are a number of different patterns of microtubule arrangement (cf. Fig. 3b and c). (The intracellular, inorganic skeletons of the radiolaria will be the topic of Chap. III.3 of this book; therefore, these organisms can be neglected at this point.)

[1] Division of Protozoology, Institute of General Zoology, Free University of Berlin, FRG
[2] Dedicated to Prof. Dr. K. E. Wohlfarth-Botttermann on the occasion of his 65th birthday

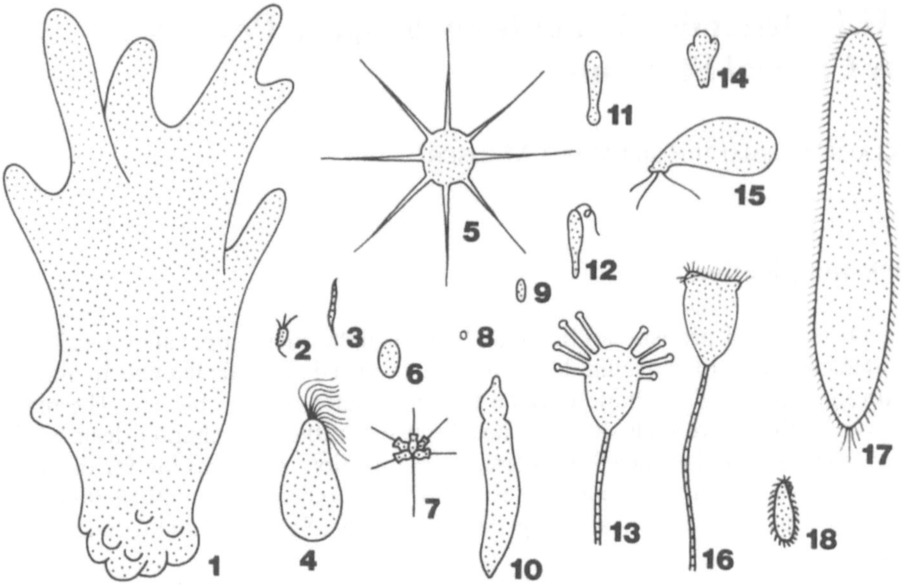

Fig. 1. Schematic synopsis of the relative dimensions of protists, drawn at the same scale (c = silicate, f = flagellate, s = sarcodine, sp = sporozoan): *1 Amoeba,* s; *2 Trichomonas,* f; *3 Trypanosoma,* f; *4 Joenia,* f; *5 Actinophrys,* s; *6 Eimeria,* sp; *7 Codonosiga,* f; *8 Microspora,* sp; *9 Myxozoa,* sp; *10 Gregarina,* sp; *11 Saccamoeba,* s; *12 Euglena,* f; *13 Discophrya,* c; *14 Entamoeba,* s; *15 Trinema,* s; *16 Vorticella,* c; *17 Paramecium,* c; *18 Tetrahymena,* c (Hausmann 1985)

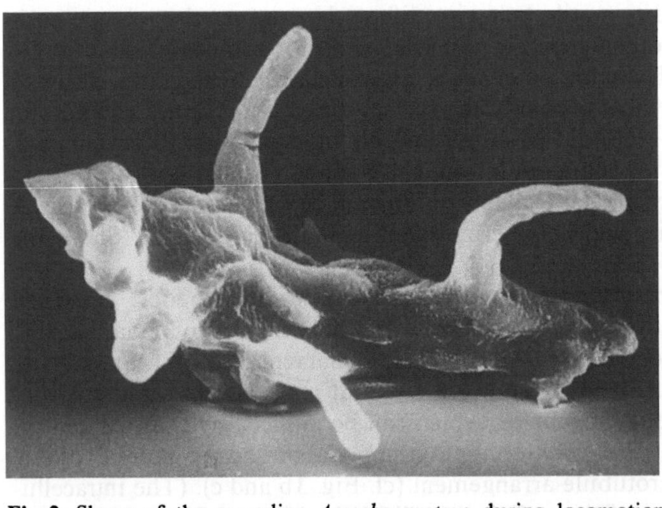

Fig. 2. Shape of the sarcodine *Amoeba proteus* during locomotion; scanning electron microscopy (SEM), × 250 (Hausmann 1985)

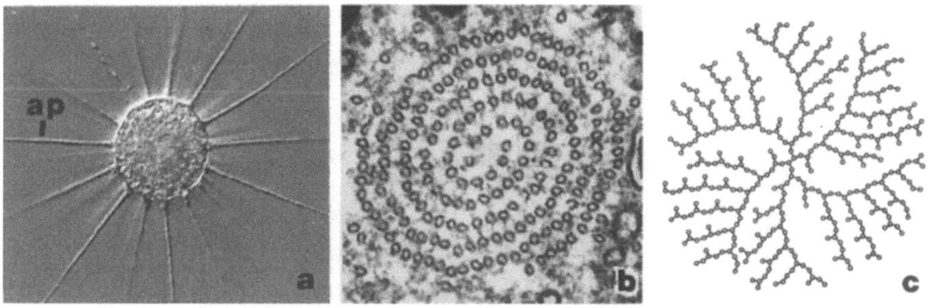

Fig. 3a–c. The sarcodine *Actinophrys sol.* **a** One main character is the presence of stiff axopods (*ap*); differential interference contrast (DIC), × 250. The axopods are stiff elements due to a bundle of microtubules; transmission electron microscopy (TEM), × 48 000. The pattern of the microtubular arrangement is different in the axopods of the diverse groups of actinopods as demonstrated for actinophryids (**b**) and nasselarids (**c**)

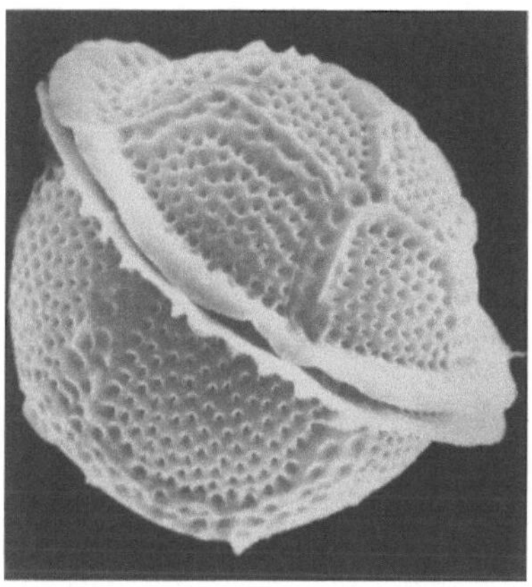

Fig. 4. Skeletal plates of the armored dinoflagellate *Gonyaulax* (SEM), × 1000 (Kindly provided by Dr. E. Small, College Park, Maryland)

Flagellates, sporozoa, and ciliates have more or less complex internal cortex structures which are responsible for the form of each cell. The simplest situation is found in some flagellates, the shape of which is controlled by the presence of cortical microtubules. This has been demonstrated, for instance, in the chrysophycean alga *Ochromonas* as removal of these microtubules leads to a loss of shape (Brown and Bouck 1973).

In cryptophycean flagellates a mosaic of thin protein plates immediately beneath the plasma membrane, the periplast, is responsible for the characteristic shape. Microtubules are not involved in maintaining the typical cell form (Dodge 1973).

The situation in euglenoid flagellates is more complicated. Here, in addition, a subcortical layer of proteinaceous material as well as certain sets of microtubules are present. The protein may be arranged in single plates. This layer enhances the rigidity of the cell body. Genera of euglenoids which show an active peristaltic movement (euglenoid movement, also called metaboly) do not have any continuity between adjacent plates. In other euglenoids the plates

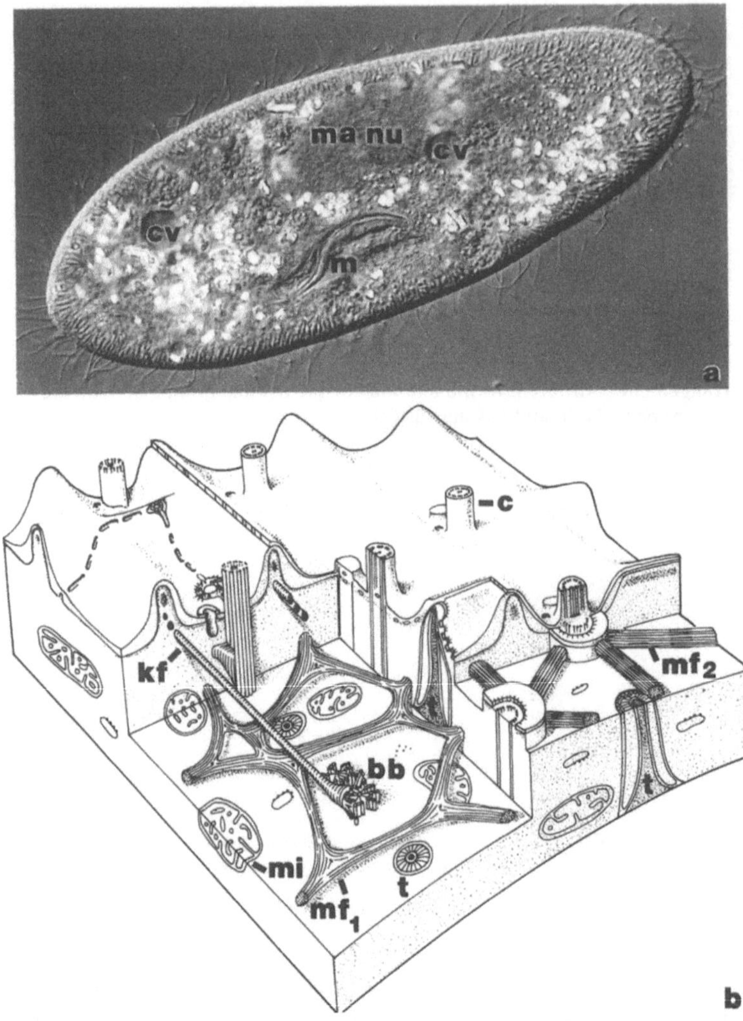

Fig. 5a, b. The ciliate *Paramecium*. **a** Habitus of a living organism (DIC), × 500; *cv* contractile vacuole; *m* mouth; *manu* macronucleus. **b** Three-dimensional reconstruction of the cortex; *bb* basal body; *c* cilium; *kf* kinetodesmal fiber; mf_1 lower microfilament system; mf_2 higher microfilament system; *mi* mitochondrion; *t* trichocyst tip (**b** Allen 1971)

are firmly connected with each other and these cells are unable to undergo metaboly (Buetow 1982).

Armored dinoflagellates (Fig. 4) also have cortical plates which result in the marked rigidity of the cells. The plates are located in flat cortical vesicles. They consist of cellulosic material (Dodge 1985).

A very complicated situation is found in the cortex of ciliates, the surface of which is not necessarily smooth but may be sculptured as it is known, e.g., for *Paramecium* (Fig. 5). The intracellular aspect is even more intriguing (Fig. 6): A mosaiclike layer of flattened vesicles (alveoli) is situated just beneath the

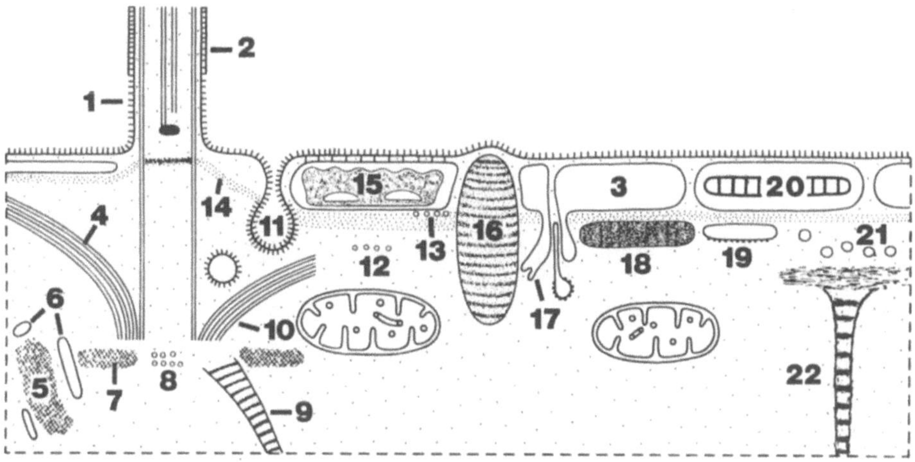

Fig. 6. Schematic synopsis of the possible structures in the cortex of ciliates: *1* plasma membrane; *2* perilemma; *3* alveoli; *4* postciliary microtubules; *5* myonema; *6* SER cisterna; *7* filamentous sheath; *8* subkinetal microtubules; *9* kinetodesmal fiber; *10* transverse microtubules; *11* parasomal sac; *12* and *13* longitudinal microtubules; *14* epiplasm; *15* calcified alveolar plate; *16* extrusome; *17* alveolocyst; *18* polysaccharide plate; *19* RER cisterna; *20* alveolar plate consisting of glycoprotein; *21* vesicles; *22* cross-striated bundles of filaments (Data from Prof. Dr. C. F. Bardele, Tübingen)

plasmalemma. The function of these alveoli is unknown. They are supposed to be involved in the regulation of the intracelluar calcium level which in turn regulates the type of beating of the cilia. In a few species, proteinaceous or calcified polysaccharide plates occur in the alveoli. These elements are interpreted as mechanical support for the cells. The term pellicle is used here to include the plasmalemma and the alveoli.

The cilia are arranged in rows (kineties). Adjacent to cilia the plasmalemma may invaginate and penetrate the alveolar system. By these so-called parasomal sacs cytotic processes are performed. Below the alveoli, and intimately connected with them, is a layer of proteinaceous material (epiplasm). Some observations suggest that the epiplasm plays an important role during morphogenesis by functioning as a matrix in which other cortical constituents are positioned.

Associated with the basal bodies of the cilia are typically three types of special root elements (Fig. 7). The first is a kinetodesmal fiber, made up of filamentous subunits, which extends toward the anterior. Secondly, a bundle of microtubules, the postciliary microtubules, points toward the rear end of the cells. The third set of root structures, the transversal microtubules, runs perpendicular to the cell's longitudinal axis toward the left side of the ciliate. The relative dimensions and the extent of complexity of these three elements are species- or at least genus-specific, but much variation is found among the ciliates.

Additional microtubules, such as longitudinal or subkinetal microtubules, may be present. Systems of filaments have been described in many species. In

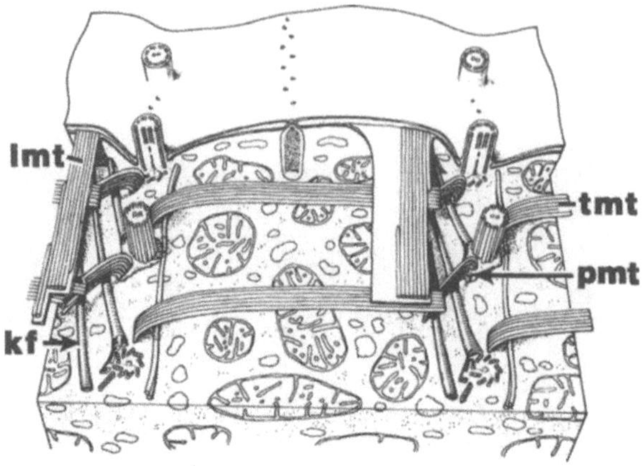

Fig. 7. Portion of the cortex of the ciliate *Tetrahymena* showing the structures which are associated with the basal bodies of the cilia. *kf* Kinetodesmal fiber; *lmt* longitudinal microtubules; *pmt* postciliary microtubules; *tmt* transverse microtubules (Allen 1967)

Paramecium the function of at least one cortical filamentous network is to separate hyaline cytoplasm from granular cytoplasm. It functions like a corset, maintaining the slipperlike cell form (Sibley and Hanson 1974).

Organelles such as extrusomes (Hausmann 1978), the pores of the contractile vacuole complexes (Patterson 1980), or the special differentiations of the cytoproct (the cell anus) (Allen 1978) are also constituents of the cortex of ciliates. The oral regions of ciliates may be thought of as a highly differentiated cortex (Corliss 1979).

Some very excellent three-dimensional reconstructions of the cortex of ciliates which show the spatial arrangement of all the different cortical elements have been made. That is true also for the apicomplexan sporozoa, i.e., the gregarines and coccidia. At least the motile stages of these parasitic protists have a defined cell form and there is a pellicle system with cortical microtubules and filaments (Hammond and Long 1973, Scholtyseck 1979). Little is known of the function of specific components.

One rare example, in which the function of some components of the cortex is known, is the highly contractile ciliate *Stentor*. Bundles of cortical, nonactin filaments can contract within milliseconds and shorten the cell body, whereas the elongation of the cell takes place within some seconds by the action of ribbons of microtubules. They stretch the organisms again to the original length by microtubule-microtubule sliding activity driven by dyneinlike elements (Huang and Pitelka 1973). Similar situations have been described for other heterotrich ciliates, e.g., *Eufolliculina* (Mulisch and Hausmann 1981).

1.3 Extracellular Cortex Structures

The cell shape is determined not only by intracellular, but also by extracellular skeletal elements. Green plant cells generally are stabilized by extracellular cell walls which counteract against high intracellular pressures. Cell walls are also found in many unicellular algae. Also, a number of heterotrophic protists exhibit a characteristic shape, which depends on extracellular structures, such as envelopes or loricae. Those structures may be permanent or may be developed under certain conditions or at a certain stage during the life cycle of an organism. A wide variety of extracellular elements has been described which differ in their shape and size, their components, and their mode of formation, as will be illustrated by the following examples. (The diatoms and foraminiferans are excluded here; for diatoms see Chap. III.2.)

Stalk formation is a common phenomenon among unicellular organisms. Stalks may fix the cells permanently or temporarily to a substrate and thereby prevent them from drifting. The mode of stalk formation differs in the diverse taxa of protists.

Under favorable conditions, the phytoflagellate *Poterioochromonas* may protrude a cytoplasmic tail which produces chitinous fibrils. They are arranged in a spiral pattern. The stalk grows distally within a relatively short period of time and ends in a wide cup. The cortical microtubules in the tail may be involved in the arrangement of the chitin fibrils (Schnepf et al. 1975). After completion of the stalk, the cytoplasmic tail is resorbed. The flagellate is not permanently fixed inside its stalk; if disturbed, it may leave it rapidly.

The heliozoan *Clathrulina elegans* undergoes a complex life cycle (Fig. 8): the sessile cell produces flagellated swarmers by multiple division. The swarmers transform into amoeboid organisms and then into heliozoan forms with stiff axopods. The swarmers build a special, tubelike pseudopodium with an inner protrusion. The protrusion contains a bundle of microtubules. Stalk material is secreted into the space between the pseudopodium and the protrusion, which thereby acts as a mold for the generated stalk. When the stalk material has hardened, the pseudopodium as well as the inner protrusion are retracted. The cell body of the heliozoan secretes an extracellular perforated capsule (Bardele 1972).

In ciliates, most stalked cells are found among the peritrichs and the suctoria. The swarmers of the sessile peritrichs secrete the material of the extracellular stalk, which consists of amorphous or filamentous elements, at the scopula. The scopula is a special area at the posterior end of the cell, characterized by a number of short, immobile cilia and the "pores." These pores are invaginations of the plasma membrane, one beside each cilium, penetrating the normal alveolar system of the cortex, like the parasomal sacs. To produce the stalk, vesicles containing the stalk material move to the pores, and the vesicle contents are secreted. Microtubules, which originate at the basal bodies of the scopular cilia, may be involved in the transport of the vesicles (Suchard and Goode 1982). Stalk secretion by suctoria occurs at the scopuloid, an organelle closely resembling the scopula.

In other cases the extracellular stalk material surrounds an intracellular contractile element, such as the spasmoneme of *Vorticella.* This is located in a long tubelike cell extension. It consists of a bundle of densely packed filaments. When it contracts, the cell is quickly drawn toward the substrate. Contraction does not depend on an actin-myosin system. It does not need ATP for contraction, but depends on Ca^{2+} (Inoué and Stephens 1975). The elastic extracellular stalk material appears to act as an antagonist to the spasmoneme, causing the subsequent elongation of the stalk.

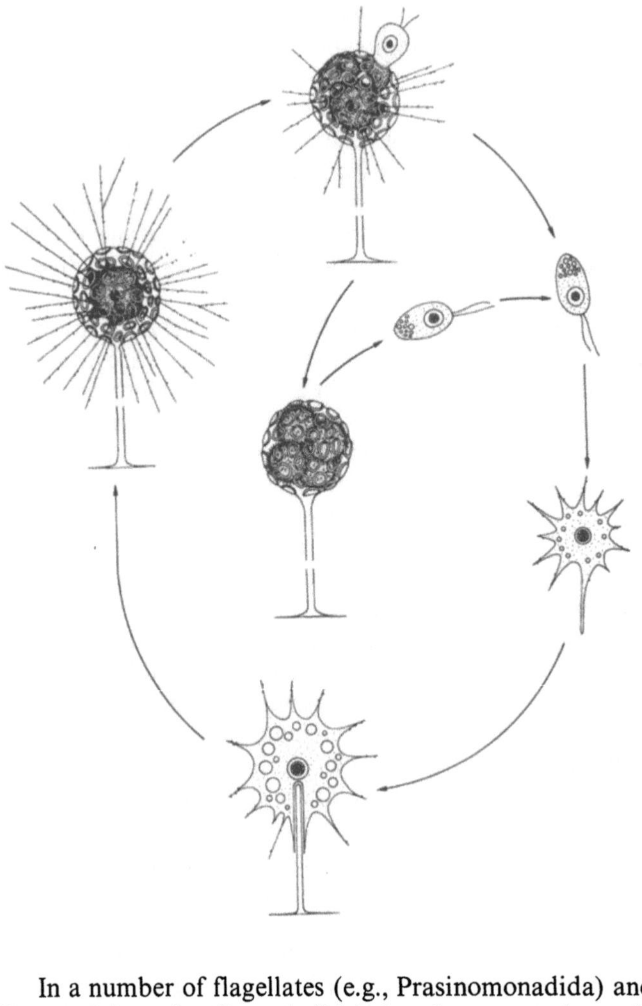

Fig. 8. During its life cycle, the sarcodine *Clathrulina* forms a solid, extracellular stalk and a perforated capsule (for explanation, see text) (Bardele 1972)

In a number of flagellates (e.g., Prasinomonadida) and sarcodines (e.g., heliozoa of the order Centrohelida) the plasma membrane is covered with spines, hairs, or scales, which may consist of organic (e.g., cellulose) or inorganic material (e.g., silica) (Fig. 9).

Scales increase the stability and provide structural support to the cells. Many are produced in dictyosomes, and each appear to be enclosed in a vesicle. During reproduction of the cell, the complete scale is discharged to the outside by exocytosis (e.g. Mignot and Brugerolle 1982). The scales are often arranged in a certain pattern which derives from movements of the cell during scale secretion.

Many species of the testate amoeba (e.g., subclass Testacealobosia and Gromiida) are characterized by tests. The test has one or two openings (pseudostomes) for the extension of pseudopods. Test material may be in the form of adhering foreign bodies or xenosomes (e.g., sand grains) (Fig. 10) and idiosomes, elements produced within the cells. Some species build their tests with one type of these elements, others may use both.

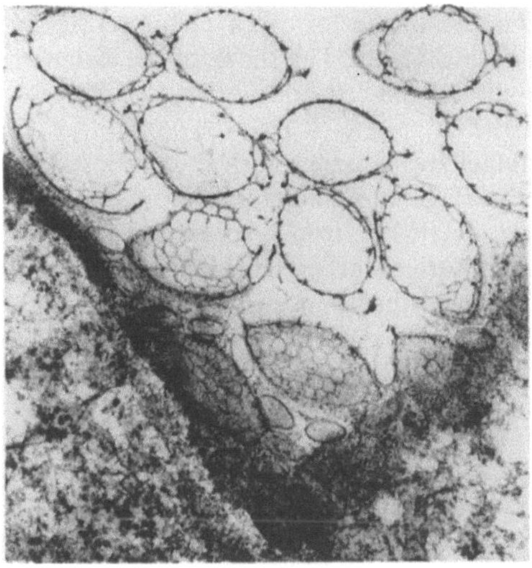

Fig. 9. Extracellular scales of the sarcodine *Cochliopodium* adjacent to the plasma membrane (TEM), × 6000 (Hausmann 1985)

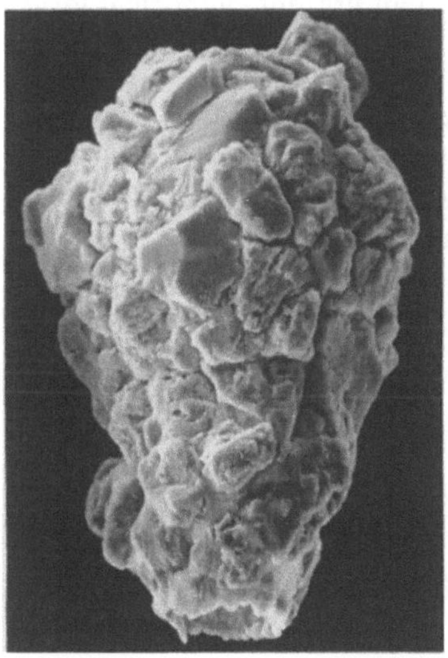

Fig. 10. The test of *Difflugia* is made of grains of sand (REM), × 280 (Hausmann 1985)

Members of the Arcellina (testate amoebae), especially the terrestrial species, are surrounded by resistent tests with only one small pseudostome. During binary fission, a small drop of cytoplasm protrudes from the pseudostome. The naked portion of the cell, which soon becomes shaped like the future shell, discharges numerous idiosomes by exocytosis. The idiosomes may be distributed more or less randomly: as the drop of cytoplasm grows, new idiosomes are deposited in the space which appears between the older ones. *Euglypha* produces idiosomes of different shape and size and these are discharged in a certain sequence to be arranged in a regular pattern (Netzel 1972).

The colonial chrysomonadid *Dinobryon,* has a lorica of cellulose fibrils which is produced by the daughter cell after division. The shape of the lorica of *Dinobryon* is the result of changes in shape of the cell body as well as of complex rotatory motions of the cell (Herth 1979).

Some of the filter-feeding choanoflagellates produce loricae composed of interconnected silica needles. A number of flagellate species live in organic loricae which are impregnated with various metals like iron (Pringsheim 1946) or manganese (Dunlap et al. 1983). The metals are absorbed from the surrounding medium by the lorica wall. The function of this absorbance is unknown.

There are many species of sessile ciliates, especially among peritrichs, spirotrichs, and suctoria, the swarmers of which produce species-specific loricae. In most examples studied, the lorica is secreted by exocytosis of vesicles at fixed sites of the swarmer (e.g., at the pores of the scopula in peritrichs), thereby resembling stalk secretion in these groups. The shape of the lorica is closely related to the behavior of the cell during lorica formation: for instance, the swarmer of the peritrich ciliate *Thuricola folliculata* first fixes at the substrate with its posterior end. Then protein is secreted from the scopula, which hardens as it makes contact with the surrounding water. By extension and contraction of the cell body, the protein is molded to an upright cylindrical lorica (Gonzalez 1979).

The tintinnids construct a wide variety of loricae, the shape of which depends on the stage in the life cycle and on environmental conditions. In *Favella ehrenbergii,* for example, smooth or spirally sculptured loricae may be formed and these may also have a variable number of "necks" (Laval-Peuto 1981). The biological function of this variability is mostly unknown.

The heterotrich sessile folliculinids are a group of ciliates with a complicated life cycle (Fig. 11). As indicated by their trivial name "bottle animalcule,"

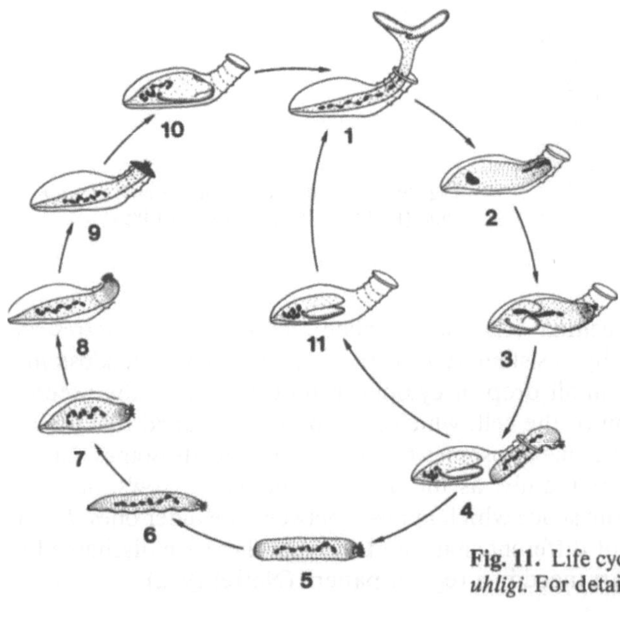

Fig. 11. Life cycle of the ciliate *Eufolliculina uhligi.* For detailed explanation, see text

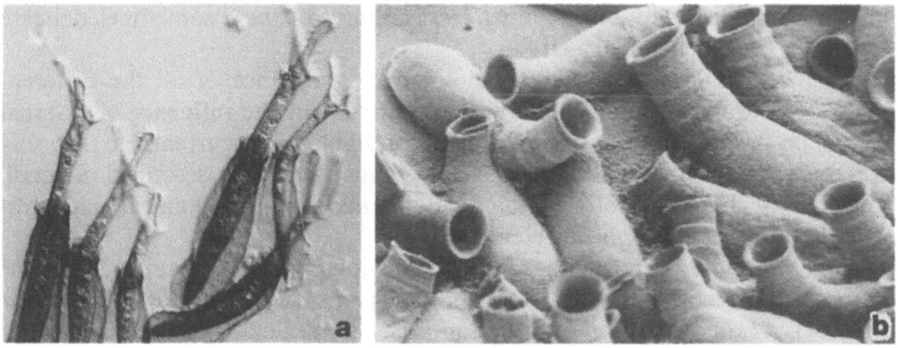

Fig. 12. **a** Living *Eufolliculina uhligi* (DIK), × 80. **b** Bottle-shaped loricae of *E. uhligi* (SEM), × 120. (Mulisch et al. 1983)

most members of this group live in a bottle-shaped lorica (Fig. 11, *1*). The division of the trophic cell results in an astome swarmer, which leaves the old lorica (Fig. 11, *2–4*). It eventually settles on the substrate (Fig. 11, *5*) and produces a new lorica in a certain sequence of events: the swarmer of *Eufolliculina uhligi* first flattens laterally and secretes a basal plate (Fig. 11, *6*). Next, lorica material is discharged from the entire cell surface giving rise to a bulgy "ampulla" (Fig. 11, *7*). Then the cell extends and rotates around its longitudinal axis, while its anterior end produces the spirally sculptured "neck" of the lorica (Fig. 11, *8*). Finally, the apical part of the cell becomes mushroom-shaped and secretes a small "brim" around the opening of the neck (Fig. 11, *9*). After lorica formation, the swarmer transforms into the trophic stage by resorbing its anterior cilia and by developing an oral apparatus with two peristomial wings (Fig. 11, *10*) (Mulisch and Hausmann 1983).

The lorica of *E. uhligi* consists mainly of chitin fibrils which are embedded in a matrix of mucopolysaccharides, pigments, and other components such as proteins and metals (Mulisch et al. 1983). Thus, a wall texture is created which can be compared in some respects with reinforced concrete. As in many other ciliates, the lorica material of *E. uhligi* is transported to the outside via exocytosis. This is different in other protists, where extracellular material is said to be produced directly at the plasma membrane (Brown 1982). However, *E. uhligi* appears to have no special organelle which would be comparable to, for example, the scopula of peritrichs. The sites of exocytotic activity as well as the cell shapes change continuously during lorica construction (Fig. 12).

1.4 Concluding Remarks

As shown by these few examples, extra- and intracellular skeletal structures are widely distributed among protists. Nevertheless, the chemical composition of these elements is mostly poorly understood. One reason for this lack of information is the difficulty of obtaining mass cultures of these sometimes delicate organisms for biochemical analyses.

Even less is known about the biological significance of all the different shapes of extra- and intracellular structures or about the influence of external factors on their formation (e.g., temperature, light, water current, etc.). In several cases the development of the skeletal elements has been described in detail, but there is almost no information available about the mechanical characteristics of the diverse supporting elements.

Cell morphology and cell behavior are obviously closely correlated with the shape of the shells, stalks, and loricae, which in turn influence the cell shape (Fig. 11). Again, almost no information exists about details of these reciprocal effects and their regulating mechanisms. We are only at the very beginning of understanding the morphogenetic factors which are responsible for the composition and the arrangement of the highly complex intracellular skeletal elements, especially in the ciliates (Aufderheide et al. 1980).

Despite the fact that we have almost no experimental data about the mechanical properties of the cortical structures in protists, some general statements can be made. Internal and external plates and scales of various location in the cortex are obviously responsible for a certain rigidity of the cells (Figs. 4 and 9). Tests and loricae seem to be very resistent to mechanical damages (Figs. 10 and 12).

Bundles of filaments and microtubules play an important role in shaping of the protistan cell body in general (Figs. 2 and 3). Moreover, as demonstrated by certain flagellates and sarcodines, microtubules are involved in shaping special transient cell extensions serving as stabilizing elements which might operate, in addition, as guidance structures for a proper arrangement of the secreted stalk material around or into an organic mold (Fig. 8).

The resulting stalks are seldom massive. In numerous cases a hollow stalk will be created. Often the stalk-forming filamentous material is arranged in spirals. Thus, the stalks have a very strong stability, but show also a certain degree of elasticity.

The most fascinating situation is that found in the ciliates' cortex. At present, no general concept exists on how the different structures might cooperate and interact in life. We can only assume some functions. The ciliary root structures (Fig. 7), for example, serve in all likelihood as anchors to guarantee the proper and firm position of the cilia. The two different sets of microfilaments in the cortex of *Paramecium* (Fig. 5) might be of static and dynamic nature: the lower one acts, as mentioned above, as a corset to separate ecto- and endoplasm. The higher one could have contractile properties and may be responsible for the slight body contractions which can be observed in nature. At least, both systems of filaments are an intriguing demonstration how bracings are used in nature to maintain rather complicated cortex and surface structures. In this context also the alveoli should be mentioned. Besides their presumed physiological function as calcium stores they might also play a role in physical respects: they underlie the plasma membrane as large flat units probably reinforcing this membrane. The sometimes several tens of micrometers thick layer of epiplasm just beneath the alveoli (Fig. 6, *14*) might support the skeletal function of these vesicles.

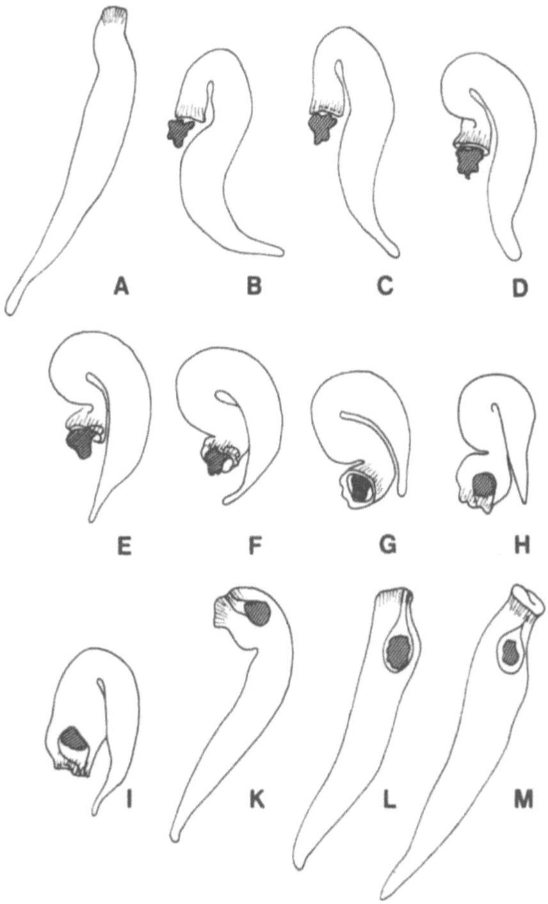

Fig. 13. Demonstration of a pronounced body flexibility in the ciliate *Homalozoon vermiculare* during food ingestion, taken from a 46s-long film sequence, × 100 (Kuhlmann et al. 1980)

If we speculate about the mechanical properties of the different external and internal cortex structures of protists, we should keep in mind, that besides a certain grade of stability very often simultaneously a fairly high degree of elasticity must exist to guarantee the cell's flexibility (Fig. 13).

Acknowledgments. With support from the Deutsche Forschungsgemeinschaft, Bonn-Bad Godesberg, West Germany. The authors thank Dr. N. Hülsmann, University of Berlin and Dr. D. J. Patterson for critical reading of the manuscript.

References

Allen RD (1967) Fine structure, reconstruction and possible functions of components of the cortex of *Tetrahymena pyriformis*. J Protozool 14:553–565

Allen RD (1971) Fine structure of membranous and microfibrillar systems in the cortex of *Paramecium caudatum*. J Cell Biol 49:1–20

Allen RD (1978) Membranes of ciliates: ultrastructure, biochemistry and fusion. In: Poste G, Nicolson L (eds) Membrane fusion. Elsevier, Amsterdam, North-Holland, pp 657–763

Aufderheide KJ, Frankel J, Williams NE (1980) Formation and positioning of surface-related structures in protozoa. Microbiol Rev 44:253–302
Bardele CF (1972) Cell cycle, morphogenesis, and ultrastructure in the pseudoheliozoan *Clathrulina elegans.* Z Zellforsch 130:219–242
Brown DL, Bouck B (1973) Microtubule biogenesis and cell shape in *Ochromonas.* II. The role of nucleating sites in shape development. J Cell Biol 56:360–378
Brown RM (ed) (1982) Cellulose and other natural polymer systems. Plenum, New York
Buetow DE (ed) (1982) The biology of *Euglena,* vol III. Academic Press, London
Corliss JO (1979) The ciliated protozoa. Characterization, classification and guide to the literature. Pergamon, New York
Dodge JD (1973) The fine structure of algal cells. Academic Press, London
Dodge JD (1985) Atlas of dinoflagellates. Farrand, London
Dunlap JR, Walne PL, Bentley J (1983) Microarchitecture and elemental spatial segregation of envelopes of *Trachelomonas lefevrei* (Euglenophyceae). Protoplasma 117:97–106
Dustin P (1984) Microtubules, 2nd ed. Springer, Berlin Heidelberg New York
Gonzales J (1979) La loge du cilie peritriche *Thuricola folliculata* (O. Müller 1786) Structure, nature chimique et formation. Protistologica 15:487–493
Hammond DM, Long Pl (eds) (1973) The coccidia. University Park, Baltimore
Hausmann K (1978) Extrusive organelles in protists. Int Rev Cytol 52:197–276
Hausmann K (1985) Protozoologie. Thieme, Stuttgart
Herth W (1979) Behaviour of the chrysoflagellate alga, *Dinobryon divergens,* during lorica formation. Protoplasma 100:345–351
Huang B, Pitelka DR (1973) The contractile process in the ciliate *Stentor coeruleus.* J Cell Biol 57:704–728
Inoué S, Stephens RE (eds) (1975) Molecules and cell movement. Raven, New York
Kuhlmann S, Patterson DJ, Hausmann K (1980) Untersuchungen zu Nahrungserwerb und Nahrungsaufnahme bei *Homalozoon vermiculare,* Stokes 1887. 1. Nahrungserwerb und Feinstruktur der Oralregion. Protistologica 16:39–55
Laval-Peuto M (1981) Construction of the lorica in ciliate tintinnina. In vivo study of *Favella ehrenbergii:* Variability of phenotypes during the cell cycle, biology, statistics, biometry. Protistologica 17:249–272
Mignot JP, Brugerolle G (1982) Scale formation in chrysomonad flagellates. J Ultrastruct Res 81:13–26
Mulisch M, Hausmann K (1981) Struktur und Ultrastruktur von *Eufolliculina spec.* Schwärmer und sessiles Stadium. Protistologica 17:285–312
Mulisch M, Hausmann K (1983) Lorica construction in *Eufolliculina* sp. (Ciliophora, Heterotrichida). J Protozool 30:97–104
Mulisch M, Herth W, Zugenmaier P, Hausmann K (1983) Chitin fibrils in the lorica of the ciliate *Eufolliculina uhligi:* ultrastructure, extracellular assembly and experimental inhibition. Biol Cell 49:169–178
Netzel H (1972) Morphogenese des Gehäuses bei *Euglypha rotunda* (Rhizopoda, Testacea). Z Zellforsch 135:63–69
Patterson DJ (1980) Contractile vacuoles and associated structures: Their organization and function. Biol Rev 55:1–46
Pringsheim EG (1946) On iron flagellates. Philos Trans R Soc Lond B Biol Sci 232:311–342
Schnepf E, Röderer G, Herth W (1975) The formation of the fibrils in the lorica of *Poterioochromonas stipitata:* tip growth, kinetics, site, orientation. Planta (Berl) 125:45–62
Scholtyseck E (1979) Fine structure of parasitic protozoa. Springer, Berlin Heidelberg New York
Sheterline P (1983) Mechanisms of cell motility: molecular aspects of contractility. Academic Press, London
Sibley JT, Hanson ED (1974) Identity and function of a subcortical cytoskeleton in *Paramecium.* Arch Protistenkd 116:221–235
Stockem W, Hoffmann HU, Gawlitta W (1981) Funktionell-morphologische Grundlagen der amoeboiden Bewegung. Verh Dtsch Zool Ges 1981:71–84
Suchard SJ, Goode D (1982) Microtubule-dependent transport of secretory granules during stalk secretion in a peritrich ciliate. Cell Motil 2:47–71

III.2 Morphogenetic Forces in Diatom Cell Wall Formation

ANNA-MARIA M. SCHMID[1]

2.1 Introduction

Diatoms are armored with a rigid cell wall which consists of two thecae overlapping in the girdle region and resembling a bipartite box (Fig. 1). Cell division occurs by binary fission parallel to the valvar plane. Each daughter cell inherits one part of the mother cell wall, which in both cells becomes the epitheca, while the hypotheca is newly created (Figs. 2, 7, 8). This causes a decrease in size for some of the daughter cells during successive divisions (Fig. 2). At a certain stage this is compensated by sexual reproduction, resulting in a large auxospore, within which the initial valves are laid down (Fig. 3). The

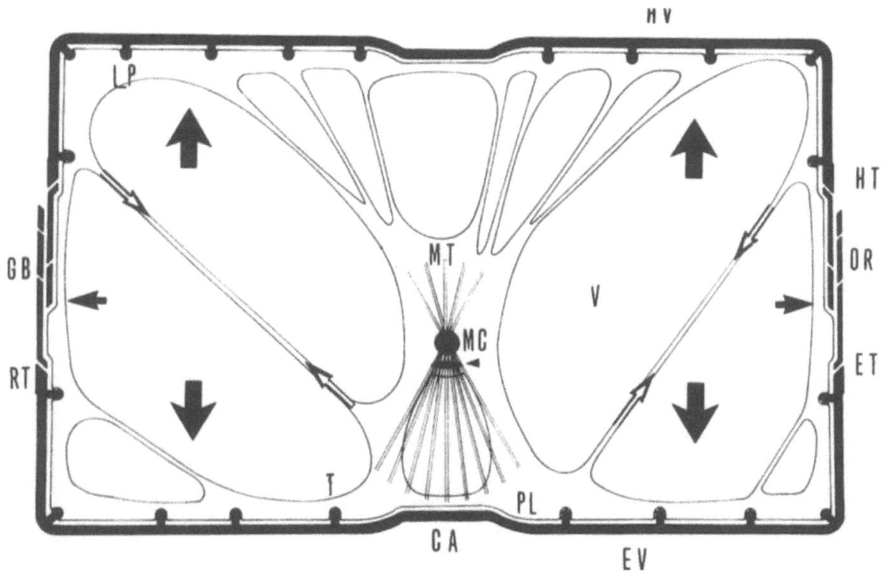

Fig. 1. *Coscinodiscus wailesii.* Schematic cross-section through the interphase cell shows a bipartite wall: epitheca (*ET*) composed of epivalve (*EV*) and three girdle bands (*GB*); hypotheca (*HT*) composed of hypovalve (*HV*) and two girdle bands. Internal turgor exerts in the flat cells a larger pushing force onto the valves (*large arrows*) than onto the girdle (*small arrows*). Sliding apart in the overlapping region (*OR*) is prevented by the tension of the guying strands (*white arrows*). The protoplast is anchored to the wall at the labiate processes (*LP*). Valve edged strengthened with ring ties (*RT*). *CA* Central area; *PL* plasmalemma; *T* tonoplast; *V* vacuole; *MT* microtubules; *MC* microtubule center; *arrowhead* spindle precursor

[1] Institut für Pflanzenphysiologie, Universität Salzburg, A-5020 Salzburg, Austria

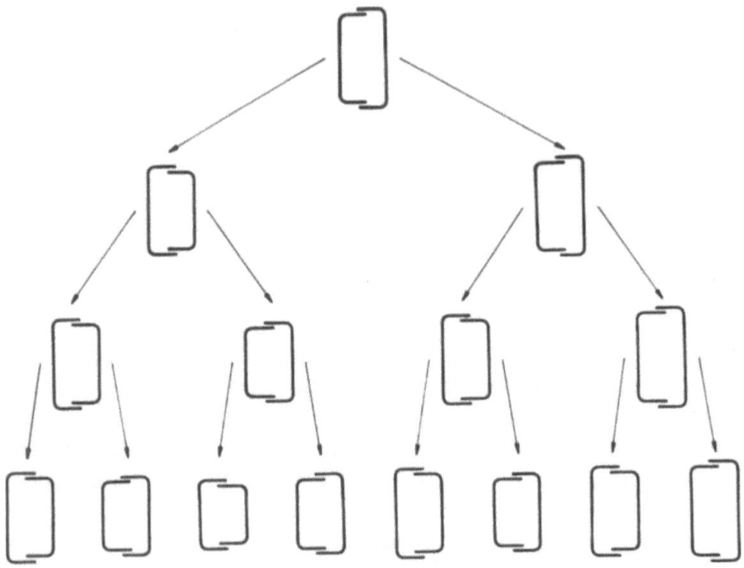

Fig. 2. Size decrease of diatom cells during successive divisions

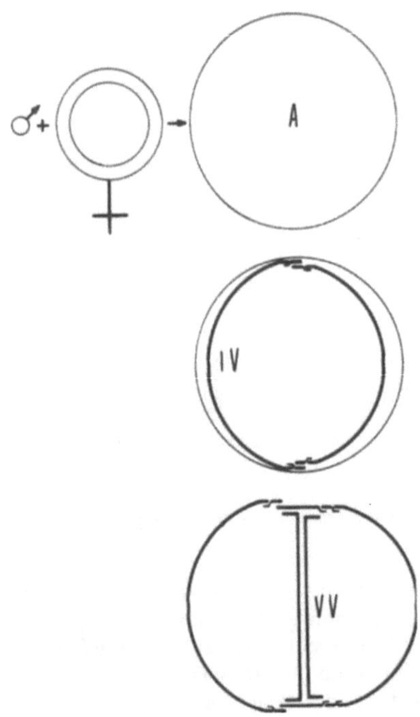

Fig. 3. Sexual reproduction resulting in an auxospore (*A*), within which the initial valves (*IV*) are laid down; vegetative valves (*VV*) after first cell division

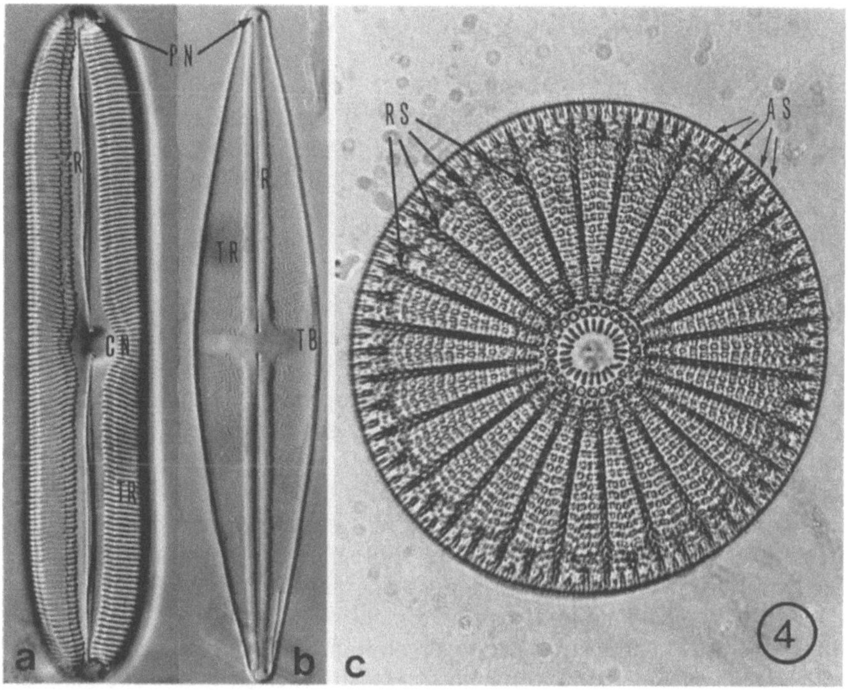

Fig. 4a–c. Main supporting valve structures in diatom valves. **a** *Pinnularia:* Raphe ribs (*R*), central nodule (*CN*), polar nodules (*PN*), and thick transapical ribs (*TR*). **b** *Stauroneis:* Raphe ribs (*R*), central transversal band (*TB*) a strengthening structure to compensate for the fine transapical ribs (*TR*), polar nodules (*PN*). **c** *Arachnodiscus:* Radiating shores (*RS*) focusing in a strongly silicified central area; angle shores (*AS*) at the margin

initial valves are rounded and do not exhibit the surface pattern characteristics of the species. The species-specific surface form of the vegetative valves is created with the first division (Fig. 3). Each individual part of the wall is composed of an inorganic opaline silica component that is enveloped by an organic coat (ref. in Schmid et al. 1981). The enormous variety and complexity of valve shapes and particularly their patterns induces the question of "how" and "why" they are formed in such a remarkable way; and what determines their outline, their topography, and their pattern? With respect to overall frustule pattern, why is there always a geometric center from which it fans out, forming a ring, or central area, in *Centrales,* while in *Pennales* it lies along an axial rib, which in a number of genera is penetrated by a slit, the raphe (Fig. 4)?

The understanding of the interaction of morphogenetic forces involved in diatom valve formation requires knowledge about the function of the mature wall. This, however, is still very limited.

2.2 Possible Functions of the Diatom Cell Wall

Diatoms are "turgor plants" because they have a higher internal osmotic value than their environment. Turgor or "hydraulic" systems need either an internal (cytoskeleton) or external (cell wall) structural device to maintain nonspherical cell shapes (e.g. Bereiter-Hahn 1977). Nonspherical, naked protists exhibit in their cortical cytoplasm a highly elaborate arrangement of microtubules and microfilaments (see Chap. III.1). The sparsity of a cortical cytoskeleton in diatom interphase cells suggests that the cytoskeletal function of shape control is provided partly by the rigid cell wall, permitting the cytoskeleton to be reduced and thus saving energy (Bonik 1978, 1979). The calculations of Darley (1974) and Raven (1983) seem to support this hypothesis. The siliceous exoskeleton would be operational however, only as a finished product. During morphogenesis, therefore, cytoskeletal elements must be present in the cell cortex, as indeed is the fact (Sects. 2.4.2, 2.4.3).

Considering the range of evolutionary strategies these organisms may have developed, one would expect a wide variety in form and pattern of the frustule resulting from various mechanical and physiological adaptations expressed in phylogenetic lineages. This scientific understanding of the origin of diatom morphological diversity adds a special dimension to their aesthetic appeal. Fracture experiments indicate that form and thickness of supporting valve structures match exactly the requirements of locally exerted stress. The main elements in *Pennales* are the raphe ribs and the central and polar nodules (Fig. 4); from there the transapical ribs run along tension lines to the valve edge (Krammer 1981). In *Centrales* the shores (RS + AS, Fig. 4), or alternatively, sandwiched honeycombs, radiate from the central ring or area along stress trajectories to the margin, which is usually strengthened with "angle shores" (Fig. 4). It seems quite probable that most of the structural patterns in diatoms can be deduced from the stress-strain constitutional law.

Plant cells are usually enveloped by firm cellulosic walls which counterbalance the turgor by tension. This cannot account, however, for a diatom cell wall, which is simply a rigid bipartite box (Fig. 1). The pressure exerted on the overlapping zone probably does not cause enough friction between the overlapping girdle bands to prevent the two thecae from sliding apart. This is conclusive, particularly from the research of Stosch (1954) and Stosch and Drebes (1964). How great, if at all, the role of organic material is in connecting the two thecae is not yet clear. Considering the tendency of a protoplast to round up, due to surface tension, the pushing force (F_p) of a diatom cell at a given turgor (p) and a given height (h) of the box, increases with valve diameter (2r) (Fig. 1):

F_v = force acting upon the valve

$F_v = p \cdot r^2 \cdot \pi$

$F_{c/2}$ = force acting upon the girdle band region

$F_{c/2} = p \cdot r \cdot \pi \cdot h$

$$F_p = \frac{F_v}{F_{c/2}} = \frac{p \cdot r^2 \cdot \pi}{p \cdot r \cdot \pi \cdot h} = \frac{r}{h}. \tag{1}$$

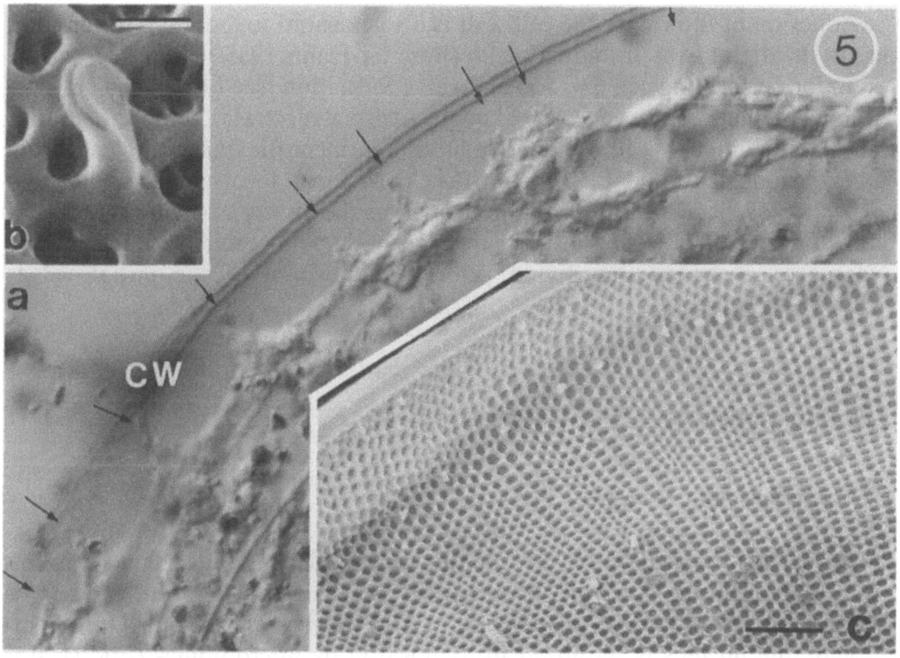

Fig. 5a–c. *C. wailesii:* Permanent attachment sites between protoplast and cell wall. **a** LM micrograph: *arrows* mark adhesion points at the valve margin; *CW* cell wall. **b** Internal tube of the labiate process; note scapuliform shape. SEM; *bar* = 1 µm. **c** Distribution of labiate processes: one ring subjacent to the valve edge, one ring at the face-mantle junction, several LPs scattered. SEM; *bar* = 10 µm

Equation (1) suggests that diatom thecae are held together by a guying device. The tension-transducing parts of the cytoskeleton have to be anchored via the plasmalemma to the cell wall. Special anchors are present in the valves for distributing forces to the rigid wall (Figs. 1 and 5). In *Centrales* they are manifested as the labiate process (LP) and the strutted process (SP), and in the *Pennales* as raphe ribs with their central and polar nodules, and fibulae if present. Gently plasmolyzing the cells shows these parts to be intimate adhesion sites (Fig. 5) (Schmid 1984a, b, 1986b).

As an example of the internal guying, *Coscinodiscus wailesii* possesses a ring of LPs at the valve edge, just above an external ring tie, and one ring at the face mantle junction (this one is missing in the domed initial valves), and several LPs scattered over the valve face (Figs. 1 and 5). At their proximal end the LPs are scapuliform (Fig. 5b), which is an ideal shape for anchoring cellular "muscles." Their role as anchors for tension is (1) indicated by the interphase arrangement of the cytoplasmic strands (Fig. 1); (2) concluded from the fact that the hemispherical initial valve of *C. wailesii* possesses only one ring of LPs at the valve edge (unpubl.); and (3) concluded from the fact that particularly *Centrales* detach from their anchors, leave their walls, and form spherical protoplasts under the influence of cytoskeletal destroying drugs (ref. in Schmid 1984b).

Adhesion of the cell to the cell wall is also essential to a diatom for another reason. In diatoms as in animal cells (Bereiter-Hahn 1985, Chap. I.1), tensile forces for performing cell division and valve formation (see below) can only develop when their cell surface makes contact with a substrate. In the case of diatoms, it is the mother cell wall, acting as an abutment for transmitting shearing forces. Raphe, LP, and SP function as "permanent focal contacts"; on disconnection of the cytoplasm from these anchors the ability to divide ceases (unpubl. observ. on nine species of seven genera). This could account for the auxospore's incapacity to divide prior to deposition of the initial valves (Fig. 3).

2.3 Preconditions of Valve Formation

2.3.1 Mitosis and Cleavage

Under normal conditions valve formation is preceded by mitosis and cytokinesis (ref. in Schmid et al. 1981), which involve nuclear migration and turgor change, and thus determine valve pattern symmetry and shape. During interphase the nucleus is stabilized in its position by microtubules (Fig. 1). For mitosis the nucleus migrates into the girdle region, with the microtubule center (MC) at the leading front (Lauterborn 1896), probably to trigger the formation of the cleavage furrow, which is initially unilateral in *Centrales* and perhaps also in *Pennales* (Schmid and Volcani 1983; Schmid 1984b; and unpubl.). Preprophase is characterized by turgor increase and thus volume increase in the pervalvar direction (Tschermak-Woess and Hasitschka-Jentschke 1958) and formation of the last girdle band of the hypotheca (Fig. 1); whereas in prophase turgor reduction begins and continues until cleavage. Marine forms can thereby plasmolyze (Höfler 1963; Stosch and Drebes 1964). Although the mechanism is still unknown, cleavage is certainly facilitated by this "active slackening" (Stosch and Drebes 1964). The main function of the turgor reduction might be to prevent opening of the cell wall due to destruction of the pervalvar cytoskeletal guying. The plasmalemma is sensor as well as effector for turgor changes in algae (Kirst 1985). In *Coscinodiscus* and *Thalassiosira,* the cleavage furrow incises from the middle of the overlapping part of the girdle, thus the last girdle band of the hypotheca, which is formed just prior to mitosis, might play an important role as a reference line for defining the cell equator. That the height of the girdle is usually kept constant although the number of bands may vary (Schmid and Volcani 1983), supports this conclusion. The position of the cleavage furrow results in an adhesion of *both* sibling cells to the hypotheca (Fig. 7 and unpubl.). The last deposited girdle band frequently bears specific, structural devices, such as lamellae (Figs. 14, 21, 27 in Schmid 1984a) which enhance adhesion and may help prevent the two thecae from sliding apart accidentally during cell division.

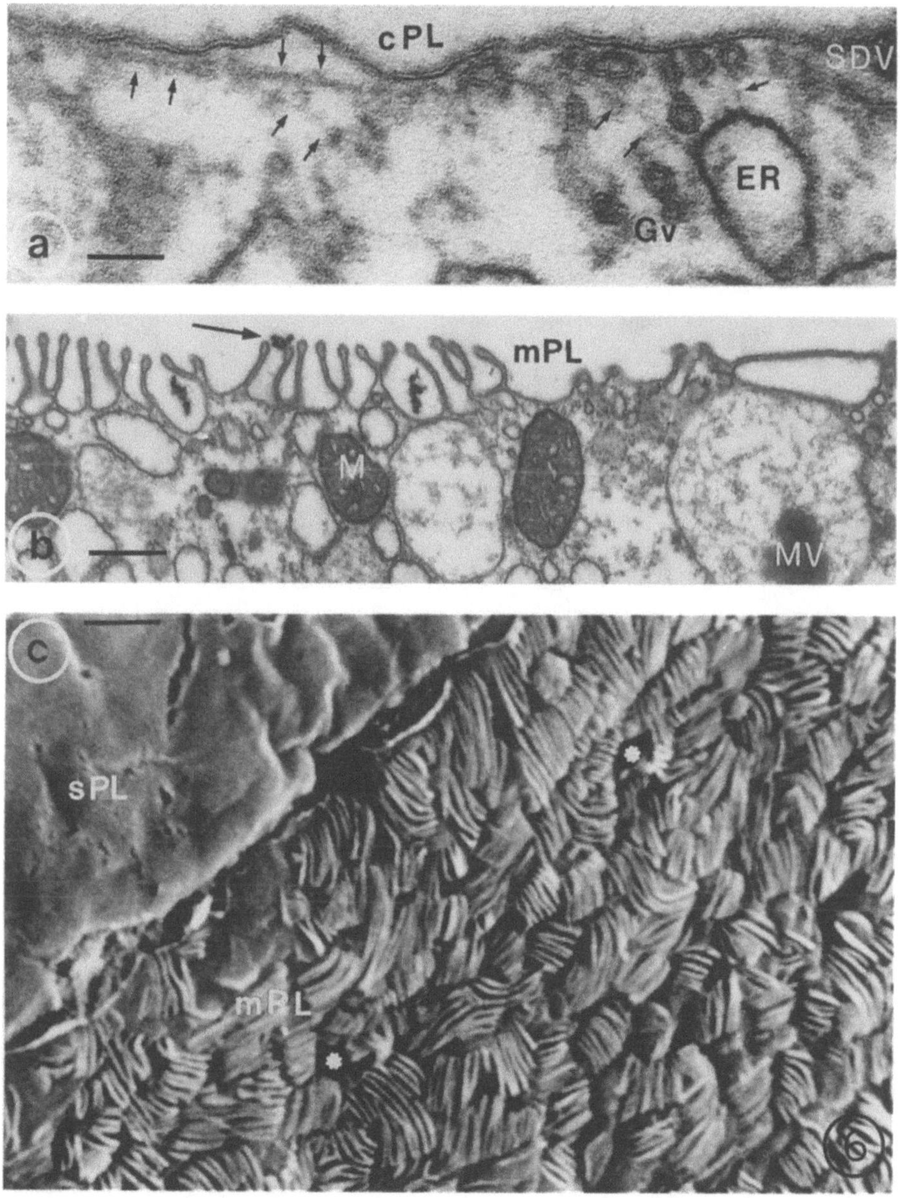

Fig. 6a–c. *C. wailesii.* **a** Plasmalemma (*cPL*) created during cleavage. Fibrous lamina interconnected with microfibrils (*arrows*). *ER* endoplasmic reticulum; *GV* Golgi vesicles prior to fusion with *SDV*. Fixation: Ga/Os/KFeCN, *bar* = 0.07 µm. **b** Metabolic active plasmalemma (*mPL*) in profile. *Arrow* points to exocytosed material. *M* Mitochondrion; *MV* mucilage vesicle. Ga/Os/KFeCN, *bar* = 0.3 µm. **c** Plasmalemma (*sPL*) molding organic layer of girdle, delimited at the valve edge against the metabolic active PL (*mPL*) that was torn during valve lifting along the transient adhesion zone. *Asterisks* mark LP position. Fixation: Ga/Os/CPD, SEM, *bar* = 1.5 µm

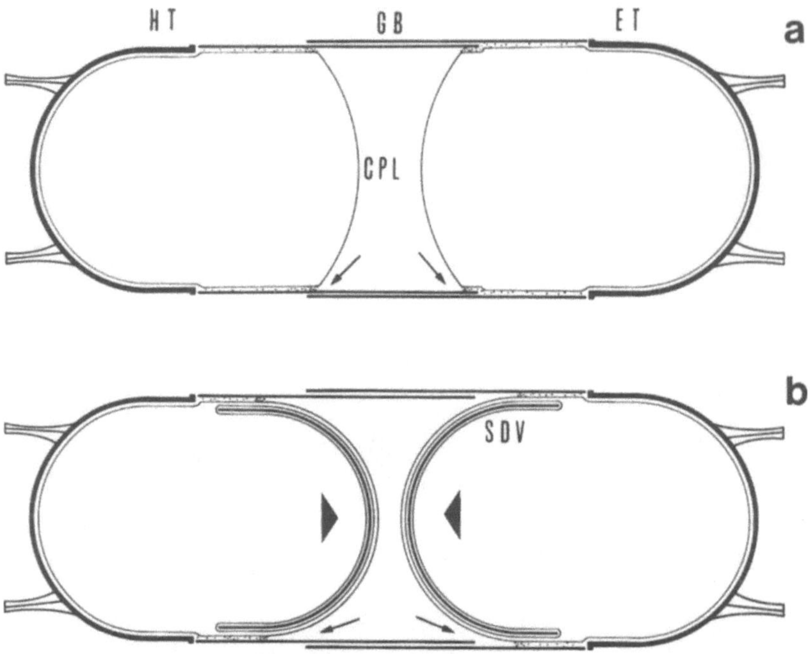

Fig. 7a, b. *Stephanopyxis turris:* Molding of the hemispherical valve topography. **a** Immediately after cleavage: both cells adhered to girdle bands of the hypotheca (*arrows*), ± flat surfaces (*cPL*) because of turgor reduction and adhesion. **b** After base layer formation: partial disconnection of the PL from the adhesion sites (*arrows*) and slight swelling (*arrowheads*) cause rounding of the molding surface due to surface tension and thus prestabilization of the forming valve (*SDV*). Girdle bands of the mother cell wall do not slide apart until valve release. (After Fig. 5f, h from Stosch and Drebes 1964; and unpubl. observ.)

2.3.2 The Molding Surface: The Plasmalemma

Three types of plasma membranes are distinguishable in *Coscinodiscus* (Fig. 6): (1) the plasmalemma created during cleavage (cPL) which seems to have no other function than molding the siliceous part of the valve (Fig. 6a) (Schmid 1984c, 1986a, b). (2) The plasmalemma created upon completion of the siliceous part of the cell wall (sPL), forming the organic layer that seals the cytoplasm against the siliceous chamber wall (Fig. 6c) (Schmid 1984c, 1986a, b). (3) The metabolically active plasmalemma (mPL) which is ruffled into a distinctive pattern of microplicae and is clearly delimited relative to those cytoplasmic surfaces involved in morphogenesis (Fig. 6b, c).

The shape of the cleavage surface in many instances directly determines the shape of the future valve. It depends on the turgor, the tension of the new plasmatic layer, and whether the mother cell adhered to all of its girdle bands prior to cleavage (*Coscinodiscus*), or only to the last and the first ones (*Thalassiosira, Stephanopyxis*). In a number of cases the cleavage surface is premolded prior to or during valve deposition. This occurs either by postcytokinetic swelling

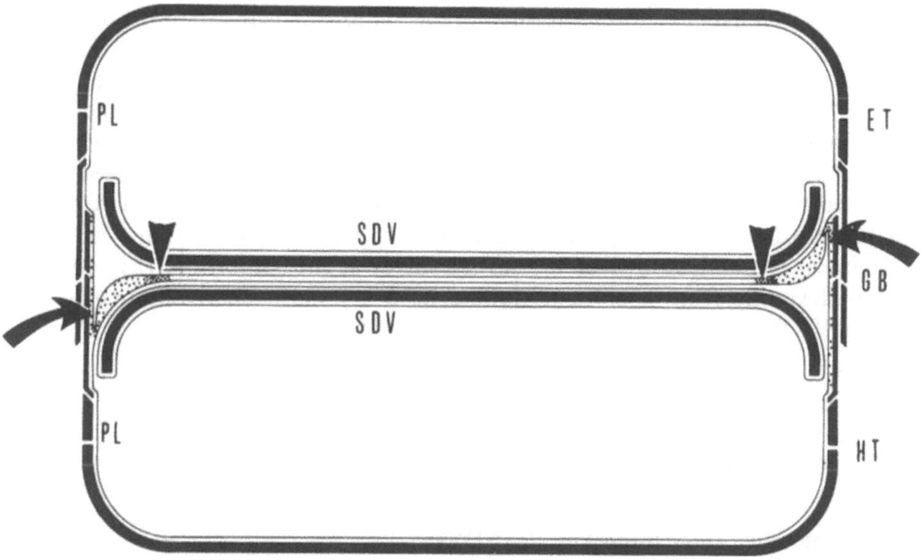

Fig. 8. *Thalassiosira eccentrica:* Stabilization of the flat valves. Plasmalemma outfold and new spotlike connections (*arrows*) occur between the siblings in the margin during the last stage of valve formation. Silicification along the tensioned plasmalemma causes stability and formation of alternating spines. *Arrowheads* Circumferential adhesion site; *dots* adhesive and incompressible mucilage

(Stosch and Drebes 1964) and by partial detachment from the adhesion sites (and thus slight swelling) (Fig. 7), or by the nucleus (Stosch et al. 1973), or by adhesion and local contractions, which could involve secretion of fluids which resist compression (ref. in Schmid 1984a). Contact between the sibling surfaces causes their shape to be mutually influenced. If adhesion is circumferential at the margin, then the mutual effects depend on the tension of the circumscribed area: high tension creates a flat surface as in *Thalassiosira* (Fig. 8), low tension causes complementary undulations or ridges and grooves (ref. in Schmid 1984a). The two cells then behave as one entity, that is as a two-celled tissue during morphogenesis. This is realized in all diatoms where $F_v/F_{c/2}$ is larger than 1, and in all freshwater diatoms. In contrast valve formation as in *Stephanopyxis* can only occur in marine diatoms when $F_v/F_{c/2}$ is smaller than 1 (see Eq. 1).

2.3.3 The Mold for the Valve Outline

Since the valves are formed within the mother cell wall, it is quite apparent that their outline is predetermined by the outline of the latter (Schmid 1985, 1986b). Thus, the question of morphogenetic determination is reduced to the initial valve formation. In *Pennales* shape determination to some extent already occurs during the growth of the auxospore: a silicified ring is deposited around the equator of the zygote, fixing the girth of the future cell. The ring is more tension-resistant than the zygote membrane. Thus, the cell expands poleward, adding one ring after the other in telescope fashion and creating a flexible abutment for initial valve formation (ref. in Stosch 1982). A *Centrales* zygote swells into a "hydrosphere" (Fig. 3), probably as a result of disintegration of the cytoskeleton which occurs during meiotic division (Schmid 1984b). A layer of small siliceous scales subjacent to the organic envelope provides some degree of reinforcement. Circumferential adhesion of the protoplast to the auxospore cover in the prospective cell equator, together with a contraction of the cell, determine shape and outline of the initial valves at the same time (Fig. 3). All forms deviating from a circular outline need a molding perizonium (ref. in Stosch 1982).

2.4 Valve Formation

2.4.1 The Silica Deposition Vesicle (SDV)

In diatoms closely apposed to the cPL, the SDV is a fundamental compartment (ref. in Schmid et al. 1981, Chap. III.4), particularly for controlling the chemistry of silica deposition (Mann 1983). This is evident from cell walls in which silica deposition is restricted to certain regions (Duke and Reimann 1977; Schnepf et al. 1980). The SDV enlarges by addition of small Golgi-derived vesicles (ref. in Schmid 1986a, b). Silica is amorphous without any inherent form. It grows into shape only when fashioned by a rigid mold. Biomembranes are not very rigid, however. Therefore, external molding forces must act upon the SDV, by applying either pressure or tension (Fig. 9).

2.4.2 The Role of the Nucleus, Microtubule Center, and Microtubules

The initiation site for the SDV (pattern center) in both centric and pennate diatoms (ref. and discussion in Schmid 1986b; Crawford and Schmid, 1986) – is dependent on the nucleus and its microtubule center (MC). This idea first proposed by Stosch et al. (1973) is testable. In *Centrales*, e.g., *Attheya* (Schnepf et al. 1980), the SDV is present prior to completion of cleavage in an excentric position as also is the nucleus. Their comigration to the prospective valve center can be inhibited: valve formation, however, continues, but the pattern center is excentric (Schmid 1979b, 1984b). Reversion of cleavage prior to ring closure (Schmid and Volcani 1983) results in a binucleate cell, with a single "internal valve" bearing *two* pattern centers (Schmid 1980, 1984b, unpubl.). The role of the MC and its microtubules in valve formation was clearly shown for *Pennales* (e.g. Pickett-Heaps et al. 1979; for exceptions see ref. in Schmid 1986b), while in *Centrales* the situation is more obscure Sect. 2.4.3). MT inhibitors primarily affect the raphe in *Pennales*, and distribution and shape of strutted and labiate processes in *Centrales*, with in fact no *direct* effect on the valve pattern (Schmid

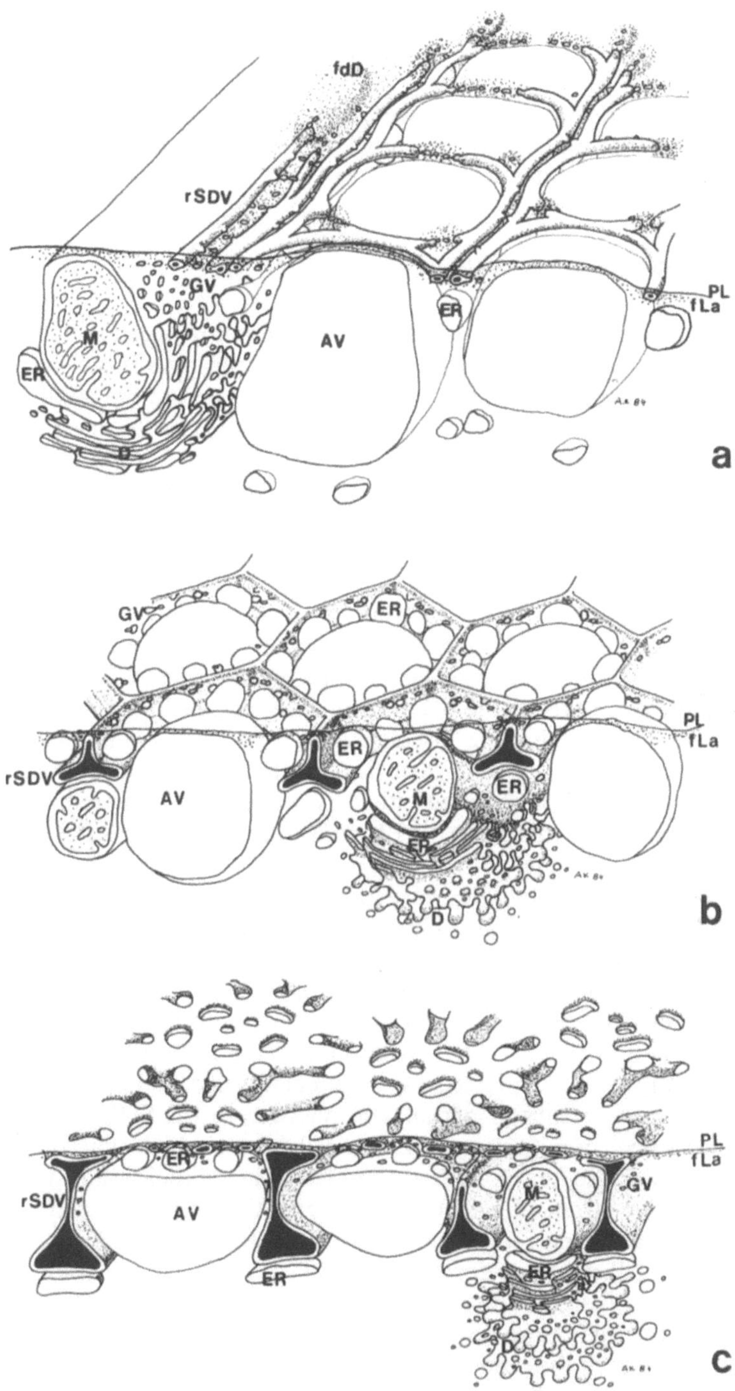

Fig. 9a–c. *C. wailesii:* Schematic drawing of valve development. **a** Development of the base layer = first horizontal differentiation; **b** development of the areolae walls = vertical differentiation; **c** development of the outer layer = second horizontal differentiation. *AV* Areolae vesicles; *D* dictyosome; *ER* endoplasmic reticulum; *GV* Golgi vesicles; *fdD* fibrous dense domains; *fLa* fibrous lamina; *M* mitochondrion; *PL* plasmalemma; *rSDV* reticulate silica deposition vesicle (Schmid 1984c, 1986a)

1980, 1984b, 1986b). Thus, microtubules seem to be involved in stabilizing position and shape of the forming attachment sites, while their formation and that of the pattern, i.e., molding of the SDV, is executed by a different system.

2.4.3 The Molding System for the Valve Pattern

Molding occurs as a precisely timed, sequential interaction between the cleavage-furrow-plasmalemma (cPL) and its fibrous lamina, and microfilaments of unidentified nature (Fig. 6). Organelles or specific vesicles are used as auxiliary systems (Schmid 1984c, 1986a, b). The valve is first outlined in its entire size, then material is added normal to the base layer: in *Pennales* from both sides (ref. in Schmid et al. 1981; Edgar and Pickett-Heaps 1984a), while in the majority of *Centrales* only from the distal side (Schmid 1984a, c, 1986a, b) (Fig. 9).

From the presence of a blueprint pattern prior to the expansion of the SDV in *Coscinodiscus* it is conclusive that the cPL and its fibrous lamina carry the information for the new valve. This information is very likely altered in a sequential manner according to the geometric design to be produced at each of the particular stages. The precisely timed and spatially controlled change in adhesion between cPL and parts of the SDV or molding organelles demonstrate this clearly (Fig. 9). The fact that the "areolae vesicles" which help in molding the chambers do not adopt a pattern of close packing, but are arranged in radial rows, also indicates that they are guided either by recognition sites in the cPL and mediated by microfilaments or by the ER which occupies their distal gaps, or by some combination of these.

In *Pennales,* microfilament bundles were found to be involved in the expansion of the SDV (Pickett-Heaps et al. 1979; Edgar and Pickett-Heaps 1984a); in *Coscinodiscus* a rather discontinuous system of fibrous material (fdD) seems to guide and stabilize the SDV across the cPL through the distal gaps of the areolae vesicles where the ER gives way to it and attaches to its proximal side until valve release (Fig. 9a). The entire distal face of the SDV adheres to the cPL, and horizontally it is compressed by the areolae vesicles. Formation of chamber side walls is by means of partial disconnection of the SDV from the cPL and the subsequent immigration of ER into this area which may inhibit lateral expansion and cause hexagonal arrangement of the vertical walls due to horizontal compressive stress (Fig. 9b). Vertical growth is terminated by detachment of the areolar vesicles from the cPL, which allows the SDV to expand again laterally and adhere as before to the cPL (Fig. 9c); premature detachment of the areolar vesicles causes the outer layer to lie directly on the base layer (unpubl.). At the same time ER vesicles collect distally to the areolar vesicles and mold the pores of the outer layer (Fig. 9c). The ER then detaches and finally the pores are filled with a very delicate lattice whose mesh size has the same dimensions as those of distal Golgi cisternae (Schmid 1984c, 1986a, b). The role of the ER is similar to that in chrysophyte scale design (ref. in Romanovicz 1981). It also resembles the situation found in pollen wall formation, where it causes the patterned destination of the colpi (ref. in Dickinson 1970). The role of the ER can partly be taken over by mitochondria, which also function in shaping raphe canals (ref. in Edgar and Pickett-Heaps 1984b; Schmid 1986b).

Specific cytoplasmic fibrous domains, the "raphe fiber," and "striated structures" are involved in the formation of raphe, LP, and SP, that is, the future attachment sites, which are at the same time the future ports to the outer world: the raphe in connection with locomotion (ref. in Edgar and Pickett-Heaps 1984b), the LP (Schnepf et al. 1980; ref. in Schmid 1986b) as possible predecessor of the raphe (Edgar and Pickett-Heaps 1984b), and the SP (Fig. 4 by D. Schulz in Schmid 1986b), penetrated subsequently by chitin fibrils (Herth 1979) that serve as colony-forming and floating devices. MTs were found below the raphe fiber, but not yet in connection with LP and SP formation; their sensitivity to microtubule inhibitors, however, is similar to that of the raphe (Schmid 1980, 1984b, unpubl.). Raphe ribs (Edgar and Pickett-Heaps 1984a), SPs, and LPs show birefringence with polarization optics, though in neither one could a reflection be obtained with Debey-Scherrer X-ray determination. Thus, birefrigence,is probably the result of a specifically oriented deposition of the opaline silica, which might be in accordance with the prospective anchor function (Kirchner and Schmid unpubl.).

2.4.4 Mechanisms for Mechanical Stabilization of the Valve

Simple models demonstrate the increase in stress resistance of a chambered valve during the course of its formation, starting with the base layer, on addition of side walls and an outer layer. The stability of the cell is achieved only at the very last stage of wall formation, provided that silicification occurs along a tensioned molding surface. During early stages of valve formation, the turgor seems to remain very low as is conclusive from LM observations (Stosch et al. 1973) and EM fixation experience. In *T. eccentrica* both siblings adhere, to each other and to the girdle bands of the mother cell's hypotheca during base layer formation. Then turgor increases gradually during the formation of the chamber side walls; the volume increase causes the girdle bands of the mother cell to slide somewhat apart and the cell which lies in the mother's epitheca becomes detached from the girdle bands of the hypotheca. In *Stephanopyxis* detachment is not accompanied by sliding of the thecae. Turgor change during the last stage of valve formation is generally more dramatic, this is evident in contractions of the sibling cells, giving rise to an increased turgor. In locally restricted unsilicified areas this leads consequently to a bulge of cytoplasm, resulting in the formation of setae and horns (Stosch et al. 1973; Schnepf et al. 1980). The chamber pattern of *T. eccentrica* is radial in the base layer, while at the distal side it appears to be tangential. The marginal spines seem to reflect the tangential lines of tension. During formation of the outer layer, the plasma membranes of the interconnected siblings were observed to fold out at the margin, using the girdle bands of the mother cell wall as substrate for what might be interpreted as "crawling motion" to establish new spotlike cell-cell contacts at the margin (Fig. 8). This event, in concert with a pervalvar shift of the interconnected surfaces, gives the impression that the cells are stretching their cPL surfaces into an optimal position. When the new cPL contacts are made, the outfolds along the girdle bands are retracted and, due to the tensile stress, the cPL is lifted tentlike at the connecting points. At this stage final silicification occurs (Schmid 1984a, c). Form and position of spines in other genera suggest rather conclusively that what occurs in *Thalassiosira* is a widely used mechanism for valve reinforcement. Broader adhesion zones than in *T. eccentrica* result in spathulate, interlocking spines which inhibit cell separation after valve completion, and thus lead to colony formation, as e.g., in *Fragilaria* or *Melosira*. The transient adhesive connections are colchicine- and APM-sensitive. This results in a deletion of spines and in a modified pattern of areolae which is accompanied by a reduction of the mechanical strength of the valve (Schmid 1984b).

2.4.5 The Organic Coat and Valve Release

The last stage of valve formation is also associated with a drastic change in biochemical production. In both *C. wailesii* and *T. eccentrica* the entire Golgi apparatus of the cells switches to produce a new type of vesicle (Fig. 10). They fuse with the proximal part of the SDV, and discharge their content into it, thus contributing organic coat material and probably a new membrane for the new plasmalemma (Schmid 1984c, 1986a, b).

As observed in *C. wailesii*, valve release seems to be a normal exocytotic process (Fig. 11). The cPL and the distal part of the SDV fuse above the valve center, and then, as concluded from serial sections, are pulled back over the valve edge into the cell. Simultaneously, the cell elongates in pervalvar direction, which also causes the mother thecae to slide somewhat apart in the girdle region (Schmid 1986a, b).

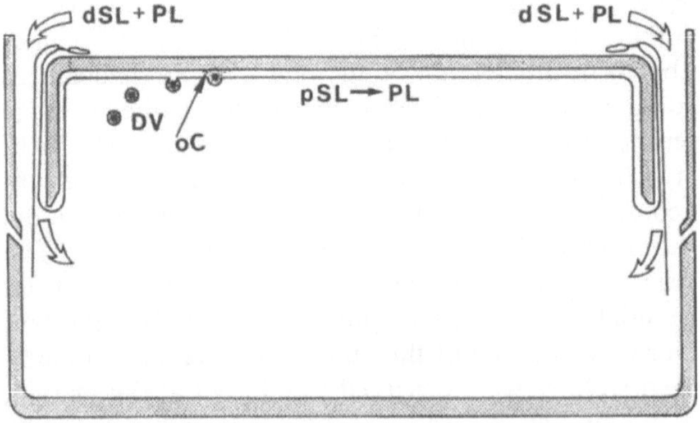

Fig. 10. *Coscinodiscus wailesii:* Organic coat (*oC*) deposition by dense Golgi vesicles (*DV*). Vesicle membrane incorporation into proximal silicalemma causes a gradual change into the new plasmalemma. Valve release occurs by membrane retraction over the valve edge (Schmid 1986b)

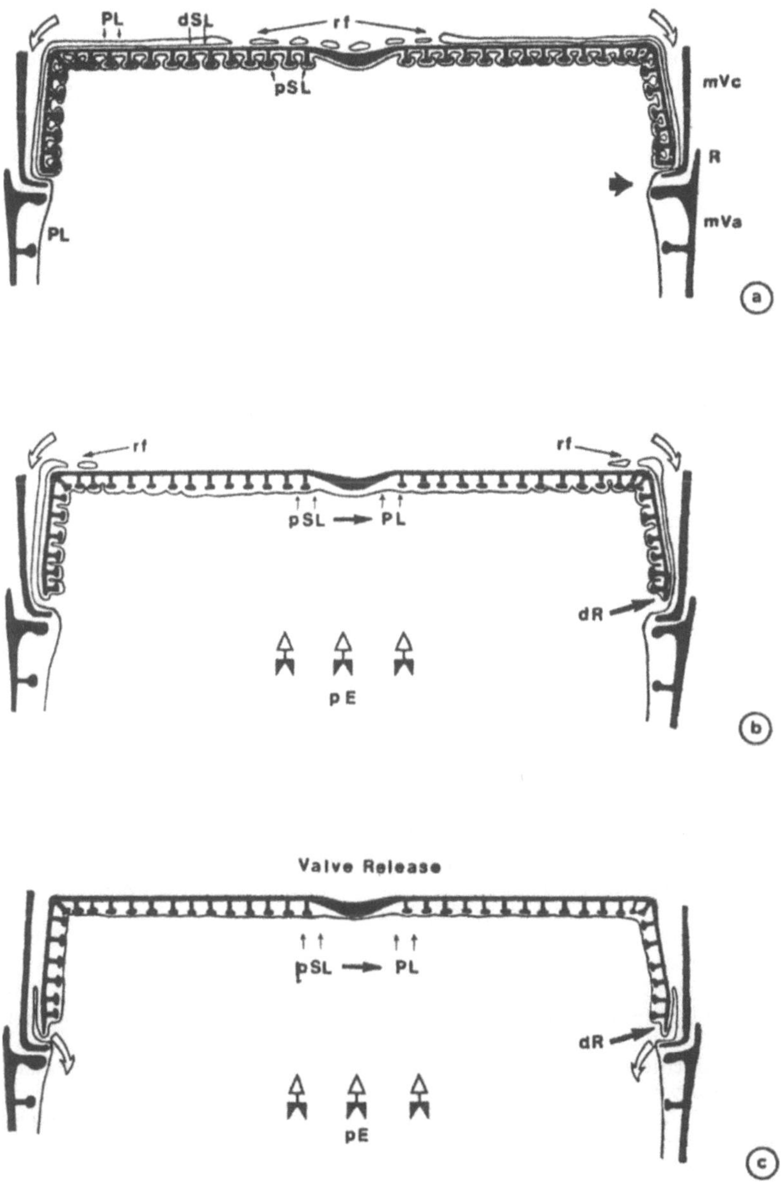

Fig. 11 a – c. *Coscinodiscus wailesii:* Valve release (Schmid 1986 a)

2.5 Conclusions

Information on diatom morphogenesis is still very limited. The basic principles of morphogenetic mechanisms are possibly the same in the whole group despite small variations in different diatom species. The main forces involved in shaping of valve surface and outline are minute changes in turgor and thus tension of the cPL, the molding surface, and local adhesions combined with compression resistant mucilage and its local swelling. Transient, interdigitating connections between sister cells and/or patterned adhesion to the mother's cell girdle both serve as an orientation reference and as anchors for tensing the molding surface. This large-scale shaping can be followed to some extent in living cells; however, pattern creation cannot be seen with the LM. This pattern formation process is a feedback mechanism in which timing plays an important role. Nothing is known about how time is measured, nor how, and in which form, instructions are given.

During morphogenesis the sibling cells are two naked protoplasts within the mother cell wall. During this stage the plasmalemma and the cytoskeleton function as the cell skeleton. This function seems to be transferred to the valve during its formation by action of the MC-MT-spindle-precursor complex, forming the future anchor and orientation sites in a precisely tuned sequence.

Acknowledgements. I am grateful to Prof. Dr. Bereiter-Hahn, to Prof. Dr. Steucek, and to Dipl.-Ing. E. Pückert for valuable discussions. Parts of the experimental work have been supported by the Austrian Fonds zur Förderung der Wissenschaftlichen Forschung and the Max Kade Foundation.

References

Bereiter-Hahn J (1977) Die Zelle, ein Pneu? In: Pneus in Natur und Technik. Inst Leicht Flächentragwerke, IL 9:152–157

Bereiter-Hahn J (1985) Architecture of tissue cells. The structural basis which determines shape and locomotion of cells. Acta Biotheor 34:139–148

Bonik K (1978) Die Entstehung der Kieselalgen – ein stammesgeschichtliches Modell. I. Die Entwicklung der Schale. Natur Museum 108:267–273

Bonik K (1979) Die Entstehung der Kieselalgen – ein stammesgeschichtliches Modell. II. Die Konsequenzen der Schalenbildung. Natur Museum 109:1–9

Crawford RM, Schmid AM (1986) Ultrastructure of silica deposition in diatoms. In: Leadbeater BSC, Riding R (eds) Biomineralization in lower plants and animals (Syst Assoc Ser No 30), pp 291–314

Darley WM (1974) Silicification and calcification. In: Stewart WDP (ed) Algal physiology and biochemistry. Blackwell Scientific, Oxford, pp 655–675

Dickinson HG (1970) Ultrastructural aspects of primexine formation in the microspore tetrad of *Lilium longiflorum*. Cytobiologie 1:437–449

Duke EL, Reimann BEF (1977) The ultrastructure of the diatom cell. In: Werner D (ed) The biology of diatoms. Botanical monographs, vol 13. Blackwell Scientific, Oxford, pp 65–109

Edgar LA, Pickett-Heaps JD (1984a) Valve morphogenesis in the pennate diatom *Navicula cuspidata*. J Phycol 20:47–61

Edgar LA, Pickett-Heaps JD (1984b) Diatom locomotion. Progr Phycol Res 3:47–88

Herth W (1979) The site of β-chitin fibril formation in centric diatoms. II. The chitin forming cytoplasmic structures. J Ultrastruct Res 68:16–27

Höfler K (1963) Zellstudien an *Biddulphia titiana*. Protoplasma 56:1–53

Kirst GO (1985) Osmotische Adaptation bei Algen. Naturwissenschaften 72:125–132

Krammer K (1981) Zur Deutung einiger Schalenstrukturen bei pennaten Diatomeen. Nova Hedwigia 35:75–105

Lauterborn R (1896) Untersuchungen über Bau, Kernteilung und Bewegung der Diatomeen. Engelmann, Leipzig
Mann S (1983) Mineralization in biological systems. Struct Bonding 54:126–174
Pickett-Heaps JD, Tippit DH, Andreozzi JA (1979) Cell division in the pennate diatom *Pinnularia*. IV. Valve morphogenesis. Biol Cell 35:199–203
Raven JA (1983) The transport and function of silicon in plants. Biol Rev 58:179–207
Romanovicz DK (1981) Scale formation in flagellates. In: Kiermayer O (ed) Cytomorphogenesis in plants. Cell biology monographs, vol. 8. Springer, Berlin Heidelberg New York, pp 27–62
Schmid AM (1979a) Influence of environmental factors on the development of the valve in diatoms. Protoplasma 99:99–115
Schmid AM (1979b) Wall morphogenesis in diatoms: the role of microtubules during pattern formation. Eur J Cell Biol 20:125
Schmid AM (1980) Valve morphogenesis in diatoms. A pattern related filamentous system in pennates and the effect of APM, colchicine and osmotic pressure. Nova Hedwigia 33:811–847
Schmid AM (1984a) Tricornate spines in *Thalassiosira eccentrica* as a result of valve modelling. In: Mann DG (ed) Proc 17th Int Diat Symp Koeltz, Koenigstein, pp 71–95
Schmid AM (1984b) Wall morphogenesis in *Thalassiosira eccentrica*: comparison of auxospore formation and the effect of MT-inhibitors. In: Mann DG (ed) Proc 7th Int Diat Symp Koeltz, Koenigstein, pp 47–70
Schmid AM (1984c) Valve morphogenesis in diatoms. In: Diatoms I. Shells in Nature and Technics. Comm IL 28:300–317
Schmid AM (1985) *Centronella reichelti* VOIGT – a very unusual diatom in the surface sediments of the Grabensee. In: Danielopol E, Schmidt R, Schultze E (eds) Contributions to the Palaeolimnology of the Trumer Lakes and the Lakes Mondsee, Attersee and Traunsee. Österr Akad Wiss: 65–78
Schmid AM (1986a) Wall morphogenesis in *Coscinodiscus wailesii* Gran et Angst. II. Cytoplasmic events of valve morphogenesis. In: Ricard M (ed) Proc 8th Int Diat Symp, pp 293–314
Schmid AM (1986b) Organization and function of cell structures in diatoms and their morphogenesis. In: Ricard M (ed) Proc 8th Int Diat Symp pp 271–292
Schmid AM, Volcani BE (1983) Wall morphogenesis in *Coscinodiscus wailesii* I. Valve morphology and development of its architecture. J Phycol 19:387–402
Schmid AM, Borowitzka MA, Volcani BE (1981) Morphogenesis and biochemistry of diatom cell walls. In: Kiermayer O (ed) Cytomorphogenesis in plants. Cell biology monographs, vol. 8. Springer, Berlin Heidelberg New York, pp 63–97
Schnepf E, Deichgräber G, Drebes G (1980) Morphogenetic process in *Attheya decora* (Biddulphiineae). Plant Syst Evol 135:265–277
Stosch HA von (1954) Die Oogamie von *Biddulphia mobiliensis* und die bisher bekannten Auxosporenbildungen bei den Centrales. Rapp Comm VIIIe Congr Int Bot Sect 17:58–68
Stosch HA von (1982) On auxospore envelopes in diatoms. Bacillaria 5:127–156
Stosch HA von, Drebes G (1964) Entwicklungsgeschichtliche Untersuchungen an zentrischen Diatomeen. IV. Die Planktondiatomee *Stephanopyxis turris* – ihre Behandlung und Entwicklungsgeschichte. Helgol Wiss Meeresunters 11:209–257
Stosch HA von, Theil G, Kowallik KV (1973) Entwicklungsgeschichtliche Untersuchungen an zentrischen Diatomeen. V. Bau und Lebenszyklus von *Chaetoceros didymum* mit Beobachtungen über einige andere Arten der Gattung. Helgol Wiss Meeresunters 25:384–445
Tschermak-Woess E, Hasitschka-Jenschke G (1985) Über das interphasische Kern- und Zellwachstum. Öst Bot Z 105:382–408

III.3 The Cytoskeletal and Biomineralized Supportive Structures in Radiolaria

O. R. ANDERSON[1], M. L. MOSS[2], and R. SKALAK[3]

3.1 Introduction

Pelagic Sarcodina, such as radiolaria and planktonic Foraminifera, are floating organisms that usually obtain nutrition by opportunistic predation. Consequently, they produce an elaborate peripheral halo of cytoplasmic structures to increase surface area and enhance prey apprehension. Radiolaria are identified taxonomically by the geometry of their siliceous skeletons (Fig. 1) and the presence of numerous, long, stiffened axopodia radiating outward from a clearly differentiated central spheroidal cytoplasmic mass called a central capsule. In many species, the axopodia are further elaborated to produce membranous sheaths (ca. 100 nm thick) enclosing algal symbionts held in the extracapsular cytoplasm. Based on electron microscope assessments of the number and distribution of axopodial strands emerging from the surface of the central capsular wall, and their calculated surface area; and the mean symbiont surface area enclosed by cytoplasmic sheaths, an estimated total outer membrane surface area for a typical large spongiose skeletal radiolarian is $1.4 \times 10^6\ \mu m^2$. This is undoubtedly a conservative estimate as additional membranous envelopes and weblike rhizopodial elements are often produced within the axopodial array. When prey is captured, moreover, large masses of ensheathing pseudopodia, called coelopodia (Anderson 1983), are elaborated to enclose the prey. The remarkable diversity of extracapsular cytoplasmic structures, their motility by cytoplasmic streaming, and the large membrane surface area enclosing these structures raise intriguing questions about the cellular biological interpretations of these phenomena. There is increasing evidence that the sometimes elaborate siliceous skeletal frameworks secreted by the radiolaria provide anchorage and surface support for an extensive pseudopodial system in many species (Anderson 1983, 1986). Some current knowledge of the cytoskeletal structures within the extracapsular cytoplasm is presented and the ontogenesis and functional biology of the biomineralized skeleton is discussed in relation to our present understanding of radiolarian cellular mechanics and physiological ecology.

[1] Biological Laboratories, Lamont-Doherty Geological Observatory of Columbia University, Palisades, New York 10964, USA
[2] Department of Anatomy and Cell Biology, College of Physicians and Surgeons, Columbia University, New York, New York 10032, USA
[3] Bioengineering Institute, Columbia University, New York, New York 10027, USA

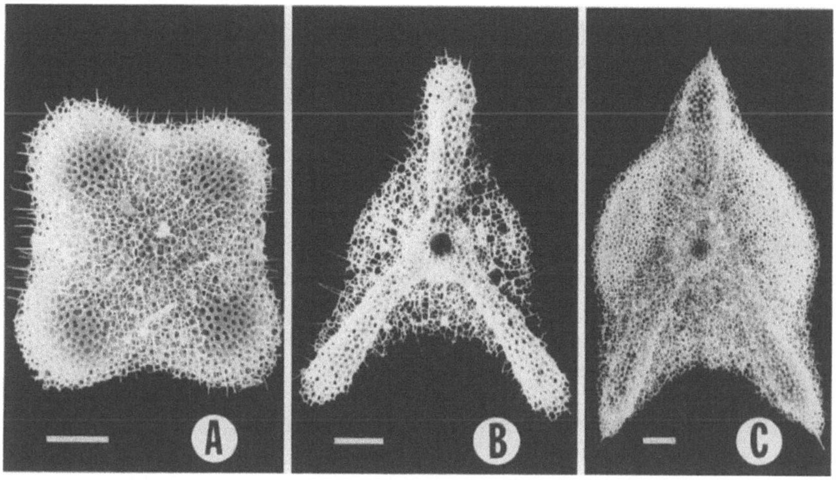

Fig. 1. Comparative scanning electron microscope views of cleaned skeletons of *Spongaster tetras tetras* (**A**) and *Euchitonia elegans* at a late stage of maturation, stage 5 (**B**), and when fully mature, stage 6 (**C**). *Bars* = 50 µm

3.2 Cytoskeletal Organization of the Axopodia

Substantial information on the organization of the central bundles of microtubules (axonemes) of radiolarian axopodia and their organizing centers near the nucleus within the central capsule has been assembled from transmission electron microscope evidence (Cachon and Cachon 1971; Anderson 1976, 1983, p 100). The axoneme is a complex three-dimensional array of microtubules exhibiting a cross-sectional image of branching spiral arms attached to one or more central microtubules, or a set of centrally located hexagonal rings (Fig. 2, cf. Chap. III.1 Fig. 3). When this cross-sectional image is extended axially in

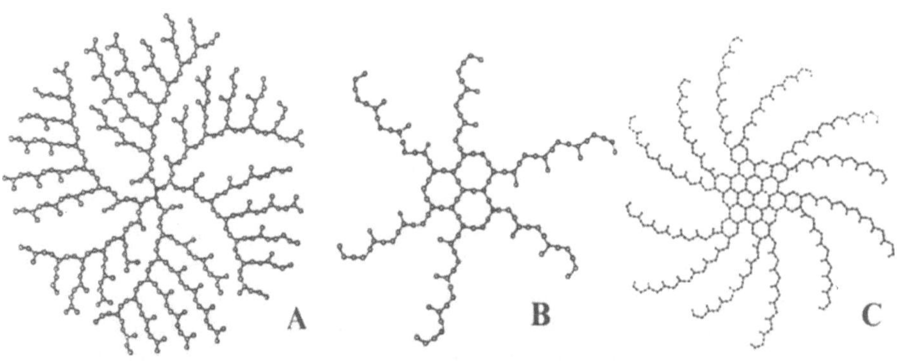

Fig. 2. Cross-sectional images of microtubular arrays forming the axoneme of radiolarian axopodia of Spumellaria (**A**) and Nassellaria (**B, C**) (After Cachon and Cachon 1971)

three-dimensions it is clear that the axoneme constitutes a substantial supportive superstructure within the axopodium. Microfilaments are usually abundant in the cytoplasm surrounding the axoneme and are especially prominent in the cytoplasm of flowing rhizopodia and coelopodia surrounding captured prey. Biochemical and electron microscope evidence (Anderson 1978) supports the conclusion that rhizopodial motility is mediated by microfilament contractile activity. A halo of shorter axopodia near the central capsule of some spherical spumellaria exhibit rhythmic contractile activity. This pulsating motion may enhance physiological exchange processes by increasing the flow of fluids near the central capsular surface.

3.3 Biomineralization and Skeletal Morphogenesis

The radiolarian skeleton is secreted within the cisterna of a cytoplasmic sheath called a "cytokalymma" (Anderson 1983), a thin cytoplasmic envelope formed by cytoplasmic streaming and expansion of the rhizopodia. As the form of the skeleton is species-specific, it is undoubtedly under genetic control, and represents a form of fixed biological information. Thus, once the skeletal geometry has been established, the siliceous scaffolding can serve as an ordering framework, directing the flow of cytoplasmic streaming along its surfaces and ensuring an orderly disposition of rhizopodia around the central cell mass. The possible stereognostic and ordering role of the skeletal framework during skeletogenesis has been discussed in other publications (Anderson 1983, 1986).

There are two fundamental patterns of skeletal morphogenesis known at present: (1) bridge growth, and (2) rim growth (Anderson and Swanberg 1981; Anderson 1981, 1983, 1986). Bridge growth is a form of crosslinking development that produces open lattices usually with large polygonal pores. Rim growth is characterized by circular or nearly circular pores of small diameter compared to the breadth of the surrounding wall material. Pores are produced by secretion of silica upon the rims of large, sometimes polygonal, pores that become reduced in diameter and have a circular border.

3.3.1 Analysis of Growth Phases

A quantitative method of describing the sequence of skeletal growth stages and the degree of integration of successive phases of growth, based on skeletal mass, has been used to analyze growth of the quadrangular skeleton of *Spongaster tetras tetras* (Fig. 1 A) and the trigonal skeleton of *Euchitonia elegans* (Fig. 1 B, C) (Anderson and Bennett 1985). Based on scanning electron microscope evidence, the growth of *E. elegans* begins as a small spongiose triangular skeleton, followed by elongation of the brachia (arms) at each vertex of the triangle (Fig. 1 B), and culminates in a trigonal skeleton with a spongy patagium between the brachia (Fig. 1 C). *S. tetras* by contrast commences growth as a small pillow-shaped, nearly discoidal, spongiose skeleton that becomes more quadrangular by extensive growth on two perpendicular axes. The surface spines appear very late in development after the major mass of the spongiose skeleton has been deposited.

The two major phases of skeletal growth in *E. elegans* (brachial elongation and patagial growth) overlap during morphogenesis. By contrast, the two phases of growth in *S. tetras* (expansion of the quadrangular skeleton and ornamentation with surface spines) exhibit little overlap during morphogenesis. By weighing the skeletons of *E. elegans* at each stage of growth and examining their developmental trends, a mathematical expression for the extent of overlap of the patagial phase with the brachial phase of growth has been developed (Anderson and Bennett 1985). The weight of the trigonal part of the skeleton at the inception of brachial growth is designated Δm_1 and the final weight of the trigonal part of the skeleton is designated Δm_2. The ratio $(\Delta m_1/\Delta m_2)$ is designated as the fractional mass increment and represents the proportion of total silica deposited in the trigonal part of the skeleton at the time patagial growth commences. By subtracting this ratio from 1.0, we obtain the mass phase lead (φ_m). This is an estimate of the degree of overlap between patagial growth and brachial growth:

$$\varphi_m = 1 - (\Delta m_1/\Delta m_2).$$

For *E. elegans*, $\varphi_m = 0.45$. This indicates that patagial growth leads into the trigonal skeletal growth phase at a point where 45% of the trigonal skeletal mass remains to be deposited (the trigonal skeleton is 55% completed). For *S. tetras*, the mass phase lead is approximately 0 as $\Delta m_1 = \Delta m_2$. The spiculate final stage of development does not overlap appreciably with the major development of the quadrangular skeleton. As the mass phase lead equation uses skeletal mass to assess the extent of overlap of growth phases, it is applicable to living and fossil specimens and may provide useful data in making taxonomic distinctions in addition to the conventional practice of using mature skeletons for species discriminations.

3.3.2 Finite Element Analysis

Further quantitative analyses of the relative rates of growth of parts of the skeleton at successive stages of development have been made using a computer-based finite element method (FEM) adopted from engineering research and recently applied with considerable success to mammalian skeletal growth (e.g. Moss et al. 1985). This method permits visualization and quantitative assessment of the direction of growth and concomitant displacement of points within identified segments of the skeleton during successive stages of maturation. Each successive stage is compared to an initial (reference) stage of growth; usually, but not necessarily, the earliest stage that can be identified. It is possible to choose any stage within the growth sequence and make all successive comparisons in relation to it. The method requires identification of a set of points (nodes) at fixed positions on the skeleton (Fig. 3A). These nodes must be

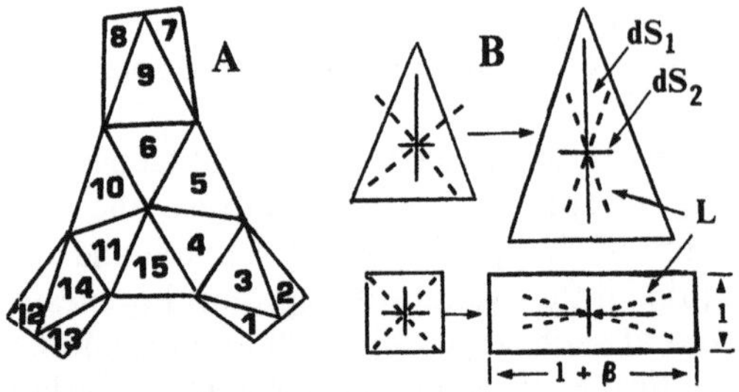

Fig. 3. Finite elements. An early stage of growth (stage 2) in *E. elegans* (**A**) has been discretized into 15 triangular elements using 16 nodes situated at the vertices of the triangles. Each element is assigned a number for convenience in identifying it during FEM analysis. During growth (*arrows*) of an element (**B**), nonuniform deformation is assessed by the change in the angle of the inclined axes (*dashed lines, L*) or by the ratio of 1: $1 + \beta$ as described in the text. The principal strain axis (dS_1) marks the direction of major deformation. The minor strain axis (dS_2) is always perpendicular to it and indicates the direction of minimum deformation

clearly identifiable at each successive stage of skeletal development. They serve as marker points to permit assessment of spatial changes in the skeleton associated with growth. The nodes are connected by straight lines to create discrete geometric sectors, called finite elements, that partition the skeleton into identifiable zones to be analyzed for changes during growth. Elements with curvilinear sides may also be used. When the analysis is two-dimensional, the elements are polygons, usually simply triangles (Fig. 3). Each triangle represents a reference zone whose change in area and principal direction of extension during growth is assessed *independently* of all other finite elements. In other words, one of the advantages of the finite element method is the invariant description of each element as a separate unit. All changes in direction of growth and associated deformation of the element during successive stages of growth are related only to the individual element and are assessed independently of changes within surrounding elements. Hence, it is possible to follow the growth parameters of each of the individual finite elements independently; the system is external-reference-frame invariant. All analyses are made with reference to changes within a given finite element. By examining the changes in the total set of elements, however, an overall assessment of growth can be made (see also Chap. I.3).

3.3.2.1 FEM Descriptors

The following descriptors are used in finite element analysis. A pair of orthogonal axes (principal axes) positioned, for convenience, at the centroid of each element, is used to geometrically represent the directions and magnitudes of growth within each finite element (Fig. 3B). The longer axis (maximal strain axis) labeled dS_1 is oriented in the direction of major deformation (extension) of the element. Its magnitude indicates the extent of increase relative to the initial state of the element. The shorter axis (minimal strain axis), labeled dS_2, represents the orientation of minimal growth of the element. It can be shown that this minimal growth direction is always perpendicular to the maximal growth direction, hence, the two axes are plotted orthogonally.

Since the skeleton grows by accretion of silica, the use of the engineering terms, strain and shear, are applied only in a descriptive sense. But, the cytoplasm enclosing and depositing the silica adjusts by streaming growth; hence, the terms can be applied to the cytoplasmic sheath in more than a figurative sense. At present, however, we do not fully understand the complex adjustments in the cytoplasm accompanying growth, thus it is not possible to quantitatively describe the deformation of the periskeletal cytoplasmic sheath.

The orientation of the maximal strain axis relative to the initial state yields information on how the geometry of the element is changing with respect to the direction of major stretch of the growing finite element. Rotation of the maximal strain axis after transition from one stage to the next indicates a change in direction of maximal stretch of the element. The relative increase in length of the maximal strain axis at a given stage relative to the initial stage is expressed as an extension ratio λ_1. Where $\lambda_1 = \varepsilon_1 + 1$, and ε_1 is the strain expressed as:

$$\varepsilon_1 = \frac{dS_1 - dS_{01}}{dS_{01}},$$

where dS_{01} and dS_1 are the line element lengths at the initial and final stages, respectively. Thus, ε_1 is a ratio of the magnitude of change in length of the line segment to its initial value. It may be thought of as a fractional increment in length of the maximal strain direction. The extension ratio ($\lambda_1 = \varepsilon_1 + 1$) multiplied by 100 describes the final length as a percent of the original length. Thus, for example, if $\varepsilon_1 = 0.5$ and $\lambda_1 = 1.5$, then the percent incremental change is 50% and the final length is 150% of the initial length. A similar descriptor can be calculated for the minimal strain axis. The difference in the length of the final and initial state of the minimal strain segment is divided by the initial value and added to 1. This yields the minimal strain extension ratio λ_2. The product of the two extension ratios ($\lambda_1 \cdot \lambda_2$) yields the area ratio which represents the areal increase of the element during growth from the initial stage to the stage being analyzed. The product $\lambda_1 \cdot \lambda_2$ is the ratio of the final area to the initial area.

For the purposes of this brief discussion, one additional descriptor, the shear strain, will be used. This coefficient represents the extent of shear deformation or change of shape of a finite element after growth. If all parts of the element change uniformly, then there is no shear strain. That is, there is uniform expansion of the element toward a larger but identical shape with respect to the initial stage. This occurs when $\lambda_1 = \lambda_2$. If, however, there is a difference in the maximal and minimal growth relative to the initial state, then some shear deformation (shape change) occurs. A geometric interpretation of shear strain is presented as a means of introducing a quantitative measure for this descriptor. More rigorous mathematical explanations can be found elsewhere (e.g. Skalak et al. 1982; Moss et al. 1985). To indicate the extent of shear strain, a second set of orthogonal axes (dotted lines, Fig. 3B) are superimposed upon the principal axes and oriented at a 45° angle. If growth is uniform in all directions, then there is no angular distortion of the second set of axes. The two line segments remain oriented at an angle of 90° to each other. If growth is not uniform in all directions, then the included angle of the second set of axes will change from 90°. The degree of variation of the angle from 90° provides a measure of the shear strain. The maximum shear strain angle may be defined as $\gamma = \pi/2 - \Theta$, where Θ is the angle between the two axes after deformation. Clearly, if $\Theta = 90°$ after deformation, $\gamma = 0$. If the angle Θ deviates from 90°, the magnitude of γ will increase concomitantly, indicating increasing shear strain within the element. The angle γ may be computed by $\gamma = \pi/2 - 2 \tan^{-1} \lambda_2/\lambda_1$). The value of γ indicates the relative extent of difference in the extension ratios of the maximal and minimal axes. If both axes change uniformly during deformation of the finite element, then $\gamma = 0$.

The ratio (λ_1/λ_2) itself may be used as a shape change indicator in the expression:

$$\beta = (\lambda_1/\lambda_2) - 1.$$

The shape parameter β (times 100) is the percentage by which growth is greater in the maximum (λ_1) direction compared to the minimum (λ_2) direction. A square (Fig. 3B) oriented along the principal directions will grow into a rectangle whose sides have the proportions $1:(1+\beta)$.

A maximum shear strain component (E_s) is computed as follows:

$$E_s = 1/2\,(E_{11} - E_{22}),$$

where

$$E_{11} = \frac{(dS_1)^2 - (dS_{01})^2}{(dS_{01})^2}$$

and

$$E_{22} = \frac{(dS_2)^2 - (dS_{02})^2}{(dS_{02})^2}.$$

This shear strain (E_s) is a measure of the amount of shear in an element and will be reported as an indicator of the extent of nonuniform deformation during growth of an element.

3.3.2.2 FEM Results

Six growth stages of *E. elegans* (Anderson and Bennett 1985) were analyzed using FEM: (1) an early triangular stage without brachia; (2) inception of brachial growth; (3) elongation of brachia (Fig. 4A); (4) onset of patagial growth and continued brachial elongation (Figs. 1A and 4B); (5) maturation of the brachia to near maximal length with concurrent strong development of the patagium

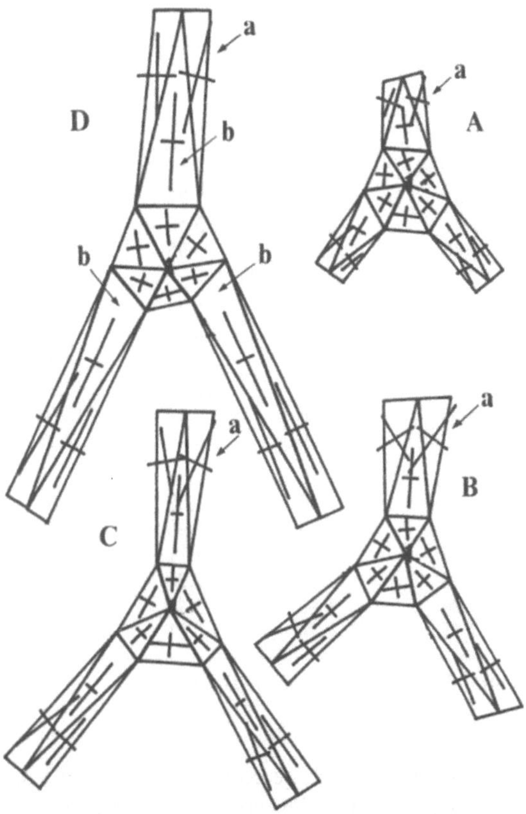

Fig. 4A–D. Diagrams of the finite elements and strain axes in the final four stages of maturation (stages 3–6) of *E. elegans*. Stages **C** and **D** correspond, respectively, to stages **B** and **C** in Fig. 1. There is clear evidence of lateral deformation in the peripheral elements of the anterior brachium (*a*) as indicated by the lateral rotation of the strain axes. The axial elements of the brachia (*b*), especially in the mature stage (**D**), show marked axial elongation. The principal strain axis is oriented parallel to the long axis of the triangular element

(Fig. 4C); and (6) full development of the patagium and final maturation of the brachia (Figs. 1B and 4D). Only the trigonal central part of the skeleton (exclusive of the patagium) was analyzed in relation to its two-dimensional profile. The succession of growth stages in two dimensions is diagrammed in Figs. 3 and 4. Fifteen triangular finite elements were constructed within the profile of the trigonal skeleton as shown in Fig. 3A. All successive stages of development were analyzed in relation to this initial set of elements. To obtain sufficiently detailed views of each growth stage, scanning electron microscope images were used of representative growth stages of different specimens. Hence, our data do not represent the growth of an individual specimen. But, we have taken care to select exemplars of each growth stage that are as representative as possible of the sequence of stages we have observed by light microscopy. Given the limitations of possible interorganismic variability in our sample, we directed our observations to some major growth characteristics that appear to be representative of ontogenesis in *E. elegans*. The 15 elements are classified into categories for convenient reference. (1) Brachial elements are those within the brachia, and are further identified as axial elements (3, 9, and 14) that occur within the central region of each brachium; and peripheral elements (1, 2, 7, 8, 12, and 13) at the periphery of each brachium. (2) Somal elements (4, 5, 6, 10, 11, and 15) are within the discoidal central body of the skeleton.

Table 1. Area ratios for finite elements categorized by position

	Growth stages				
	2	3	4	5	6
Brachial axial element no.					
3	1.02	1.81	3.45	2.72	7.63
9	1.66	2.41	3.64	3.53	8.98
14	0.91	1.86	2.89	3.55	7.54
Brachial peripheral element no.					
1	0.94	2.29	5.22	7.07	10.66
2	1.04	2.24	4.94	6.21	9.75
7	3.03	4.80	12.46	13.78	17.38
8	2.18	2.98	9.20	10.45	13.33
12	1.72	3.03	6.96	8.29	12.68
13	1.01	1.83	4.99	6.21	9.27
Somal element no.[a]					
4	1.36	1.34	1.50	1.38	1.59
5	1.70	1.52	1.83	2.48	2.96
6	1.40	1.29	1.26	1.11	3.24
10	1.56	1.48	1.89	2.36	3.03
11	1.00	1.26	1.47	1.48	1.66
15	1.84	2.15	2.34	3.71	1.70

[a] Somal elements are those within the central discoidal region of the skeleton.

The area ratios for the finite elements of each stage compared to stage 1 are presented in Table 1. As is expected, the brachial elements exhibit the greatest arêa ratios among all of the elements, especially in stages 3 to 6 when major brachial elongation occurs. The use of the finite element method allows us to quantify these changes and highlight the relative growth of each area. It is of interest to note that the area ratios of the peripheral elements of the brachia are, in general, larger than the area ratios of the axial elements. This is understandable as the peripheral elements of the brachia exhibit both elongation and lateral expansion as the brachia mature. This effect is clearly reflected in the orientation and magnitude of the principal strain axes (Figs. 3 B and 4) that orient laterally in many of the peripheral elements during stages 3 to 5. This is consistent with scanning electron microscope evidence of major thickening of the brachia during this growth period. It is interesting to note that the anterior brachium (containing elements 7–9) exhibits the greatest lateral development as indicated in part by the rotation of the principal strain axis. The two posterior brachia show considerable difference in their lateral development. The left brachium (containing elements 12–14) shows more lateral growth than the right brachium (containing elements 1–3). This is evident when the magnitude and orientation of the principal strain axes within the peripheral elements of the two brachia are compared. In general, the FEM data showing strong elongation of the axial elements within the brachia, and concurrent broadening of the peripheral elements, coincides with scanning electron microscope evidence that the brachia grow by extension of axial spines that become crosslinked and

Table 2. Maximum shear strain and shear angles[a] for elements categorized by position

	Growth stages				
	2	3	4	5	6
Brachial axial element no.					
3	0.09 (10)	0.27 (44)	0.23 (57)	0.55 (72)	0.18 (70)
9	0.15 (26)	0.18 (41)	0.24 (61)	0.45 (73)	0.11 (63)
14	0.22 (22)	0.22 (40)	0.27 (57)	0.33 (67)	0.16 (67)
Brachial peripheral element no.					
1	0.03 (3)	0.16 (36)	0.09 (44)	0.09 (52)	0.10 (65)
2	0.07 (9)	0.15 (34)	0.10 (44)	0.08 (46)	0.10 (63)
7	0.08 (25)	0.09 (40)	0.04 (41)	0.04 (48)	0.04 (53)
8	0.03 (8)	0.02 (6)	0.01 (10)	0.03 (30)	0.03 (41)
12	0.16 (30)	0.16 (44)	0.07 (43)	0.09 (55)	0.08 (64)
13	0.15 (17)	0.10 (19)	0.04 (24)	0.07 (43)	0.08 (55)
Somal element no.					
4	0.11 (17)	0.18 (25)	0.13 (22)	0.47 (52)	0.21 (34)
5	0.04 (8)	0.09 (16)	0.15 (28)	0.17 (40)	0.04 (12)
6	0.10 (16)	0.07 (10)	0.11 (16)	0.36 (38)	0.06 (23)
10	0.16 (26)	0.09 (15)	0.10 (21)	0.19 (41)	0.07 (24)
11	0.18 (20)	0.09 (13)	0.18 (28)	0.25 (36)	0.08 (14)
15	0.12 (23)	0.08 (20)	0.07 (19)	0.08 (32)	0.22 (36)

[a] Shear angle (degrees) is presented in parentheses following each maximum shear value.

extended laterally by bridge growth producing the spongiose peripheral skeleton of the brachia.

Further evidence of differences in growth among the 15 elements is presented in Table 2 showing the maximum shear strain and shear angle of each element categorized according to location within brachial or somal regions of the skeleton. A distinct peak in shear strain occurs at growth stage 5 for axial elements 3, 9, and 14; and for somal elements 4, 5, 6, 10, and 11. Based on electron microscope evidence (Anderson and Bennett 1985), growth stage 5 exhibits heavy patagial development between the anterior brachium and each posterior brachium concurrent with major elongation of the three brachia. This extensive nonuniform expansion is reflected in the peak values of the shear strain values for the elements at stage 5. The axial elements and the opposed somal elements (e.g., the pair of elements 3 and 4), as well as the elements at the sides of the discoidal region (elements 5 and 10), also have high shear strain values at stage 5. By contrast, the shear strain coefficient for the peripheral brachial elements (1, 2, 7, 8, 12, and 13) does not peak at stage 5. Indeed, it is less in stage 5 than some of the preceding stages. This is attributed to the extensive lateralization of the brachia underway at this time that partially compensates for the strong extensional growth, and thus begins to bring the shape of the mature peripheral elements closer to their initial configuration. In general, at stage 6, shear strain declines for most of the elements as peripheral growth nearly restores the overall shape of the elements to their original form. Element 15, however, positioned in the axil between the posterior brachia exhibits a peak in shear strain at

Table 3. Maximal and minimal extension ratios for growth stage 6

	Finite elements														
	1	2	3	4	5	6	7	8	9	10	11	12	13	14	15
Extension ratios															
Maximal	6.90	6.30	6.51	1.73	1.92	2.20	7.21	5.38	6.11	2.15	1.46	7.41	5.45	6.13	1.84
Minimal	1.55	1.55	1.17	0.92	1.54	1.50	2.41	2.48	1.47	1.41	1.14	1.71	1.70	1.23	0.93

the final stage. This may be attributed to nonuniform changes in conformation associated with the divergent growth of the brachia and the moderate peripheral growth deposited during patagial development. It is interesting to note, moreover, that at stage 5, elements 5, 10, and 15 grow about twice as much as 4, 6, and 11. Values for the maximal (λ_1) and minimal (λ_2) extension ratios at stage 6 compared to stage 1 are presented for the 15 finite elements in Table 3.

3.3.2.3 Limitations and Implications of FEM Analysis with Radiolaria

Radiolarian skeletons grow by accretion as opposed to volumetric growth. Hence, deformation of a finite element does not represent uniform deformation of all points within the continuum of the element as assumed in the continuum mechanics theory of FEM analysis. Nonetheless, while recognizing this limitation, FEM can be used validly to quantitatively and graphically analyze patterns of deformation associated with growth of the finite elements. The numerical results in Tables 1 and 2 represent the equivalent uniform growth. Extension of the method to three dimensions will undoubtedly yield additional information and provide a more complete description of changes accompanying maturation. It is clear, however, that even with the two-dimensional analysis, the quantification of changes during growth and insights gained about form changes generated by the FEM analysis complement and support descriptions of growth patterns made by microscope observations and analyses of growth phases based on changes in mass of the skeletons (Anderson 1983; Anderson and Bennett 1985). These quantitative analyses of growth may also be useful in comparative morphogenetic studies among species and as an aid in identifying those stages of skeletal growth where correlated cytoplasmic events may be of interest. It is becoming increasingly clear that although the radiolaria are single-celled organisms, some aspects of their microanatomy and cellular physiology (e.g. Anderson and Botfield 1983) and the intricacy of their skeletal morphogenesis are remarkably complex.

Acknowledgments. We express appreciation to Drs. K. Sen, L. Moss-Salentijn, and other members of the Bioengineering Institute of Columbia University who have given advice and assistance during the FEM analyses. This work was supported by grants OCE 84-08137 from the Biological Oceanography Division of the National Science Foundation, and HD-14371 and HD-19446 of the National Institutes of Child Health and Human Development.

References

Anderson OR (1976) A cytoplasmic fine-structure study of two spumellarian radiolaria and their symbionts. Mar Micropaleontol 1:81–89

Anderson OR (1978) Light and electron microscopic observations of feeding behavior, nutrition, and reproduction in laboratory cultures of *Thalassicolla nucleata*. Tissue Cell 10(3):401–412

Anderson OR (1981) Radiolarian fine structure and silica deposition. In: Simpson TL, Volcani BE (eds) Silicon and siliceous structures in biological systems. Springer, Berlin Heidelberg New York, pp 347–380

Anderson OR (1983) Radiolaria. Springer, Berlin Heidelberg New York, 255 pp

Anderson OR (1986) Biomineralization in radiolaria: silica deposition and ontogenetic origins of form. In: Leadbeater BSC, Riding R (eds) Biomineralization in lower plants and animals. Oxford Univ Press, Oxford, pp 375–391

Anderson OR, Bennett P (1985) A conceptual and quantitative analysis of skeletal morphogenesis in living species of solitary radiolaria: *Euchitonia elegans* and *Spongaster tetras*. Mar Micropaleontol 9:441–454

Anderson OR, Botfield M (1983) Biochemical and fine structure evidence for cellular specialization in a large spumellarian radiolarian, *Thalassicolla nucleata*. Mar Biol 72:235–241

Anderson OR, Swanberg NR (1981) Skeletal morphogenesis in some living collosphaerid radiolaria. Mar Micropaleontol 6:385–396

Cachon J, Cachon M (1971) Le système axopodial des Radiolaires Nassellaires. Arch Prostitenkd 113:80–97

Moss ML, Skalak R, Patel H, Sen K, Moss-Salentijn L, Shinozuka M, Vilmann H (1985) Finite element method modeling of cranio-facial growth. Am J Orthod 87:453–472

Skalak R, Dasgupta G, Moss M, Otten E, Dullemeijer P, Vilmann H (1982) Analytical description of growth. J Theor Biol 94:555–577

IV. Mechanical Factors Determining Plant Cell Morphogenesis

IV.1 Mechanical and Hydraulic Aspects of Plant Cell Growth

DANIEL J. COSGROVE[1]

1.1 Introduction

As any student beleaguered with plant anatomy or morphology will attest, plant cells come in a myriad of shapes and sizes. In vascular plants, each member of this cell menagerie can usually be traced back through development to small, nearly isodiametric cells of the meristem. This chapter will review the physical basis by which this transformation takes place. Because of space limitations, many important aspects of cell morphogenesis will be omitted. For detailed discussion of biochemical and cytological aspects of plant cell morphogenesis, readers are referred to recent reviews and collections of articles on this subject (Darvill et al. 1980; Dugger and Barnicki-Garcia 1984; Kiermayer 1981; Maclachlan and Fèvre 1982; Robinson and Quader 1982; Taiz 1984). Likewise, it is beyond the scope of this chapter to consider specific influences of environmental agents and hormones on cell shape and size. Readers interested in this aspect of plant growth should consult relevant reviews and other sources (Cleland 1981; MacMillan 1980; Scott 1981; Wareing 1982; Weisenseel and Kicherer 1981).

The physical control of plant cell shape and size may be divided into three parts. For our purposes, a plant cell may be thought of as a pressurized sack of fluid, surrounded by the plasma membrane and restrained from outward expansion by a rigid wall. Hence, the structural, mechanical, and biochemical properties of the primary cell wall are of paramount importance in controlling cell shape and these will be reviewed first. The rate of cell expansion is a separate issue from the directionality of growth and will be examined next. Finally, when cells enlarge, the increase in volume arises mostly from uptake of water. Water absorption, turgor maintenance, and solute uptake are essential for cell expansion, and under some conditions may place physical limitations on the growth rate of plant cells. This aspect of plant growth will be examined in the last section.

1.2 Directionality of Cell Growth

It is not unusual for the internal hydrostatic pressure (turgor pressure) within plant cells to build up by osmosis to values of 0.3 to 0.8 MPa. Such high pres-

[1] Department of Biology, The Pennsylvania State University, University Park, Pennsylvania 16802, USA

sures are possible because plant cells are surrounded by a rigid wall which withstands and restrains the expansive tendencies of the pressurized cell. By stretching and thereby tightening the cell wall, the fluid pressure gives the cell added mechanical rigidity and structural stability; it provides what has been aptly termed a "hydraulic skeleton" (Wainwright 1970). The consequence of this arrangement, however, is that plant cells can change shape during growth only by a controlled, irreversible *expansion* of the cell wall. Unlike most animal cells, plant cells as a rule cannot change shape by contraction. The few exceptions to this rule, e.g., stomatal guard cells, pulvini, and contractile roots, will not be covered in this chapter.

1.2.1 Patterns of Expansion

For plant cells to grow, the wall must increase its surface area and it does so by two major patterns. *Tip growth* is a localized distention of the cell wall; it occurs during the growth of root hairs, pollen tubes, phloem fibers, and other invasive cells. In these tubular cells, increase in surface area is mostly limited to the dome-shaped tip where dictyosomes deposit wall material. Tip growth can be maintained either by *anisotropic* surface expansion, in which the longitudinal and transverse extension rates differ, or by isotropic expansion, in which the directional character of growth arises from a steep descending gradient in the rate of surface expansion on the dome (see Fig. 1). According to Green (1969), the latter is the most common means for tip-growing cells.

The second growth pattern is *nonlocalized expansive growth,* which entails an increase in surface area throughout most or all of the cell wall. This pattern of growth is found in the majority of cells which make up growing stems, roots, and leaves. Expansive growth may be isotropic, as when a sheet of rubber expands uniformly in all planar directions, or it may be anisotropic. Anisotropic growth is typical of elongating roots, stems, and other structures, and arises because of the structural bias of the wall to expand preferentially in a certain direction.

1.2.2 Wall Architecture

The primary wall of vascular plants is composed of crystalline microfibrils of cellulose embedded in a highly hydrated matrix composed of various mixlinked polysaccharides, protein, and other substances. The microfibrils are

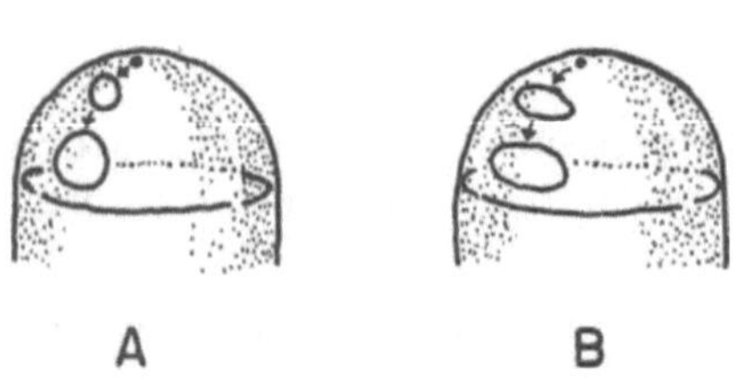

Fig. 1 A, B. Two modes of surface expansion in a hemispherical growth zone. In **A** the local pattern of surface expansion is everywhere isotropic, so a small area enlarges without distortion. In **B** the area is distorted transversely as it expands. (After Green 1969)

bonded to the matrix, and thereby indirectly to each other, through noncovalent bonds, including hydrogen bonds and Van der Waals forces (Frey 1986). Studies with the electron microscope sometimes show that the primary wall consists of multiple layers of closely packed, parallel microfibrils deposited in alternating directions (Roland and Vian 1979). In other cases, multi-ply construction is not so evident in primary walls, but is more pronounced in secondary walls.

Because of their appearance and mechanical properties, plant walls are often compared to fiber-reinforced composites such as fiberglass, in which stiff glass fibers are embedded in an amorphous matrix (Wainwright et al. 1976). In such composites, the rigid fibers increase the mechanical strength of the material. When they are arranged in an oriented fashion, the fibers may also impart anisotropic mechanical behavior to the composite. For example, in a composite in which the fibers are arranged parallel to each other, the deformation, or strain, which occurs when the material is stressed is greatest in the direction perpendicular to the aligned fibers. Shear interactions are weakest in this direction and strongest in the direction parallel to the fibers.

In plant cells, anisotropic expansion has been explained in a similar fashion. The concept goes as follows. When cellulose microfibrils are deposited on the inner surface of the wall in a *nonrandom* pattern, they impart to the wall a preferred direction of expansion. The directionality of growth (i.e., the ratio of transverse to longitudinal expansion) is a function of the stress pattern in the wall and the microfibril arrangement. When the interfibrillar network is weakened or loosened by biochemical processes (see Sect. 1.3), the wall yields anisotropically to the wall stresses generated by cell turgor. In this way, a "hoop-reinforced" cell may elongate to form a cylinder, despite the tendency of turgor pressure to round out the cell.

Major support for this concept comes from studies which have examined the anisotropy of the *Nitella* wall with optical and mechanical methods. With polarized light microscopy one may measure the average orientation of the crystalline microfibrils. The directionality of cell expansion correlates with microfibrillar orientation, i.e., when the microfibrils are oriented transversely the wall expands preferentially in the longitudinal direction (Green 1958). When cells are treated so as to cause random microfibril deposition, growth becomes isotropic and the cell expands spherically (Richmond et al. 1980). Studies of isolated wall specimens have likewise demonstrated good correlation between mechanical anisotropy of the wall and directionality of growth (Probine 1963; Probine and Barber 1966; Taiz et al. 1981).

1.2.3 Multinet Growth

The foregoing mechanical view of wall expansion has a corollary: that microfibrils are realigned during growth. Both of these ideas are encountered in the *multinet growth hypothesis* advanced by Roelofson (1965). According to this concept, the wall is constructed by apposition of successive layers of microfibrils, which become passively realigned during cell enlargement. Originally, multinet growth was used to interpret the appearance of the wall as seen in early studies with the electron microscope (Roelofson and Houwink 1953). These micrographs showed that microfibrils of the inner face of the wall lie dispersed in a direction transverse to the long axis of the elongating cell, whereas at the outer surface the microfibrils are aligned nearly longitudinally. Multinet growth attempted to explain the change in microfibril orientation through the wall thickness as the result of passive realignment during expansion.

More recent studies, however, have shown that a cross-lamellate structure, analogous to plywood, is perhaps more typical of plant cell walls (reviewed in Boyd 1985; Roland and Vian 1979). On the basis of such observations, some authors have dismissed multinet growth entirely. In doing so, they have emphasized the lack of a smooth gradient in microfibril angle, from transverse to axial, through the wall thickness. Nevertheless, the existence of cross-lamellate walls does not contradict a central tenet of multinet growth: that the directionality of wall expansion is a function of microfibril orientation, and that as a result of wall expansion the microfibrils are passively realigned. This thesis provides the simplest mechanical mechanism for the control of cell shape. It makes no difference whether the net microfibril orientation is attained via discrete lamellae crossing one another or via uniform apposition of microfibrils dispersed about the transverse direction.

Roland and Vian (1979) have put forward a more elaborate proposal. They suggested that the cell can control the direction of its growth by selectively expanding certain lamellae in the wall. This idea is quite different from the purely mechanical view envisioned in multinet growth. It is attractive because it would afford the cell an additional degree of freedom in directing its growth, but there seems to be little direct support for such a mechanism and the simpler mechanical view seems at present to account for the directionality of growth.

A case against the mechanical view might be made if an example were found of a cell expanding in a direction different from that predicted by the mechanical and optical properties of the wall. An illuminating example of such behavior was reported by Richmond (1983). He treated *Nitella* with the microtubule-disrupting drug, isopropyl N-phenylcarbamate, which caused cellulose to be deposited in a random orientation. (Microtubules are believed to control the direction of cellulose deposition; see Hepler 1981; Robinson and Quader 1982.) When the drug was removed, the cells attained their normal growth anisotropy before the optical anisotropy of the wall was reestablished. To explain this discrepancy, Richmond (1983) suggested that only the inner portion of the wall is the stress-bearing region, and therefore, only the inner portion controls growth. In effect, he argued that the optical and growth anisotropies did not correspond during the recovery period because the outer wall is mechanically fragmented to such an extent that it does not bear wall stress, yet it is still optically active. A critical test of these ideas would be to compare the mechanical anisotropy with the growth anisotropy of the wall during the recovery period. Richmond et al. (1980) showed a qualitative correspondence between these two properties. Unfortunately, quantitative comparison is complicated by the fact that growth requires that the wall stress exceed a critical yield point (see Sect. 1.3) and the yield point may differ for the transverse and longitudinal directions (Métraux and Taiz 1978; Probine and

Barber 1966). Moreover, methods available to gauge the mechanical properties of isolated walls may not be appropriate for evaluating the wall characteristics that govern wall yielding (Cosgrove 1986; Taiz 1984).

1.2.4 The Wall Matrix

Up to this point, the discussion of the directionality of cell growth has focused on the ordered arrangement of microfibrils. It is possible that the arrangement of matrix components may also contribute to wall anisotropy. In support of this notion, studies show that the pectic components of the wall are not arranged completely at random, but show a degree of orientation parallel to the direction of cell expansion (Hayashi et al. 1980; Morikawa et al. 1978). In another vein, Taiz et al. (1983) concluded that ethylene treatment selectively decreased longitudinal extensibility of pea stems. This effect, however, may have resulted from an ethylene-induced reorientation of microfibrils from transverse to longitudinal. There do not yet seem to be any quantitative tests of the significance of an ordered arrangement of the matrix on either the mechanical or the growth properties of the cell wall.

Despite considerable progress in chemical characterization of the cell wall (Darvill et al. 1980; McNeil et al. 1984), we still do not understand how the components of the wall fit together (Frey 1986). Which of the polymers bear the tensile stress in the wall has not yet been established. More to the point, which of the load-bearing bonds is (are) loosened during cell expansion still has not been identified (Taiz 1984). This is an essential step in our quest for understanding the control of plant growth.

To sum up this section, turgor pressure creates wall stresses which are necessary for physical expansion of the wall. The directionality of growth is determined largely by the pattern of these wall stresses and by the mechanical anisotropy introduced into the wall by the ordered deposition of cellulose and other wall polymers.

These remarks are pertinent to the determination of cell shape. The rate at which cells enlarge is another question altogether. There does not seem to be a causative correlation between the arrangement of microfibrils and the growth rate. Moreover, Richmond and Métraux (1984) showed that cellulose synthesis could be inhibited without affecting the rate of cell expansion; hence, growth rate is not coupled to microfibril deposition, at least in short-term experiments. The nature of the expansion process will be reviewed next.

1.3 Wall Loosening and Expansion

The nature of wall expansion was examined in a landmark study by Ray and Ruesink (1962) who studied the elongation of oat coleoptile segments. They found that the elongation rate responded to changes in temperature within 20 s, and that cyanide treatment or oxygen deprivation stopped growth within a few minutes. They interpreted their results as indicating that wall expansion is directly and tightly controlled by metabolic reactions, and they proposed that wall expansion is a type of chemorheological process in which critical bonds in the wall are broken by metabolically-linked reactions. After such scission, the wall was thought to expand via a viscoelastic expansion. Ray and Ruesink assumed that their treatments did not affect turgor pressure substantively, an assumption later confirmed by direct turgor pressure measurements with the pressure microprobe (Cosgrove and Cleland 1983b).

1.3.1 Physics of Wall Expansion

Numerous studies have also demonstrated that the rate of cell expansion is strongly dependent on turgor pressure, and that a minimum turgor pressure is required for growth. Lockhart (1965) formalized these observations in an analysis in which he proposed the simple relation (see also Cosgrove 1986):

$$dV/(Vdt) = \varphi (P - Y), \quad (1)$$

where $dV/(Vdt)$ is the relative rate of cell expansion, P is turgor pressure, Y is the minimum turgor required for growth (the yield threshold), and φ is a strain-rate compliance relating the effective turgor pressure $(P - Y)$ to the growth rate. Lockhart called φ extensibility. However, readers should be cautioned that this term has also been used for several other wall properties which differ from extensibility as defined in Eq. (1).

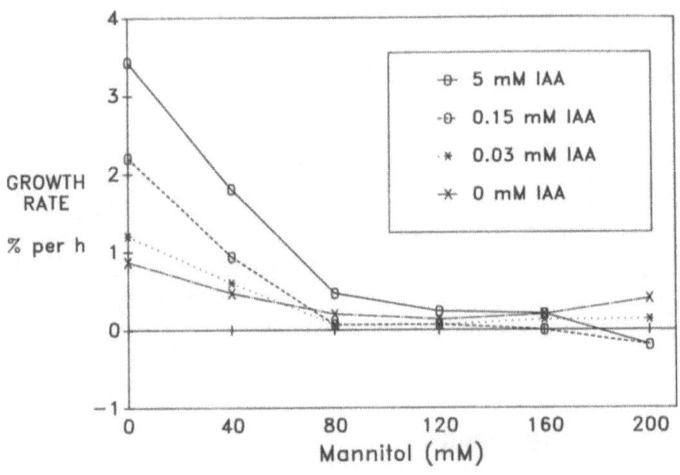

Fig. 2. Growth rate of excised oat coleoptiles as a function of the external concentration of the osmoticum. Segments were incubated in a graded series of mannitol concentrations as a means of reducing cell turgor pressure. Up to about 80 mM, the growth rate was a function of the mannitol concentration. The changes in the slope of the line with varying concentrations of indole acetic acid (*IAA*) suggest that auxin alters the wall-yielding coefficient (wall extensibility), and has little effect on the yield threshold. (After Cleland 1976)

The conventional procedure for measuring φ and Y has been to measure growth rate as a function of turgor pressure (Fig. 2). The slope of the line is taken as a measure of φ; Y is estimated from the point in the line where the slope changes to zero or nearly zero. From such assays it is apparent, for example, that auxin stimulates growth in large measure by increasing φ (Cleland 1976).

1.3.2 Stress Relaxation

One disadvantage of the conventional procedure is that growth depends not only on the wall-yielding properties, but also on the water uptake properties of the growing tissue (see Sect. 1.4). A new procedure, termed in vivo stress relaxation, has recently been devised to circumvent this problem (Cosgrove et al. 1984; Cosgrove 1985). It works by the following principle. If a growing cell or tissue is isolated from an external supply of water, then its volume cannot increase and so it cannot grow in the usual sense. However, the biochemical processes which loosen the wall continue. As a result, when stress-bearing bonds in the wall are broken, wall stress decreases, or "relaxes" as the wall polymers undergo rearrangement without changing the net size of the wall. Because turgor pressure is equal and opposite to wall stress, turgor relaxes at the same time. Given Eq. (1), then the time course for stress relaxation should take the form of an exponential decay to the yield threshold (Cosgrove 1985), with a halftime ($T_{1/2}$) given by:

$$T_{1/2} = \frac{\ln(2)}{\varphi \cdot \varepsilon}, \tag{2}$$

where ε is the volumetric elastic modulus of the cell ($\varepsilon = VdP/dV$).

Experimentally, the procedure may be carried out with the use of the pressure probe to measure turgor pressure directly. A typical stress relaxation time course for growing pea stem tissue is shown in Fig. 3. For rapidly growing tissue, the relaxation halftime is in the range of 15 to 20 min, whereas for slowly growing tissue the halftime is much longer. In studies carried out to date, various treatments which stimulate or inhibit the growth of pea stems seem to exert their effects principally through a change in φ, not through a change in Y. That is, regardless of their growth rate, pea tissue relax to the same threshold turgor of about 0.3 MPa (Cosgrove et al. 1984; Cosgrove 1985), but the halftime varies with the treatment. In other tissues, changes in the apparent threshold turgor have been noted (Hsiao et al. 1985; Matthews et al. 1984; Van Volkenburgh and Cleland 1986).

It should be emphasized that the loss of turgor pressure during the stress relaxation procedure is not due to simple evaporative water loss. Evaporation is prevented by sealing the tissue in a humid chamber. This fact can be demonstrated by measurement of tissue dimensions with sensitive position transducers (Cosgrove 1985). Thus, the loss of turgor occurs without loss of water. As one would expect, basal (nongrowing) stem segments do not undergo stress relaxation and do not lose turgor (Cosgrove et al. 1984).

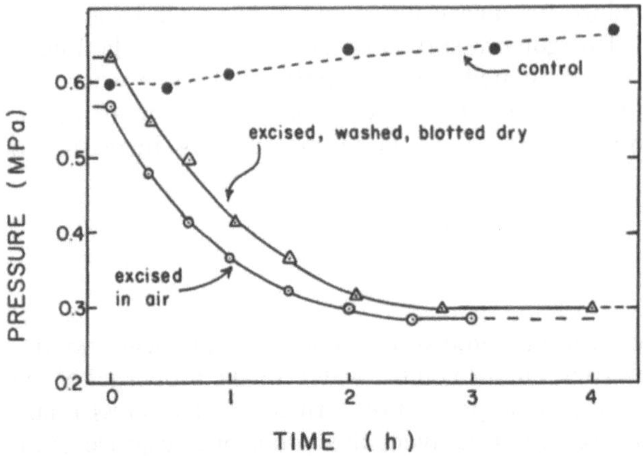

Fig. 3. Time course for in vivo stress relaxation of young pea stems. In situ turgor pressure was measured at each time point in about ten outer cortical cells in the region 4–6 mm below the apical hook, using the pressure microprobe. At zero time the growing tissue was excised and thereby deprived of an external water supply. Washing the cut surfaces to remove solutes released from the damaged cells did not appreciably alter the pattern of relaxation. As a control, water was left in contact with the cut surfaces of some segments; these segments did not undergo relaxation. (After Cosgrove 1985)

1.3.3 Molecular Models of Wall Loosening

What do φ and Y represent at the molecular level? If we imagine the wall as an array of parallel microfibrils held together by a complex of proteins, pectins, and other mixed-linked polysaccharides, then extension may be initiated when one of the load-bearing bonds that hold the microfibrils together is enzymatically broken. If the wall stress is sufficient to allow wall polymers to slip past one another, then a viscoelastic extension of the wall will ensue. By this model, the yield turgor, Y, represents the minimum wall stress needed to initiate a passive viscoelastic extension; i.e., the wall acts like a Bingham plastic, with a definite yield point. The extensibility term φ is more complex; it is a function of both the rate of wall loosening and the passive viscoelastic properties of the wall that govern the amount of extension that will occur for each wall-loosening event.

In the preceding model, the turgor dependence of cell growth derives from the step involving viscoelastic expansion of the wall. As an alternative to this model, enzymatic cleavage of load-bearing polymers may require that the wall be under tension, either to allow steric access by a cleaving enzyme to a sensitive site, or to raise the internal energy of the polymers such that bond breakage is energetically feasible. Turgor dependence would arise because the rate of enzymatic cleavage depends on wall stress and because viscoelastic expansion requires the wall to be under tension. Such a model might explain why walls of oat coleoptile segments do not become progressively weaker after cell turgor pressure is reduced below the yield threshold (Cleland and Rayle 1972). However, direct evidence for turgor-dependent enzymatic cleavage is lacking.

A mechanism termed "chemical creep" has also been proposed (Cleland and Rayle 1972; Ray and Ruesink 1962). This model assumes that wall expansion is governed by cleavage and re-formation of a set of covalent bonds. Bond cleavage initiates wall expansion, which stops when the bond re-forms. Thus, only when the wall is under tension does this wall metabolism translate into a physical expansion.

These models assume that covalent bond breakage initiates the wall expansion process. An alternative idea is that some constituent of the matrix must be added to the wall to promote expansion, without the necessity of breaking covalent bonds. For example, intercalation of matrix polymers into the wall might allow the microfibrils to separate a certain amount before the wall would again be pulled taut. Such strain-hardening of the wall would insure that expansion does not get out of control and cause cell rupture (Green et al. 1977).

Although our current knowledge does not yet allow us unequivocably to affirm one of these models of wall loosening, there are various pieces of circumstantial evidence that point to an enzymatic cleavage, perhaps of one of the matrix polymers (for reviews, see Labavitch 1981; Taiz 1984). First, turnover of hemicellulose in the wall is associated with conditions of rapid wall expansion (Darvill et al. 1978; Terry et al. 1981). Second, isolated cell wall preparations from growing tissues undergo creep (prolonged yield under constant stress) when they are placed in mildly acid pH (Cleland et al. 1983, 1986; Taiz and Richmond 1984). This creep is eliminated by boiling the wall or by treating it with proteinase. Third, antibodies prepared against wall proteins are reported to be effective in inhibiting auxin-induced cell expansion (Huber and Nevins 1981). Unfortunately, attempts to isolate a "wall-loosening enzyme" have not yet been successful. Only if this is achieved will the enzymatic hypothesis be confirmed.

1.4 Water Uptake and Turgor Maintenance

In our model of cell expansion, water uptake accomplishes two things. It increases cell volume and it maintains high turgor pressure. The first aspect is important because the increase in cell size is mostly due to water uptake. Eighty to 95% of plant cell volume is water. The second aspect is important because turgor pressure provides the force needed to expand the network of polymers and microfibrils in the wall.

1.4.1 Physics of Water Uptake

Water uptake is a passive process which requires a gradient in water potential. The rate of water uptake depends on the magnitude of this gradient and on a water transport coefficient of cell, the hydraulic conductance (L). The relevant equation for a single cell is:

$$dV/(Vdt) = L(\Delta\psi) = (A/V)\, L_p(\Delta\pi - P)\,, \qquad (3)$$

where dV/(Vdt) is the relative rate of water uptake (growth rate under steady conditions), $\Delta\psi$ is the water potential gradient, A/V is the area/volume ratio of the cell, Lp is the specific hydraulic conductivity of the plasma membrane, and $\Delta\pi$ is the gradient in osmotic pressure across the plasma membrane.

In a nongrowing cell, water uptake quickly increases turgor pressure, and thereby eliminates the water potential difference that induces the water influx. In other words, at equilibrium turgor pressure (P) equals the osmotic pressure difference ($\Delta\pi$) and $\Delta\psi$ is zero. In a growing cell, wall relaxation reduces turgor pressure, and thereby maintains the driving force for water absorption by reducing the cell's water potential below that of its surroundings. In a normal situation, stress relaxation of the wall is not observed because water absorption continually rebuilds turgor pressure and thus restores the wall stress.

During steady growth, wall yielding and water uptake are exactly balanced. This coordination comes about because turgor pressure is a linking factor which influences and is influenced by both processes (see Figs. 1 and 3). Thus, in a cell with a very low hydraulic conductance, wall relaxation may reduce turgor pressure substantially. This has two effects: It creates a water potential difference to drive water influx; and it reduces wall expansion and slows cell growth. In principle, the rate of growth may be limited either by φ or L, or by some combination of the two.

1.4.2 Restriction of Growth by Water Transport

For single, isolated cells, it is easy to demonstrate that water transport is so rapid that growth is entirely controlled by the wall-yielding properties (Cosgrove 1981, 1986). On the other hand, the water transport pathway in multicellular tissues is much longer and more complex (Molz and Boyer 1978; Silk and Wagner 1980).

The *magnitude* of the growth-sustaining water potential gradient has been a subject of controversy. Boyer (1985) has argued that it is large, even in small tissues like the soybean hypocotyl. As evidence, he cited a number of studies which show by means of thermocouple psychrometry and other methods that the water potential of nontranspiring, growing tissue is frequently about 0.3 MPa below that expected of a well-hydrated tissue. If this low water potential represented the magnitude of the growth-sustaining water potential gradient, then cell expansion in these tissues would be restricted in large part simply by the ability of the tissue to absorb water. If confirmed, this conclusion would have important implications concerning the means available to a plant to control its growth rate.

An alternative interpretation has been put forward by Cosgrove and Cleland (1983a). They have shown that the concentration of solutes in the cell wall free space of growing tissues is large enough to account for much of the reduced water potential in young stems. It should also be noted that psychrometric techniques may register abnormally low water potentials in growing tissues because of stress relaxation of the wall (Cosgrove 1981, 1986). Recent studies have documented this type of artifact (Cosgrove 1985; Cosgrove et al. 1984; Hsiao et al. 1985; Van Volkenburgh and Cleland 1986).

A recent theoretical analysis (Cosgrove 1985) showed that one way to gauge the relative important of φ and L in controlling growth is by comparison of the kinetics of wall relaxation with the kinetics of water uptake. Each process may be characterized with a halftime. The appropriate equation is given by:

$$\frac{T_{1/2}(\text{swelling})}{T_{1/2}(\text{stress relax.})} = \frac{\varphi\varepsilon}{L(\varepsilon+\pi)+\varphi\varepsilon} \approx \frac{\varphi}{L+\varphi} \quad (\text{for } \varepsilon \gg \pi)\,. \tag{4}$$

Experimental measurements show that $T_{1/2}$ for swelling of pea stems is about 1−2 min, whereas $T_{1/2}$ for stress relaxation is about 18 min (Cosgrove 1985, 1986). Such results indicate that L restricts the growth rate of pea stems by only a small amount and is not a major limitation of the expansion rate of this tissue.

1.4.3 Solute Uptake

Turgor pressure arises because cells build up an intracellular solute concentration in excess of that present outside the plasma membrane. If new solutes were not taken up or otherwise generated in growing cells, water absorption during cell expansion would soon dilute the solutes. The result would be a reduction of turgor and cessation of cell expansion.

Studies have shown that water uptake by expanding tissues is closely paralleled by solute uptake, such that the osmotic pressure of the cell sap changes very little during the course of many-fold expansion (Cosgrove 1985; Hsiao et al. 1985). These results imply that the cell regulates either its osmotic pressure or turgor pressure. In marine algae turgor regulation is accomplished by control of the activity of ion pumps at the plasma membrane (Kirst and Bisson 1979; Nuccitelli and Jaffe 1976). Although the nature of the sensory mechanism is not yet established, most models have focused on effects of turgor pressure on the electromechanical characteristics of the plasma membrane (Bisson and Gutknecht 1980; Coster et al. 1976).

In seed plants, on the other hand, solutes are delivered to expanding organs largely via the phloem. There is substantial evidence that the solute unloading mechanism is turgor-sensitive in nonexpanding tissues (Daie and Wyse 1984; Wolswinkel 1985), but little work has been carried out with expanding organs. In particular, the exact means for coordination of uptake by expanding cells and delivery by the phloem has not yet been worked out. Current biophysical models of cell expansion assume, probably incorrectly, that cells are able to maintain constant osmotic pressure despite dilution by cell expansion. What is the dynamic range of the system delivering solutes to expanding organs? What causes net solute flux to cease, or at least to diminish greatly, when cells mature? Is solute delivery to expanding organs a limiting factor for growth? To these questions we do not yet have definite answers.

In water-stressed plants, solutes accumulate in the growing region (reviewed in Morgan 1984). This response is commonly thought to be adaptive, in that it helps to maintain turgor pressure in the face of falling water potentials, and thereby helps to sustain cell expansion.

Electrogenic H^+ pumps in the plasmalemma are thought to play a key role in many transport processes (Reinhold et al. 1984; Reinhold and Kaplan 1984). They may also be important in the maintenance of the low pH necessary for wall expansion (Cleland and Rayle 1978). Recent work has linked such pumps to the control of expansive growth in young seedlings and leaves (De Boer et al. 1985; Okamoto et al. 1984), but many details remain to be worked out. Because the vacuole is the largest compartment in rapidly expanding cells, the control of tonoplast pumps also deserve particular study.

Considering its importance, the dearth of information and models on the coordination of solute uptake and expansion is surprising and needs to be remedied by further research.

1.5 Summary

For tip-growing cells, the directionality of cell expansion seems to be controlled by the pattern of deposition of wall material at the expanding tip. More generally, the deposition of aligned cellulose microfibrils gives to the wall a bias or direction of preferential expansion. The final directionality of expansion (longitudinal versus transverse extension) is determined by the pattern of wall stress and by the net orientation of microfibrils in the load-bearing region of the wall.

From a physical point of view, plant cell expansion begins with stress relaxation of the wall, followed by water uptake in response to the water potential disequilibrium created by such relaxation. The molecular nature of wall relaxation remains in doubt. Although wall-loosening enzymes are commonly thought to play a role, a likely candidate has not yet been found. Solute uptake is coordinated with cell expansion, and the activity of membrane pumps is implicated in this control, but further work is necessary to elucidate the mechanism of such coordination.

References

Bisson MA, Gutknecht J (1980) Osmotic regulation in algae. In: Spanswick RM, Lucas WJ, Dainty J (eds) Plant membrane transport: current conceptual issues. Elsevier, Amsterdam, North-Holland, pp 131–146

Boyd JD (1985) Biophysical control of microfibril orientation in plant cell walls. Martinus Nijhof/Dr Junk, Dordrecht

Boyer JS (1985) Water transport. Ann Rev Plant Physiol 36:473–516

Cleland RE (1976) The control of cell enlargement. Integration of activity in the higher plant. Symp Soc Exp Biol 31:101–115

Cleland RE (1981) Wall extensibility: hormones and wall extension. In: Tanner W, Loewus FA (eds) Encyclopedia of plant physiology, new series: plant carbohydrates II. Extracellular carbohydrates. Springer, Berlin Heidelberg New York, pp 225–276

Cleland RE, Rayle DL (1972) Absence of auxin-induced stored growth in Avena coleoptiles and its implication concerning the mechanism of wall extension. Planta (Berl) 106:61–71

Cleland RE, Rayle DL (1978) Auxin, H^+-excretion and cell elongation. Bot Mag Tokyo 1:125–139

Cleland RE, Cosgrove DJ, Tepfer M (1983) Characterization of the in vitro acid extension curves of Avena coleoptiles. Plant Physiol 72:S – 74
Cleland RE, Cosgrove DJ, Tepfer M (1986) Long-term acid-induced wall extension in an in vitro system. Planta 170:379 – 385
Cosgrove DJ (1981) Analysis of the dynamic and steady-state responses of growth rate and turgor pressure to changes in cell parameters. Plant Physiol 68:1439 – 1446
Cosgrove DJ (1985) Cell wall yield properties of growing tissues. Evaluation by in vivo stress relaxation. Plant Physiol 78:347 – 356
Cosgrove DJ (1986) Biophysical control of plant cell growth. Ann Rev Plant Physiol 37:377 – 405
Cosgrove DJ, Cleland RE (1983a) Solutes in the free space of growing stem tissues. Plant Physiol 72:326 – 331
Cosgrove DJ, Cleland RE (1983b) Osmotic properties of pea stem internodes in relation to growth and auxin action. Plant Physiol 72:332 – 338
Cosgrove DJ, Van Volkenburgh E, Cleland RE (1984) Stress relaxation of cell walls and the yield threshold for growth: demonstration and measurement by micro-pressure probe and psychrometer techniques. Planta 162:46 – 52
Coster HGL, Steudle E, Zimmermann U (1976) Turgor pressure sensing in plant cell membranes. Plant Physiol 58:636 – 643
Daie J, Wyse RE (1984) Evidence on the mechanism of enhanced sucrose uptake at low cell turgor in leaf discs of Phaseolus coccinius. Physiol Plant 64:547 – 552
Darvill A, McNeil M, Albersheim P, Delmer DP (1980) The primary cell walls of flowering plants. In: Stumpf PK, Conn EE (eds) The biochemistry of plants. Academic Press, London, pp 91 – 161
Darvill AG, Smith CJ, Hall MA (1978) Cell wall structure and elongation growth in Zea mays coleoptile tissue. New Phytol 80:503 – 516
De Boer AH, Katou K, Mizuno A, Kojima H, Okamoto H (1985) The role of electrogenic xylem pumps in K^+ absorption from the zylem of Vigna unguiculata: the effects of auxin and fusicoccin. Plant Cell Environ 8:579 – 586
Dugger WM, Barnicki-Garcia S (1984) Structure, function, and biosynthesis of plant cell walls (Symposium in Botany, University of California, Riverside). American Soc Plant Physiologists, Rockville, MD
Frey SC (1986) Cross-linking of matrix polymers in the growing cell walls of angiosperms. Ann Rev Plant Physiol 37:165 – 186
Green PB (1958) Structural characteristics of developing Nitella internodal cell walls. J Biophys Biochem Cytol 4:505 – 515
Green PB (1969) Cell morphogenesis. Ann Rev Plant Physiol 20:365 – 394
Green PB, Bauer K, Cummins WR (1977) Biophysical model for plant cell growth: auxin effects. In: Jungreis AM, Hodges TK, Kleinzeller A, Schultz SG (eds) Water relations in membrane transport in plants. Academic Press, London, pp 30 – 45
Hayashi R, Morikawa H, Nakajima N, Ichikawa Y, Senda M (1980) Oriented structure of pectic polysaccharides in pea edipermal cell walls. Plant Cell Physiol 21:999 – 1005
Hepler PK (1981) Morphogenesis of tracheary elements and guard cells. In: Kiermayer O (ed) Cytomorphogenesis in plants. Springer, Berlin Heidelberg New York, pp 327 – 347
Hsiao TC, Silk WK, Jing J (1985a) Leaf growth and water deficits: biophysical effects. In: Baker NR, Davis WD, Ong C (eds) Control of leaf growth, Soc Exp Bio Sem, vol. 27. Cambridge University Press, Cambridge, pp 239 – 266
Hsiao TC, Silk WK, Diedenhofen U, Matson C (1985b) Spatial distribution of osmoticum and potassium and their deposition rates in the primary root of corn, Zea mays. Plant Physiol 77:S-35
Huber DJ, Nevins DJ (1981) Wall-protein antibodies as inhibitors of growth and of autolytic reactions of isolated cell wall. Physiol Plant 53:533 – 539
Kiermayer O (1981) Cytomorphogenesis in plants. Springer, Berlin Heidelberg New York, p 439
Kirst GO, Bisson MA (1979) Regulation of turgor pressure in marine algae: ions an low-molecular weight organic compounds. Aust J Plant Physiol 6:539 – 556
Labavitch JM (1981) Cell wall turnover in plant development. Ann Rev Plant Physiol 32:385 – 406

Lockhart JA (1965) An analysis of irreversible plant cell elongation. J Theor Biol 8:264–275
Maclachlan G, Fèvre M (1982) An overview of cell wall biosynthesis. In: Lloyd CW (ed) The cytoskeleton in plant growth and development. Academic Press, London, pp 127–146
MacMillan J (1980) Encyclopedia of plant physiology, new series, hormonal regulation of development. I. Molecular aspects of plant hormones. Springer, Berlin Heidelberg New York, p 681
Matthews MA, Van Volkenburgh E, Boyer JS (1984) Acclimation of leaf growth to low water potentials in sunflower. Plant Cell Environ 7:199–206
McNeil M, Darvill AG, Fry SC, Albersheim P (1984) Structure and function of the primary cell walls of plants. Ann Rev Biochem 53:625–663
Métraux JP, Taiz L (1978) Transverse viscoelastic extension in Nitella I. Relation to growth rate. Plant Physiol 61:135–138
Molz FM, Boyer JS (1978) Growth-induced water potentials in plant cells and tissues. Plant Physiol 62:423–429
Morgan JM (1984) Osmoregulation and water stress in higher plants. Ann Rev Plant Physiol 35:299–319
Morikawa H, Kitamura S, Senda M (1978) Effect of auxin on changes in the oriented structure of wall polysaccharides in response to mechanical extension in oat coleoptile cell walls. Plant Cell Physiol 19:1553–1556
Nuccitelli R, Jaffe LF (1976) Current pulses involving chloride and potassium efflux relieve excess pressure in Pelvetia embryos. Planta 131:315–320
Okamoto H, Mizuno A, Katou K, Ono Y, Matsumura Y, Kojima H (1984) A new method in growth-electrophysiology: pressurized intra-organ perfusion. Plant Cell Environ 7:139–147
Probine MC (1963) Cell growth and the structure and mechanical properties of the wall in internodal cells of Nitella opaca. J Exp Bot 14:101–113
Probine MC, Barber NF (1966) The structure and plastic properties of the cell wall of Nitella in relation to extension growth. Aust J Biol Sci 19:439–457
Ray PM, Ruesink AW (1962) Kinetic experiments on the nature of the growth mechanism in oat coleoptile cells. Dev Biol 4:377–397
Reinhold L, Kaplan A (1984) Membrane transport of sugars and amino acids. Ann Rev Plant Physiol 35:45–83
Reinhold L, Seiden A, Volokita M (1984) Is modulation of the rate of proton pumping a key event in osmoregulation? Plant Physiol 75:846–849
Richmond PA (1983) Patterns of cellulose microfibril deposition and rearrangement in Nitella: in vivo analysis by a birefringence index. J Appl Polymer Sci: Appl Polymer Symp 37:107–122
Richmond PA, Métraux J-P (1984) Cellulose synthesis inhibition, cell expansion, and patterns of cell wall deposition in Nitella internodes. In: Dugger WM, Bartnicki-Garcia S (eds) Structure, function, and biosynthesis of plant cell walls. American Society of Plant Physiologists, Rockville, USA, pp 475–476
Richmond PA, Métraux J-P, Taiz L (1980) Cell expansion patterns and directionality of wall mechanical properties in Nitella. Plant Physiol 65:211–217
Robinson DG, Quader H (1982) The microtubule-microfibril syndrome. In: Lloyd CW (ed) The cytoskeleton in plant growth and development. Academic Press, London, pp 109–126
Roelofson PA (1965) Ultrastructure of the wall in growing cells. Adv Bot Res 2:69–149
Roelofson PA, Houwink AL (1953) Architecture and growth of the primary cell wall in some plant hairs and in the Phycomyces sporangiophores. Acta Bot Neerl 2:218–225
Roland JC, Vian B (1979) The wall of the growing plant cell: its three-dimensional structure. Int Rev Cytol 61:129–166
Scott TK (1981) Encyclopedia of plant physiology, new series, hormonal regulation of development. II. From the cell to the whole plant. Springer, Berlin Heidelberg New York
Silk WK, Wagner KK (1980) Growth-sustaining water potential distributions in the primary corn root. A noncompartmented continuum model. Plant Physiol 66:859–863
Taiz L (1984) Plant cell expansion: regulation of cell wall mechanical properties. Ann Rev Plant Physiol 35:585–657

Taiz L, Richmond PA (1984) Regulation of cell wall mechanical properties in vivo and in vitro. In: Dugger WM, Bartnicki-Garcia S (eds) Structure, function, and biosynthesis of plant Cell Walls. American Society of Plant Physiologists, Rockville, USA, pp 185–194

Taiz L, Métraux J-P, Richmond PA (1981) Control of cell expansion in the Nitella internode. In: Kiermayer O (ed) Cytomorphogenesis in plants; cell biology monographs. Springer, Berlin Heidelberg New York, pp 231–264

Taiz L, Rayle DL, Eisinger W (1983) Ethylene-induced lateral expansion in etiolated pea stems: The role of acid secretion. Plant Physiol 73:413–417

Terry ME, Jones RL, Bonner BA (1981) Soluble cell wall polysaccharides released from pea stems by centrifugation. I. Effect of auxin. Plant Physiol 68:531–537

Van Volkenburgh E, Cleland RE (1986) Wall yield threshold and effective turgor in growing bean leaves. Planta 167:37–43

Wainwright SA (1970) Design of hydraulic organisms. Naturwissenschaften 57:321–326

Wainwright SA, Biggs WD, Currey JD, Gosline JM (1976) Mechanical design in organisms. Edward Arnold, London, p 423

Wareing PF (1982) Plant growth substances 1982. Academic Press, London, p 683

Weisenseel MH, Kicherer RM (1981) Ionic currents as control mechanism in cytomorphogenesis. In: Kiermayer O (ed) Cytomorphogenesis in plants. Springer, Berlin Heidelberg New York, pp 379–400

Wolswinkel P (1985) Phloem unloading and turgor-sensitive transport: factors involved in sink control of assimilate partitioning. Physiol Plant 65:331–339

IV.2 Plant Cytomechanics and Its Relationship to the Development of Form

PHILIP M. LINTILHAC[1]

2.1 Introduction

What is the relationship between cellular mechanics and the development of form in organisms?

We see from the chapters in this book that the cell is rich in ultrastructural machinery which performs various kinds of tasks within the cell; organizing subcellular components, contributing to directional transport, and in some cases controlling the shape and morphology of individual cells.

But how do these intricate cytomechanical processes relate to the organization of the whole organism? Are the details of molecular architecture and cellular organization merely part of the necessary machinery of the cell or do they contribute in some direct way to the overall life history of the organism?

In this chapter I hope to illustrate, by means of examples taken from the plant kingdom, how cytomechanical models may contribute to our understanding of whole organism development. Specifically, I hope to illustrate how, in plants, the evolution of ultrastructural mechanisms of stress-mechanical sensitivity at the cellular level may be one of the key elements in the development of whole plant architecture, and may in fact serve as the principal organizational link which, along with transmitted stress, lends logic and coherence to the development of organic form in plants.

2.2 The Logic of Development

2.2.1 The Role of the Genome in the Development of Form

My approach to the development of form diverges somewhat from the usual concept of developmental control which, simply stated, says that the specifications for the development of mature form are contained within the genome, and that the individual events which comprise organismal ontogeny are proscribed within the genetic code. I do not believe that this reflects an accurate or useful picture of the developmental process. Nor do I believe that it does justice to the inherent self-organizing abilities of biological materials (cf. Chap. I.1.7.1).

[1] Botany Department, The University of Vermont, Burlington, Vermont, 05401, USA

There is no detailed specification of developmental ontogeny within the genome, and furthermore there is no description of the completed adult (Lintilhac 1984). The widespread idea that there could be such a blueprint is in fact of rather recent origin, and dates to the elucidation of the basic syntax of the genetic code. Previously, the trend of biological thought was toward a more holistic if somewhat sketchy picture of the organism as a complex but malleable summation of individual cellular responses conspiring to channel developmental events in a stochastic way.

The idea that form itself can be traced back to some master blueprint lying on the genetic drawing board ignores the basic fact that biological form never arises de novo, but always grows out of a preexisting form. In this sense the development of form bears less resemblance to the construction of a new building on an untouched site than it does to the continual remodeling of an ancient habitation. In this regard, the developmental question that I address here is this: What provides the continuity between successive forms, maintaining the integrity and flexibility of development without having to resort to a detailed set of construction specifications?

The possibility that the genome does not contain information controlling the important developmental process of cell wall orientation does not deny its importance as the ultimate repository of all information specifying the structure of the enzymatic machinery available to the cell. Clearly the developmental potential of any organism must eventually be traced back to information encoded in its genome. The purpose of that information is to ensure the proliferation of responsive and competent cells, however, and not to proscribe future events. The genome provides the detailed information necessary for the production of an array of biochemical and macromolecular equipment which gives the cell its particular abilities to react to signals in its environment.

2.2.2 The Role of the Environment in the Development of Form

For some cells, the environment may be dominated by signals from the external world; soil, water, or air. For other cells the environment is entirely organismically generated, consisting of chemical, physical, and electrical signals of various kinds which give the organism the channels through which to manipulate cellular environments within itself. For most plant cells cellular behavior may in large part be controlled through these internal channels, with certain major triggering changes controlled by signals from the outside, such as light or temperature.

Plants, which are essentially architectural beings, offer a particularly transparent model where we can study these epigenetic control mechanisms and relate them to the basic cytomechanical machinery which underlies them (Lintilhac and Vesecky 1980). This is because the basic cellular events which underlie morphogenesis are permanently recorded in the multicellular patterns which we see at the tissue level.

2.3 The Architecture of Plant Form

Plant cells are contained within rigid boxes whose walls are composed of the biopolymer cellulose and a complex matrix of polysaccharide and protein. It is the nature of plant cell division that the new cells which result are permanently attached to each other. After division, plant cells do not move with respect to each other, although they are free to enlarge and distort in various ways. They are essentially trapped within a system of permanent walls, giving plant tissues their characteristic architecture which we see in microscopic view. This implies that the history of cell division activity within a tissue is frozen into the cell patterns which are revealed in any section, and that this history, if properly interpreted, could be used to unravel the development of form.

2.3.1 Division and Growth. The Basic Events

Plant cells divide by a process which is quite unlike that of animal cells. Being contained within a rigid box, division involves the erection of a new partition across the cell, much as a new partition wall would be erected within a room to divide it into two smaller rooms. The new partition, which is initially unattached and free to assume any orientation within the parent cell, eventually achieves a precise orientation and becomes rigidly attached to the walls of the preexisting cell.

It is at this point that the architectural analogy has its greatest validity. Just as the architect must ensure that the support walls in an architectural structure will accomodate the loads which are acting on them, so the cell must ensure that the new partition wall will effectively support its share of the forces acting through the tissue. This is accomplished mainly by orienting the new partition so that it coincides with the principal stresses acting on the parent cell (Lintilhac and Vesecky 1981).

2.3.2 Growth as a Source of Mechanical Stress

The forces with which individual growing plant cells have to deal come from growth itself (see Chap. IV.1), however, and not from external loads such as gravitational forces and wind, as they do in the architectural analogy. This becomes clear when we realize that growth in plants is itself the result of a force acting within the individual cells (Ray et al. 1973), and not simply the result of the accumulation of newly synthesized protoplasm. New synthesis does account for a small part of plant growth, but the greatest part of growth is due to water uptake, which generates high internal pressures (turgor), and high stresses in the wall. The high wall stresses are accomodated by a controlled relaxation (creep), resulting in cell enlargement and growth.

The osmotic potential of the cell's interior acting on the semirigid constraint of the cell wall can generate pressures of up to 20 atm within plant cells, forcing a creep-mediated stretching of the wall and a transmission of growth stresses from cell to cell throughout the tissue. These high internal pressures represent an enormous potential for the generation of locked-in stresses within growing tissues, stresses which could easily destroy improperly oriented walls.

We now have a picture of a plant tissue as an aggregation of hydrostatic units, water-filled balloons trapped in a system of common walls which are rigidly connected but extensible due to viscoelastic creep. The walls are strong but thin, which means that they can easily be ruptured by shearing forces. The cell's only protection against this is to make sure that new walls are installed in a plane which is free from shear, perpendicular to one of the principal stresses (Lintilhac and Vesecky 1981).

Growth of the tissue as a whole must be treated as a two-part process. The first part involves the division (partitioning) of the individual cells (and the synthesis of new protoplasm). This involves minimal actual volume growth, but provides the basic modulation of form through the creation of new cells. This is amplified in the second part of the growth process by enlargement growth, which contributes most of the volume growth and visible increase in size.

It is here that we can see the connection between individual cell behavior and the development of organic form in the whole plant. The individual cells, reacting to the forces acting on them, divide in particular orientations, thereby contributing to the emergence of a new form. To complete our understanding of this control process we must then determine what controls the distribution of transmitted stress acting through the tissue.

2.3.3 Factors Affecting Stress Distribution in Embryonic Plant Organs

There is much to be learned about the mechanics of plant tissues. If we ignore individual cells, and sidestep the complex stress analysis which is necessary to interpret the behavior of the individual cell as a reinforced pressure vessel (which in fact it is), then we can look at whole multicellular tissues. If we are dealing with whole embryonic organs such as leaf, root, or shoot primordia, for instance, then we can consider the tissue as a continuous medium through which stress acts in conformity with all the normal rules of stress mechanics (Heywood 1969). Stresses follow the surface contours of the object, with families of trajectories acting parallel and perpendicular to the free surfaces. Tensile and compressive stresses also achieve a normal equilibrium within a tissue or organ, so that any region of overall compressive stress must be balanced against a region of tensile stress. Stresses in tissues become concentrated around notches and irregularities in the same way that they do in engineering materials, contributing to the characteristic patterning of cells that one sees in the axils of embryonic leaves (Lintilhac and Vesecky 1980).

The distribution of stress through the tissue is thus determined by two factors, the first being the location of active regions of enlargement growth, because these regions constitute the sources from which stress emanates into the surrounding tissue. The second factor is the shape of the organ as a whole since the stresses, in following the surfaces of the structure, can actually be considered to be molded by the form of the structure within which they are trapped.

2.3.4 The Role of Stress in the Generation of Form

The emergence of form in growing plant organs is a matter of adding new enlarging cells to a preexisting cellular mass, this means that the new form of the organ will be determined cumulatively by the orientations of the individual cell divisions. However, since individual partition wall orientations are keyed to the distribution of stresses in the tissue, and the stresses are in turn shaped by the overall form of the structure, it would appear that division wall orientation, growth, stress, and form are inextricably linked in a repetitive developmental cycle of control.

The outward form of the growing structure, being the only aspect which we can readily appreciate without analytical or experimental probes, tends to be the aspect which we identify as the legitimate end product of development, and therefore the object of genetic specification. In actual fact, of course, it is no more an end product than any other level in the developmental hierarchy, and so it should be no surprise to discover that form itself can also be a critical controlling element in development.

2.4 The Ultrastructural Basis of Cell Behavior

If one can identify a pivotal and central event in this ongoing evolution of form out of preexisting form, it must surely be the orientation of the partition wall during cell division, because it is here that genetic specification, such as it is, is translated into a sophisticated structural responsiveness. This means that it is the ultrastructural machinery with which the cell is equipped which bestows upon it the ability to respond to the mechanical signals in its environment. This ability to make structural decisions casts the cell in the role of a behavioral unit, rather than simply a collection of biochemicals which mediates information transfer between the genome and the development of form.

It is possible, then, to make a strong argument for the case that "form begets form," with the cell serving as the behavioral nexus through which genetic potential is realized in terms of morphogenetic decisions (Lintilhac 1984).

2.4.1 The Role of Cytomechanics in the Development of Form

The case to be made for cytomechanics, then, is this: in plant cells in particular, the cytomechanical machinery within the cell is the basic equipment which provides the cell with its extraordinary sensitivity to external structural-mechanical cues, and its ability to make appropriate decisions in terms of architectural responses. It constitutes the essential link between stored information in the nucleus, and the development of morphogenetic form, which is the salient expression of cellular behavior.

2.5 Other Responses to Mechanical Stimuli. Reaction Wood

Plants, with their architectural habit of growth, have used to great advantage the structural possibilities conferred by the individual cell's responsiveness to directional forces.

Plants show various kinds of responses to mechanical stimuli. One of the best known of these is termed "reaction-wood" formation, which includes the formation of "compression-wood" on the lower side of horizontal branches in conifers, and "tension-wood" which is formed on the upper side of horizontal branches in hardwoods (Wilson and Archer 1977). Both of these gross adaptive responses result from cellular responses to the increased stress due to the gravitational loading of horizontal branches.

2.5.1 Tropic Responses

Other kinds of growth responses to mechanical stimuli involve gravistimulated responses which bring roots and stems into a vertical orientation with respect to the gravity vector (Shen-Miller and Hinchman 1974). These responses, which are of obvious adaptive significance to plants, since they drive stems upward toward the light and roots down into the soil, are still not fully understood; although here too, it is clear that there must be some cellular sensor which interprets gravitational signals and triggers a specific response resulting in one side of the axis growing faster than the other and bringing it back to the vertical.

Clear-cut growth responses to such stimuli as wind, vibration (Quirk and Freese 1976), and even mechanical rubbing (Jaffe 1977) have been demonstrated in a number of plants. These generally result in an inhibition of overall elongation growth resulting in a shorter, stouter axis.

The mechanisms by which these different responses act are varied. In some cases mechanical stimulation appears to result in the localized release of hormonelike compounds which can cause either an inhibition or a stimulation of growth on one side of an axis. One thing they have in common is that the response is generally manifested in terms of changes in the relative rate of enlargement growth of the individual cells. In this respect they are different from the partition-wall orientation mechanisms which underlie the generation of form itself. Nonetheless, they serve to illustrate the varied and sensitive mechanisms which plants use to accomodate to the architectural problems which arise during growth.

2.6 Meiosis as a Mechanically-Induced Process

In the last section of this chapter I would like to illustrate how a major developmental change in plants may also be explained in terms of a sophisticated architectural response which is mediated by cytomechanical discrimination of stress-mechanical stimuli.

In plants the process of meiosis proceeds in a slightly different context from meiosis in animals. In plants there is no "germ-line" of gonadal tissue from which meiotic cells differentiate. Any cell can be a meiotic cell if it happens to be in the right place at the right time. Since all the somatic cells of the plant are genetically identical, and since no line of cells is predestined to undergo meiosis, the plant must have some method for the selection of a cell or a group of cells and reliably triggering them to undergo meiosis. Various hypotheses can be proposed to explain this, but a close examination of the developmental logic of the situation reveals constraints which might not be obvious at first.

The most obvious hypothesis would be that selection of the cell or cells which undergo meiosis is genetically programmed. But this is impossible, since all the cells are genetically identical to begin with. If any cell's genome contains a statement which says: "I am the one" then all the cells of the plant contain the same statement, and therefore selection of meiotic cells cannot come from the genome. Similarly, one cannot invoke a genetically determined sensitivity of a few cells to some external messenger such as a hormone, because, at least initially, all cells would possess the same sensitivity.

Clearly then, the first discriminant has to be some environmental selector which can stimulate a few cells to behave differently from their cohorts even though they are initially genetically identical. Once this initial differentiative behavior is established then genetic mechanisms can come into play, allowing the cells to read different portions of an identical genome which would equip them with different intracellular machinery. This amounts to a genetic amplification of differences which are imposed from the outside.

What is the nature of the external discriminant? I propose that here again plants have used their cytomechanically based architectural sensitivity to build a stress-mechanical triggering device which serves to isolate and release a few cells into a meiotic mode. This device is the sporangium.

2.6.1 The Sporangium as a Stress-Focusing Device

The sporangium is one of the universal structural constants of the plant kingdom. Wherever meiosis occurs in a multicellular setting it is defined by this unique and yet varied structure. The sporangium is typically a club-shaped or bottle-shaped mass of cells within which a single cell or group of cells (the sporogenous cells) undergo meiosis to give rise to the sperm or eggs, or their precursors (Fig. 1).

The shape of the sporangium can be as varied as bottles can, and meiosis can be induced in one or many cells, but it is always characterized by a number of features. First of all, meiosis is always internal, which is to say it always takes

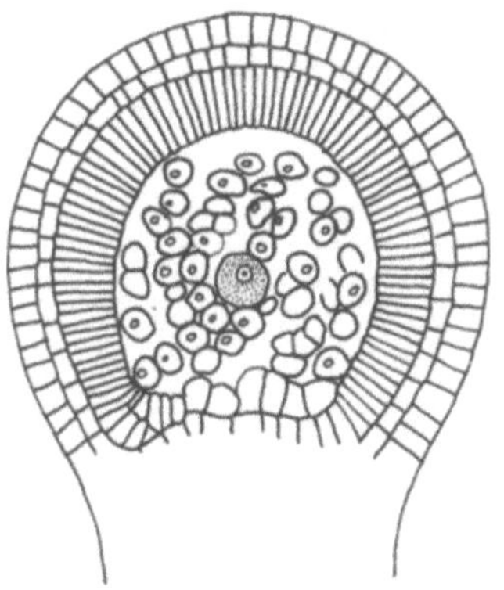

Fig. 1. *Selaginella erythropus* megasporangium (Goebel 1905), showing a highly structured sporangial wall and a single, centrally located viable megaspore amongst other residual megaspores in various stages of degeneration

place within the body of the bottle, rather than in the surface layers of cells which comprise the bottle wall. Secondly, when many cells are induced to undergo meiosis they begin the process almost simultaneously. Thirdly, in cases where only a single cell is induced, it is always precisely located at the geometric center of the sporangium; and lastly, at the time when the discrimination of the meiotic cells is still not evident, before any visible differences exist between the cells of the sporangium, there is always good mechanical coupling between the cells of the sporangial wall and those of the sporogenous tissue which it encloses. In other words, there is a continuity of walls from the cells of the sporangial wall into the center of the sporangium. (Bottle wall and contents are not yet differentiated and are mechanically coupled through a system of common walls.)

Considered as a stress-mechanical structure the sporangium can be interpreted as follows. The cells comprising the walls of the sporangium represent a potential source of growth stresses, which because of the fact that they are growing in a tier around the perimeter of the structure will generate compressive stress which propagates tangentially (circumferentially). Now, since the sporangium as a whole must be considered to be in stress-mechanical equilibrium, it follows that the interior of the structure must be subjected to a uniform radial tension which acts across the center of the sporangium. This amounts to a condition of "locked-in stress", a frequently useful, and sometimes dangerous condition which is familiar to engineers dealing with tempered or prestressed materials.

2.6.2 Isotropic Stress as a Developmental Effector

Plants have used the ability to generate and accumulate internal stresses as a way of creating a unique condition of isotropic (three-dimensionally focused) tension which can be applied to a cell or group of cells which is otherwise undistinguished except for its geometric location within the center of the structure (Fig. 2). These focused tensile forces can be used as the environmental signal which acts as the initial discriminant, triggering otherwise identical cells to enter a divergent developmental pathway which takes them toward meiosis.

Fig. 2. The sporangium as a stress-focusing device. Outer sporangial wall cells generate circumferential compression which is translated into tensile forces acting across the interior of the sporangium

This explains the lack of instances of superficial meiosis, which would be impossible with a mechanism such as this, which requires the spherically symmetrical loading which can be brought to bear on a deep-seated group of cells. It also explains the necessity for mechanical continuity through the structure, since without mechanical continuity the tensile stresses set up in response to the superficial compression could never be transmitted to the cells lying at the center of the structure.

The stalked sphere, or bottlelike shape of sporangia, clearly relates to the potential which this shape offers in terms of focusing stresses on a strictly defined group of cells. In cases where the bottle is considerably elongated, such as in the developing anther sacs of flowering plants, the number of cells which lie at the focus of the induced tensile forces may be considerable. In other cases, the number of cells at the focus may be restricted. In many cases only a single cell, precisely located at the center of the sporangial mass is triggered to undergo meiosis.

The surprising universality shown by sporangiate structures from all the groups of multicellular plants thus reflects the stringent mechanical conditions which must be met for them to fulfill the task of isolating and channeling a number of otherwise undistinguished cells into the meiotic developmental pathway. It becomes clear, in fact, that sporangia are not simply convenient receptacles for meiotic products; they are precisely structured stress-mechanical devices which use the forces released during growth to control the mechanical environment of a few cells, presenting them with the unique signals upon which the initiation of the meiotic process depends.

This developmental scenario, although only hypothetical at this point, is supported by a variety of circumstantial evidence, mostly relating to those consistent features of the sporangiation process which are conserved throughout the multicellular plants. It is also supported by the problems encountered in explaining the differentiation of single cells within a clonal cell mass by means of gene switching or hormone targeting; a difficult proposition in plants, where there is no permanent germ-line and no reliable way of focusing diffusing chemical morphogens on a single cell.

There is an economy and fitness to a stress-mechanical targeting system, which relies on the inherent architectural sensibilities of plant cells, and which can accurately identify and trigger cells solely on the basis of their location within a developing organ. Furthermore, the requirements placed on the genome are considerably reduced, since the cell itself becomes the active decision-making unit. The genome is no longer required to provide the details of developmental sequencing, but is only called upon to amplify events which have already been called into play at the cellular level.

Cytomechanics, the ability of the cytoplasm to transmit, generate, and respond to mechanical stimuli, lies at the heart of this interplay between the cell and its environment, and defines the cell as an architecturally reactive unit.

Any experimental approach to these developmental problems is complicated by technical difficulties, however. The ideal approach to these questions would be to find a nondestructive method of mapping the stresses in a living plant organ, allowing one to compare the in vivo stress conditions directly to the cell wall patterns. Even the largest growing apical meristems are smaller than the smallest implantable strain sensors, however, and photoelastic methods cannot be applied because the high intrinsic birefringence of the cellulosic walls swamps out any small photoelastic effect that might be present. This means that only indirect methods can be used to determine the actual state of stress in a living structure.

In my laboratory we have tried to develop an in vitro model which allows us to apply known mechanical loads to tissues growing in sterile culture, where

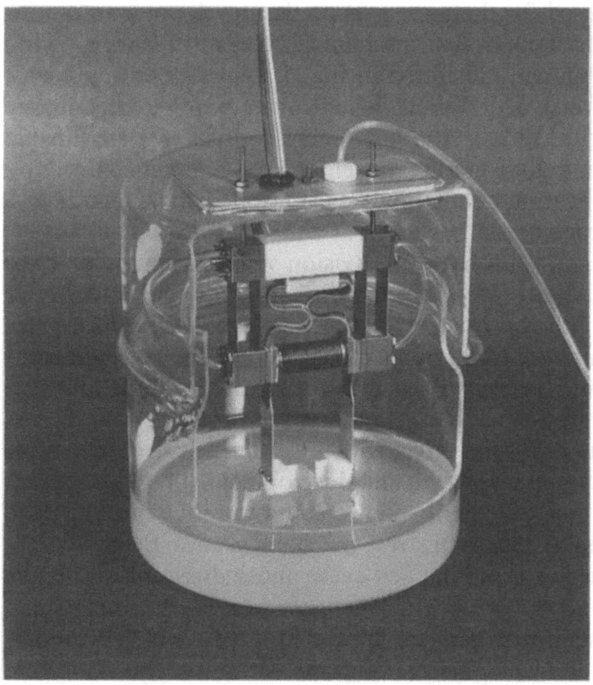

Fig. 3. Cut-away view of a culture system for applying controlled mechanical forces to tissues growing in sterile culture. Compressive forces generated by a centrally located metal bellows are transmitted to the tissue by strain-gauged arms hanging down into the culture medium. External feedback controls maintain the load at a constant level and provide for variable control of the applied load level

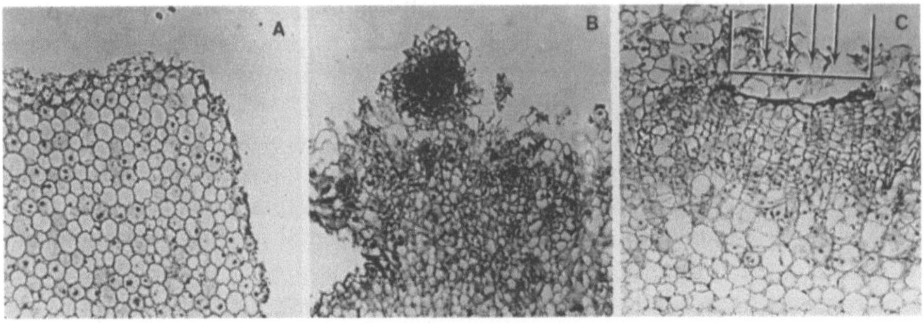

Fig. 4 A–C. Transverse sections of cultured and uncultured explants of *Nicotiana* pith, ×100. **A** The initial condition of the explant before transfer into sterile culture. **B** and **C** Aspects of an equivalent explant after 5 days in a standard defined medium. **B** Unrestrained callus proliferating from one corner of the explant; **C** another area of the same section, growing under continuous compressive stress applied through a narrow Teflon interface (*arrows*). Note radially aligned files of cells extending from the original pitch core to the interface surface

they are free from the various stress-mechanical unknowns which are at work in the intact plant, and where the exact nature of the mechanical stimuli acting on dividing cells can be predicted more easily.

Our work has been focused for the most part on the development of an instrument for the application of controlled mechanical forces to tissues growing in sterile culture. This instrument, illustrated in Fig. 3, allows us to apply continuously controlled compressive or tensile loads to any consolidated tissue which can be grown in sterile culture. The instrument consists of a sterilizable forcing frame, suspended in a glass culture vessel, and an external control system which permits the application of continuously controlled static load levels, or cyclic loads whose parameters can be varied at will. This instrument has allowed us to explore the relationship between division wall orientation and applied load in an unequivocal setting, where the directionality and intensity of stresses being transmitted into the tissue are known. We have shown, for instance, that in tobacco pith tissue freshly explanted into culture, constant compressive loads in the range of 75 KPa will suffice to reestablish parallel division in a tissue which would otherwise be dividing randomly (Fig. 4).

This result illustrates the profound effect which transmitted stress can have on the most elemental event in plant growth and development.

We hope to pursue this line of experimentation, and to refine our ability to control basic cellular process by means of stress-mechanical cues. Eventually we hope to use these techniques to bring the cell cycle under direct manipulative control without the need to resort to blocking agents and chemical intervention. Further research along these lines is very much needed if we are to fully understand the epigenetic aspect of developmental control.

References

Goebel K (1905) The organography of plants, part II. Oxford
Heywood RB (1969) Photoelasticity for designers. Pergamon, Oxford
Jaffe MJ (1977) Experimental separation of sensory and motor functions in pea tendrils. Science 195:191–192
Lintilhac PM (1984) Positional controls in meristem development. In: Barlow P, Carr DJ (eds) Positional controls in plant development
Lintilhac PM, Vesecky TB (1980) Mechanical stress and cell wall orientation in plants. I. Am J Bot 67:1477–1483
Lintilhac PM, Vesecky TB (1981) Mechanical stress and cell wall orientation in plants. II. Am J Bot 68:1222–1230
Quirk JT, Freese F (1976) Effect of mechanical stress on growth and anatomical structure of red pine: stem vibration. Can J For Res 6:375–381
Ray PM, Green PB, Cleland R (1973) Role of turgor in plant cell growth. Nature 239:163–164
Shen-Miller J, Hinchman RP (1974) Gravity sensing in plants: a critique of the statolith theory. BioScience 24:643–651
Wilson BF, Archer RR (1977) Reaction wood: induction and mechanical action. Ann Rev Plant Physiol 28:23–43

IV.3 Mechanical Properties of the Cyclamen Stalk and Their Structural Basis

K. Dierks[1] and L. Hafner[2]

At the end of the 19th century among others Schwendener (1874) and Haberlandt (1909) described anatomy of plants as a mechanical system. Emphasis was given to "specifically mechanical" cells or "stereids" which, in their entirety, form the "stereom", a cellular framework to which the stiffness and bearing capacity are ascribed. The contribution of turgor to rigidity of cells with thin cell walls was disregarded in these studies, as in the later work of Mergenthaler-Paulen (1952), who calculated the moment of inner forces being half that of external forces. Since these forces are commonly supposed to be in equilibrium, they should have been equal. G. Helmcke and F. Otto initiated the present study by drawing our attention to this unsatisfactory level of knowledge.

3.1 Anatomy of the Cyclamen Persicum Flower Stalk

Stiffness and bearing resistance of vascular plant tissue result mainly from internal hydrostatic pressure and for a small part only from stiffened cell walls. Regarding its mechanical properties four tissue areas of the *Cyclamen persicum* stalk are of significance (Figs. 1 and 2). The first area is the outermost layer of a cross-section, the epidermis with subepidermal collenchyma, the parenchyma of the cortex, the vascular cylinder (xylem and phloem), and the pith in the center. A starch-containing endodermis forms the boundary between cortex and vascular cylinder (Esau 1969). No sklerenchyma is found in the stalk. Longitudinal walls of epidermis and collenchyma are considerably thickened (up to 8 μm). This multilayer construction forms a tensile envelope being submitted to pressure from within. The only lignified elements are found in the xylem and are arranged radially. Due to their central location they contribute only minimally to the bending resistance of the stalk.

[1] Institut f. Tragkonstruktionen u. wirtschaftl. Fertigung der Techn. Universität Berlin, Straße d. 17. Juni 150–152, 1000 Berlin 12, FRG
[2] Inst. f. Biologie der Techn. Universität Berlin, 1000 Berlin 12, FRG

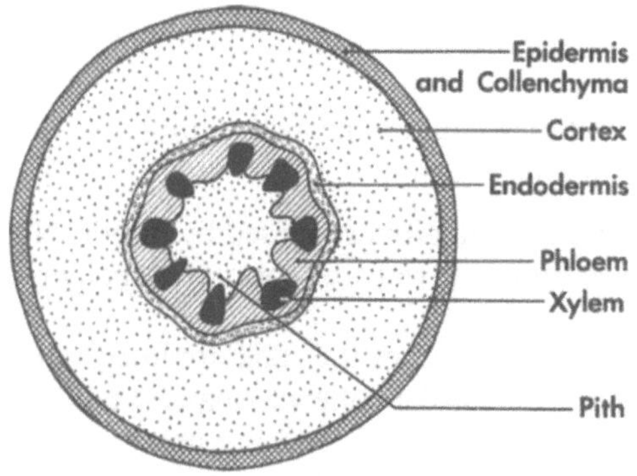

Fig. 1. Diagrammatic view of a transverse section of the stalk of *Cyclamen persicum*

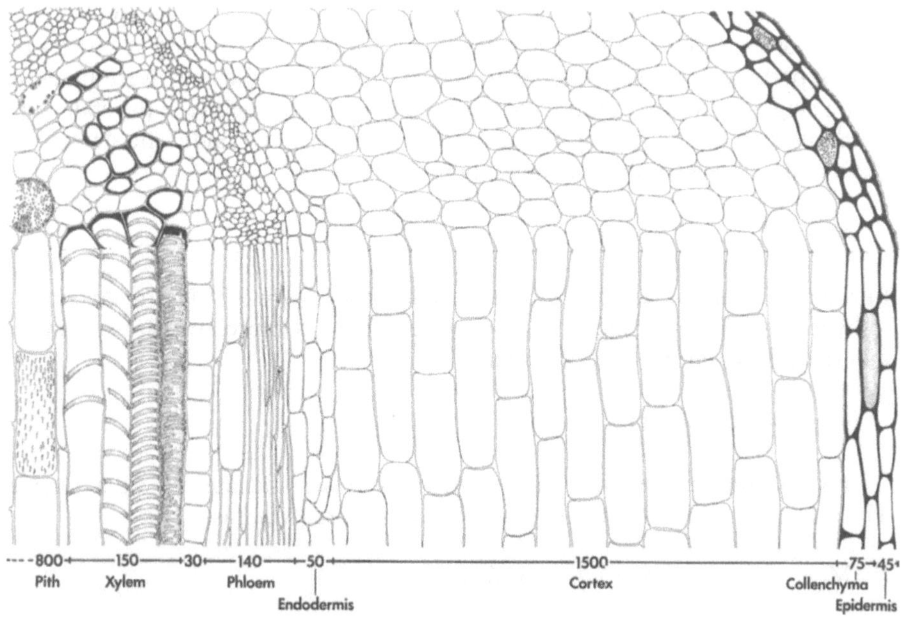

Fig. 2. Anatomy of the stalk (all measurements are in μm)

3.2 Internal Hydrostatic Pressure

The average turgor (p) of the stalk was determined, according to Moore (1974) by immersing specimens of tissue in solutions of varying osmolality, to be $p = 0.58\ N/mm^2$. The water potential of the cell sap of parenchymatous cortex cells was determined by incipient plasmolysis to be $p = 0.51\ N/mm^2$. The average value of $p = 0.55\ N/mm^2$ is used in the following calculations.

3.3 Behavior Under Ultimate Load[3]

Regarding anatomy either of two hypotheses could conceivably describe the state of inner forces of the stalk due to turgor.

A: The walls of all cells, with the exception of the pith are tensed by turgor. No tensile stress difference exists between the cell walls of cortex parenchyma and collenchyma epidermis (Fig. 4).

B: Components of turgor in the direction of the axis of the stalk are in equilibrium by tensile stresses mainly in cell walls of the collenchyma and epidermis. The cell walls of the cortex, xylem phloem, and pith are nearly stress-free (Fig. 5). The stalk behaves like a tube under internal pressure.

The hypotheses were tested by measuring the behavior of the stalk under external load.

Fresh samples of stalk, 7-cm-long, were taken to observe the behavior under the load of the *Cyclamen persicum* stalk. The samples were placed as girders on two supports with an effective span of 6 cm. In the middle they were made to bear a single-point load until they ruptured (Fig. 3). In each case the stalk ruptured midspan due to the top layers of stalk fiber buckling under the stress. The mean of the observed ultimate bending moment was

$$M_{UT} = 9.38 \pm 0.75 \text{ Nmm.}$$

Further calculations are based on a sample with $M_{UT} = 9.75$ Nmm. According to hypothesis A, rupturing is to be expected, if the compressive stress σ_c in the outermost layer equals that of turgor,

$$\sigma_c = p = 0.55 \text{ N/mm}^2.$$

In this case the tensile stress in the cell walls ceases and the cell walls buckle. The calculated ultimate bending moment at rupture (M_{UC}) is

$$M_{UC} = \sigma_c \cdot W = 0.55 \cdot 7.93 = 4.46 \text{ Nmm.}$$

Because the observed ultimate bending moment is more than twice this calculated value, hypothesis A does not describe adequately the behavior of the stalk.

According to hypothesis B, the rupturing is to be expected if the compressive stress σ_c due to the external load in the outermost layer equals that of the tensile stress σ_E resulting from turgor in the epidermis and the collenchyma.[4] Because of their softness, epidermis and collenchyma buckle under compressive

[3] An introduction into the terminology used in this chapter is given, e.g., by Mc Donagh (1984)

[4] The ring-shaped tensile stress vertical to the long axis of the stalk is of no relevance for this rupturing behavior

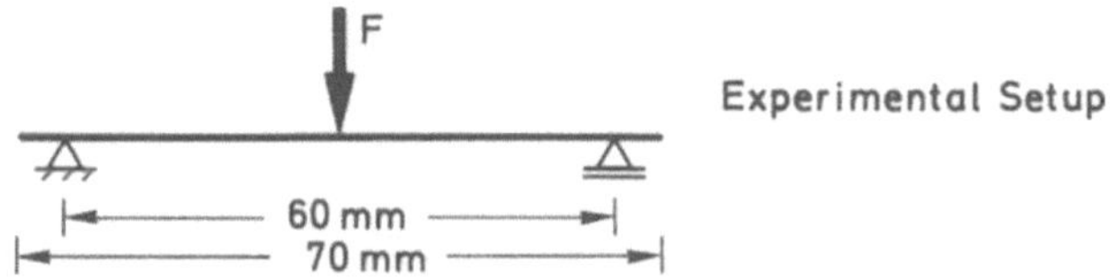

Fig. 3. Experimental setup for the determination of ultimate loads and ultimate bending moments

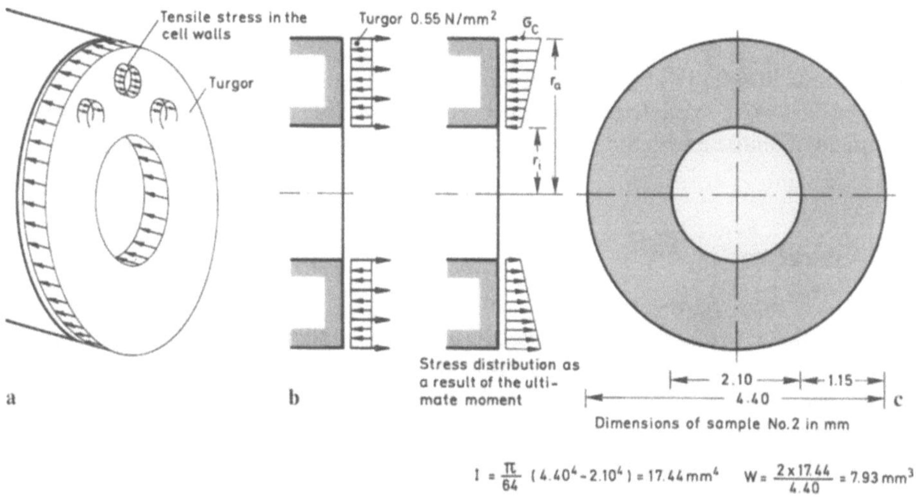

Fig. 4a–c. Stress distribution according to hypothesis A. **a** Cross-section, stresses due to turgor; **b** stress distribution due to turgor (*left*) and bending moment (*right*); **c** cross-section: *I* second moment of area (moment of inertia), *W* modulus of section

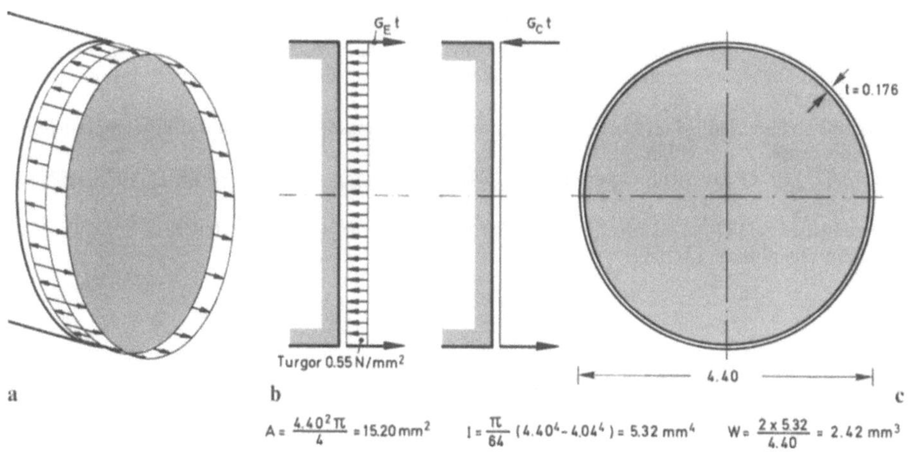

Fig. 5a–c. Stress distribution according to hypothesis B. **a** Cross-section, stresses due to turgor; **b** stress distribution due to turgor (*left*) and bending moment (*right*); **c** cross-section: *A* cross-sectional area under turgor (diameter equals arithmetic mean of stalk diameters at both ends), *I* second moment of area according to the area of epidermis and collenchyma, *W* modulus of section

stress. According to Fig. 5 the tensile stress in the epidermis and collenchyma is

$$\sigma_E = \frac{Ap}{2\pi r t} = \frac{15.2 \cdot 0.55}{4.40 \cdot 0.176 \cdot \pi} = 3.44 \text{ N/mm}^2 .$$

The calculated ultimate moment of rupture corresponding to σ_E is

$$M_{UC} = \sigma_E W = 3.44 \cdot 2.42 = 8.32 \text{ Nmm} .$$

This comes very close to the observed ultimate moment of rupture $M_{UT} = 9.38$ Nmm. The 17% difference lies within the range of the measurement in accuracy to be expected. Hypothesis B seems to be suitable for describing the firmness of shape and bending resistance of the *Cyclamen persicum* stalk.

3.4 Summary

Bending failure tests on fresh *Cyclamen persicum* stalks indicated a high degree of coincidence with the hypothesis that the firmness of shape and the bending resistance of the *Cyclamen persicum* is due to a great extent to turgor. A skin consisting of epidermis and circular collenchyma surrounds the stalk like a tube and is under tensile stress lengthwise and circularly.

References

Esau K (1969) Pflanzenanatomie. Gustav Fischer, Stuttgart

Haberlandt G (1909) Physiologische Pflanzenanatomie, 4. Aufl. Wilhelm Engelmann, Leipzig

McDonagh (1984) Mechanical science for technicians, vol 1 and vol 2. Edward Arnold, p 160 and p 232

Mergenthaler-Paulen M (1952) Die Spannungsverhältnisse in einigen zylindrischen Pflanzenorganen. Z Bot 40:363–384

Moore TC (1974) Research experiences in plant physiology. Springer, Berlin Heidelberg New York

Schwendener S (1874) Das mechanische Prinzip im anatomischen Bau der Monocotylen. Wilhelm Engelmann, Leipzig

V. Mechanical Forces Determining the Shape of Metazoan Cells

V.1 Forces Shaping an Erythrocyte

RICHARD E. WAUGH[1] and ROBERT M. HOCHMUTH[2]

1.1 Introduction

Of all cells in living systems the mammalian erythrocyte is structurally the simplest. It consists of an elastic membrane and a liquid interior that is principally a Newtonian solution of hemoglobin with a viscosity of approximately 0.6 centipoise at a concentration of 0.31 kg/l (Cokelet and Meiselman 1968). Because the interior is liquid, all of the elastic character of the cell resides in the membrane. Therefore, the shape of the cell comes from the natural shape of the membrane and the elastic response of the cell is determined by the elasticity of the membrane. Because of its simple structure and the dominance of the membrane in determining the cellular response to external forces, the erythrocyte has served as a model system for the study and characterization of the mechanical properties of biological membranes (Evans and Skalak 1979).

The molecular organization of the membrane (Fig. 1) has also been well characterized (see Gratzer 1983 for a recent review). The outer surface consists

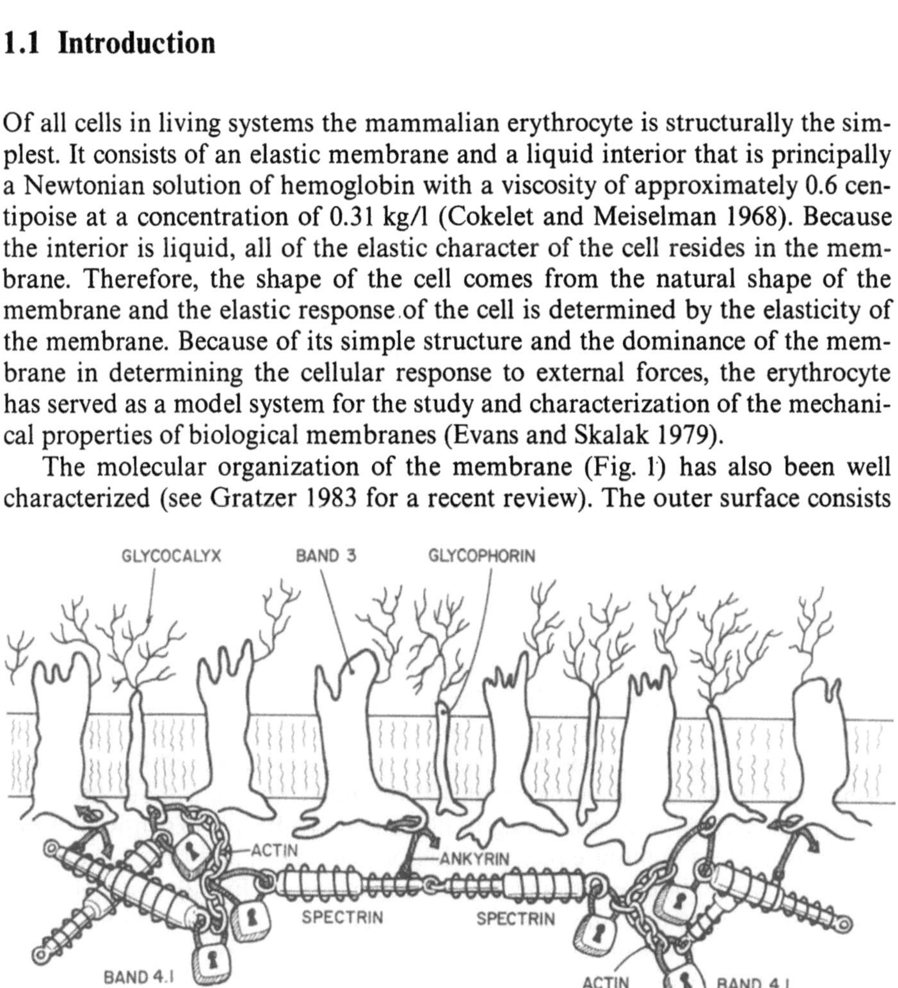

Fig. 1. Schematic organization of the erythrocyte membrane showing major protein constituents and putative organization. The illustration emphasizes the viscoelastic character of the membrane skeleton

---

[1] Department of Biophysics, University of Rochester, School of Medicine and Dentistry, Rochester, New York 14642, USA

[2] Department of Mechanical Engineering and Materials Science, Duke University, Durham, North Carolina 27706, USA

of a phospholipid bilayer containing integral membrane proteins (e.g., band 3 and glycophorin). At the inner surface of the bilayer there is a protein network, the membrane skeleton, consisting of spectrin in association with short filaments of actin. The protein, band 4.1, stabilizes the spectrin-actin interaction and also binds to the integral protein glycophorin. The protein ankyrin binds to both spectrin and the integral protein band 3, linking the membrane skeleton to the bilayer.

Because biomembranes, in general, and the red cell membrane, in particular, are only a few molecules thick, they form a continuum only in the plane of the membrane. Thus, the methods of classical three-dimensional continuum mechanics must be compressed into a two-dimensional world in which the forces within the material are expressed as "force resultants" or "tensions" (Evans 1973a; Skalak et al. 1973). Measurement of the surface force resultants and the corresponding surface deformations permits the intrinsic material properties of the membrane surface to be calculated. These surface properties are a summation of the mechanical contributions of the lamellar, molecular structures that form the membrane (Evans and Hochmuth 1977).

The elastic and viscous properties of the red cell membrane are determined by using a light microscope to observe individual red cells as they are manipulated with a small hollow glass tube (micropipette). The changes in material properties that occur under different environmental conditions and as a result of alterations in molecular structure provide important information about the molecular basis for the mechanical behavior of the membrane.

1.2 Membrane Elasticity

Although cell and membrane deformation can be quite complex in general, for a two-dimensional, incompressible, elastic material like the red cell membrane, any deformation can be expressed in terms of three fundamental, independent deformations (Evans and Hochmuth 1978; Evans and Skalak 1979). These deformations are (1) an elongation or "shear" of the membrane without either increasing the surface area or bending it, (2) dilation (isotropic expansion) of the membrane surface without either shearing or bending it, and (3) a bending of the membrane without either shearing or expanding it (Fig. 2). Each of these fundamental elastic deformations is characterized by an elastic modulus: (1) a shear modulus μ with units of N/m, (2) an area expansion modulus K with units of N/m, and (3) a bending modulus B with units of N m. In general, a larger modulus indicates a greater resistance to that particular form of deformation.

1.2.1 Shear Elasticity

For the constant area extension of a two-dimensional material, the maximum shear resultant T_s occurs along a line at 45° to the direction of extension. Shear

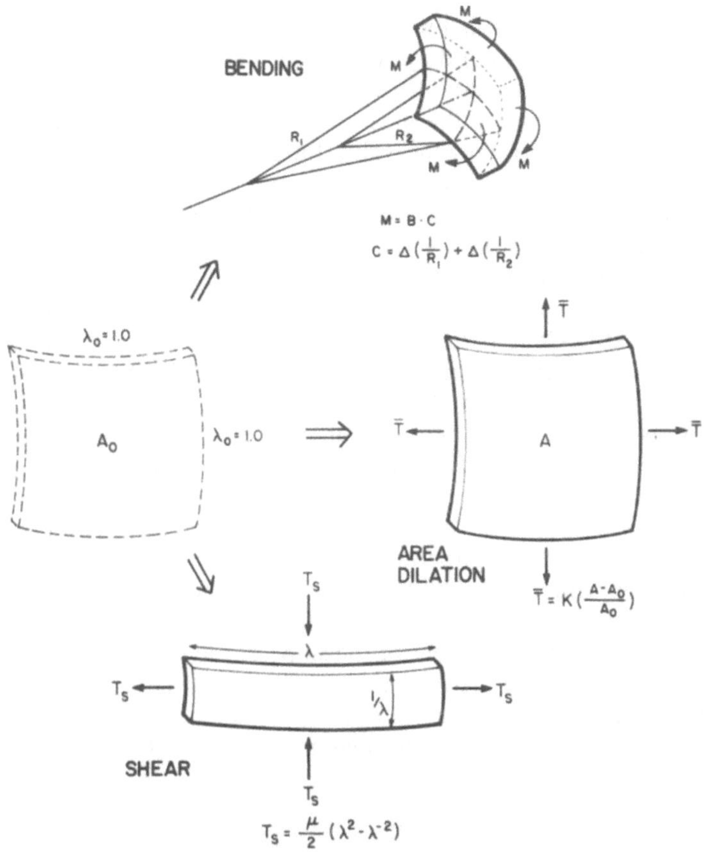

Fig. 2. Illustration of the three modes of deformation for a membrane surface. The membrane moment resultant is related to the change in membrane curvature via the bending stiffness B. The isotropic membrane force resultant is related to the fractional change in membrane area via the membrane expansivity modulus K. The extension (shear deformation) of the surface is related to the maximum shear resultant via the membrane shear modulus μ

and extension represent the same phenomenon and are characterized by a single elastic constant: the shear modulus μ. The relation between the shear resultant T_s and the extension ratio λ (extended length/original length) is (Evans 1973a; Evans and Hochmuth 1977):

$$T_s = (\mu/2)\,(\lambda^2 - \lambda^{-2}). \tag{1}$$

When the portion of the membrane at the dimple of a flaccid cell is aspirated into a pipette of radius R_p, the relation between the aspiration pressure P (relative to the pressure outside the cell) and the extension up the pipette L is (Chien et al. 1978; Waugh and Evans 1979):

$$P\,R_p/\mu = (2\,L/R_p) - 1 + \ln\,(2\,L/R_p). \tag{2}$$

From Eq. (2) it is readily seen that μ is proportional to the slope of the P vs L line. This slope can be measured accurately. The major uncertainty in the determination of a value for μ is the uncertainty in the measurement of a value for the pipette radius R_p. A 20% error in the measurement of R_p (e.g., 0.2 µm of 1.0 µm) will give about a 40% error in the value for μ. Based upon a single datum point for two different human red cells at room temperature, Evans (1973b) first obtained a value for μ of 5×10^{-3} mN/m. This value agrees with the value for μ obtained from a study of the shear deformation of point-attached red cells (Hochmuth and Mohandas 1972). Since the original measurements, many investigators have obtained values for μ ranging from about 4×10^{-3} mN/m (Evans and LaCelle 1975; Chien et al. 1978) to $6-7 \times 10^{-3}$ mN/m (Waugh and Evans 1979; Linderkamp and Meiselman 1982; Smith and Hochmuth 1982; Nash and Meiselman 1985). Recently, however, Evans et al. (1984) accurately measured the inner diameter of their pipettes from the insertion depth of a tapered microneedle that had been calibrated with a scanning electron microscope. Their value is $\mu = 9 \pm 1.7 \times 10^{-3}$ mN/m. Because this value is based on an accurate measurement for R_p, it must be given greater weight. Thus, for human red cells at room temperature, $\mu = 6-9 \times 10^{-3}$ mN/m. For a given pipette and a given sample of cells from the same donor, the standard deviation in the measurement of μ is about ±20%. This is caused either by cell-cell variability or by errors in measurement of the slope of the P vs L line. Experimental data indicate that μ, as defined by Eq. (1), is constant, although Fischer et al. (1981) suggested that the value for μ could be several times smaller at very small membrane strains. Such small strains cannot be measured with present techniques.

1.2.1.1 Molecular Basis of Shear Elasticity

It has long been thought that the shear elasticity of the membrane is due to the presence of the membrane skeleton at the inner surface of the bilayer (Evans and Hochmuth 1977; Fischer et al. 1978). This conclusion is based on the molecular organization within the bilayer and observations of protein mobility in the membrane surface (Frye and Edidin 1970; Singer and Nicholson 1972), indicating that the bilayer portion of the membrane behaves like a two-dimensional liquid. Recent measurements of shear modulus of membranes having inherited molecular disorders (Waugh 1987) indicates that, more specifically, the shear elasticity is directly proportional to the surface density of the skeletal component spectrin. This observation contrasts with observations on membranes having abnormalities associated with ankyrin or band 4.1, in which there was little or no correlation between the reduction in shear modulus and the extent of the molecular abnormality (Waugh 1987). These observations confirm previous hypotheses that spectrin is the molecule primarily responsible for elastic energy storage during membrane deformation (Waugh and Evans 1979; Stokke 1984).

The temperature dependence of the shear modulus gives an indication of the mechanism by which the spectrin stores energy during deformation. The fact that the shear modulus decreases with increasing temperature indicates that the energy storage mechanism is not a simple entropic one as is the case for most rubberlike materials (Waugh and Evans 1979). Energy storage by changes in the internal energy of the molecules (as opposed to entropy) is likely to depend on the details of the spectrin conformation. This view is consistent with the observation that heating red cells to temperatures above 47 °C increases the shear modulus irreversibly by a factor of two or more (Rakow and Hochmuth 1975; Nash and Meiselman 1985). These temperatures correspond to temperatures at which purified spectrin in solution undergoes a conformational change detectable by scanning calorimetry and circular dichroism (Brandts et al. 1977). Finally, Chabanel et al. (1985) have found an increase in shear modulus for patients having an abnormality in spectrin that reduces its self-association. This increased modulus could be due to conformational differences in the abnormal spectrin analogous to those induced by high temperature.

1.2.1.2 Metabolic, pH, and Ionic Effects

The shear modulus is also affected by environmental conditions such as ionic strength, metabolic levels, and pH. Crandall et al. (1978) observed a dramatic increase in μ below pH 6.0. The modulus increased 100-fold at pH 5.0 over a period of 6 h. Metabolic depletion of cells does not affect the shear modulus except in the presence of calcium, which causes a 15% increase in the modulus (Meiselman et al. 1978; Baker 1981). Elevated levels of intracellular 2,3-diphosphoglycerate (2,3-DPG) up to 20 mM have no effect on the shear modulus except at slightly reduced salt concentration. At 20 mM 2,3-DPG and 240 mOsmol saline, μ is reduced by about 30% (Waugh 1986). A thorough study of the effects of ionic strength on material properties has not been done.

1.2.2 Area Elasticity

An isotropic dilation of the red cell membrane is produced by aspirating a preswollen cell into a micropipette until the portion of the cell outside the pipette forms a sphere. The isotropic tension T is given by (Evans et al. 1976):

$$T = P \cdot R_p/[2 \cdot (1 - R_p/R_c)], \qquad (3)$$

where P is the aspiration pressure, R_p is the pipette radius, and R_c is the cell radius. As the aspiration pressure is increased, the area of the cell expands because of an increase in the isotropic tension in the membrane. The expansion of the surface area is measured accurately by measuring the movement up the pipette of the membrane "tongue" within the pipette (Evans et al. 1976; Evans and Waugh 1977). Extremely small area dilations can be measured in this way.

Evans et al. (1976) proposed a simple linear relation between the isotropic tension in the membrane and the relative area expansion:

$$T = K \cdot \Delta A/A_o, \qquad (4)$$

where ΔA is the increase in surface area, A_o is the original area, and K is the area expansion modulus. The value for K at room temperature is K = 450 mN/m with a standard deviation of about ±15–20% (Evans and Waugh 1977; Waugh and Evans 1979).

1.2.2.1 Molecular Basis of Area Elasticity

Unlike the resistance to shear deformation, the resistance to area dilation may have contributions from both the membrane skeleton and the bilayer. Kwok and Evans (1981) found a value for K at room temperature of approximately 150 mN/m for a bilayer vesicle made from egg lecithin. More recently, Needham and Evans (1986) have found that the modulus of the phospholipid bilayer membrane increases substantially with addition of cholesterol up to a value of approximately 800 mN/m at a 1:1 mol ratio. Based on these measurements on cholesterol-containing bilayers, the bilayer of the red cell could account for much, if not all, of the membrane resistance to area change. Evidence for a contribution from the membrane skeleton to the area modulus comes from unpublished observations of Smith and Hochmuth, who found that cells deficient in spectrin had an area modulus approximately half as large as normal. Observations of the temperature dependence of the area modulus led Evans and Waugh (1980) to hypothesize that the skeleton accounted for about one-third of the isotropic rigidity at room temperature, and that the skeletal contribution became small as the temperature approached 49 °C, where spectrin denatures. Observations by Evans and Kwok (1982) that the area compressibility of purely lipid membrane (dimyristoylphosphatidylcholine) above the phase transition is not highly temperature-dependent, is consistent with the view that the temperature dependence of the compressibility of red cell membrane is due to the presence of the skeleton. However, recent observations by Evans and Needham (1986) indicate that the temperature dependence of the membrane bilayer compressibility is more complicated than originally thought. Addition of cholesterol to a pure lipid bilayer causes a marked increase in the area modulus, but further addition of protein reduces the modulus. This is thought to be the result of a repartitioning of the cholesterol from lipid to protein regions in the membrane surface. If the partitioning of cholesterol is temperature-dependent (as is likely) then changes in temperature could produce significant changes in the compressibility of membrane that contains lipid, protein, and cholesterol, such as the erythrocyte membrane. The interpretation of experiments on the area modulus is further complicated by observations by Katnik and Waugh (1986) that the area modulus appears to depend on the magnitude of the electric field applied across the cell and by the presence of metal ions (Zn^{2+} and Cu^{2+}) in solution. Thus, full understanding of the molecular basis for the resistance of the red cell membrane to changes in surface area must await further experiments.

1.2.3 Bending Elasticity

Although small, the membrane's resistance to bending stabilizes the biconcave shape of the cell at rest, during the early phases of cell swelling (Zarda et al. 1977; Fischer et al. 1981), and during pipette aspiration (Evans 1980). The resistance to bending is characterized by the membrane bending stiffness, B, which relates the moment resultant (bending moment per unit length) M to the change in curvature of the membrane (Helfrich 1973; Evans 1974):

$$M = B(C_1 + C_2 - C_0), \tag{5}$$

where C_1 and C_2 are the principal curvatures and C_0 is the curvature of the stress-free state. Several methods have been used to estimate the value of B for the red cell. Based on the behavior of the cell during swelling Zarda et al. (1977) and Fischer et al. (1981) estimated a value of approximately 10^{-16} mNm (10^{-12} dyn cm) for a value of 6.0×10^{-3} mN/m for the shear modulus. Based on the behavior of the cell in micropipette aspiration experiments, Evans (1980) calculated B to be in the range 10^{-15} to 10^{-17} mNm. Most recently, Evans (1983) developed a method for calculating B based on the appearance of creases in the cell surface as it is aspirated into a micropipette. The magnitude of the aspiration pressure at which creasing occurs depends directly on the value for B and inversely on the cube of the pipette radius. Analysis and measurements indicate that $B = 1.8 \pm 0.2 \times 10^{-16}$ mNm (Evans 1983; Evans et al. 1984).

1.2.3.1 Molecular Basis of Bending Rigidity

The bending resistance of the membrane arises from the differential expansion and compression of the surface area of the molecular layers of the membrane as it changes curvature. This concept was introduced qualitatively by Sheetz and Singer (1974) and formulated quantitatively by Evans (1974) and Evans and Skalak (1979). Thus, the underlying molecular basis for the bending stiffness is similar to the molecular basis for area elasticity, except that the relative areas of the different surface layers and the associations between the layers play an important role.

Because techniques for measuring bending stiffness have been developed recently, there is little information about what factors affect the value of B. Evans et al. (1984) found no effect on B of increasing intracellular hemoglobin concentrations for normal or sickle cells, except at the highest hemoglobin concentrations for sickle cells, for which a 50% increase in B was reported. Nash and Meiselman (1985) reported little or no effect (<20%) of heat treatment on the value of B. In both of those experiments, the change in B was much less than the change in μ under similar conditions. In contrast, Waugh (1987) has found reductions in the bending stiffnesses of hereditary spherocytes that closely parallel both the reduction in shear modulus and the reduction in spectrin density for the abnormal membranes. Taken together, these observations suggest that the membrane skeleton makes an important contribution to the bending stiffness, but that the mechanism of its contribution is different from its contribution to shear rigidity (μ).

No studies have yet been performed to test the relationship between changes in area modulus and changes in bending stiffness. The results of such studies would certainly be interesting and would provide important information about the relative contributions of the skeleton and the bilayer to the bending rigidity and the importance of bilayer-skeletal interactions in determining the curvature elasticity of the membrane.

1.3 Membrane Viscosity

The elasticity of the membrane characterizes its resistance to deformation; the viscosity of the membrane characterizes its resistance to a rate of deformation. The rate of deformation is one of shear. Therefore, a shear viscosity is measured. When the membrane of a flaccid red cell is elongated by micropipette aspiration or fluid shear deformation and then the force of deformation is suddenly removed, the cell rapidly recovers its undeformed shape. For this recovery process, Evans and Hochmuth (1976) calculated that the viscous dissipation in the membrane is about two orders of magnitude greater than that in the internal hemoglobin solution. Thus, the membrane is the dominant source of viscous dissipation during the recovery process.

Evans and Hochmuth (1976) accounted for the dissipation in a "viscoelastic" solid membrane by the addition to Eq. (1) of a "Newton's law of viscosity" term:

$$T_s = (\mu/2)(\lambda^2 - \lambda^{-2}) + 2\,\eta\,\partial \ln\lambda/\partial t, \tag{6}$$

where t is time and η is the coefficient of surface viscosity. A characteristic time constant t_c is obtained in Eq.(6) simply dividing η by μ:

$$t_c = \eta/\mu. \tag{7}$$

The time required for the membrane of a red cell to recover its stress-free state when the force of deformation is removed is given by the integration of Eq. (6) with $T_s = 0$ (Evans and Hochmuth 1976; Hochmuth et al. 1979). A best fit of this integrated equation to the experimental data gives a value for t_c (Hochmuth et al. 1979). Typically, $t_c = 0.1$ s and, thus, if $\mu = 10^{-2}$ mN/m, $\eta = \mu\, t_c = 10^{-3}$ mN s/m.

Rapid deformation of a red cell produced either by rapid aspiration of the membrane into a pipette (Chien et al. 1978) or by "tank treading" the membrane in a fluid shear field (Tran-Son-Tay et al. 1984) indicates that "shear thinning" of the membrane may occur at high rates of deformation. Values for t_c and, presumably, η (if μ is a constant) may decrease by as much as a factor of six from their values at a very low rate of deformation (Chien et al. 1978).

1.3.1 Molecular Basis of Membrane Viscosity

The membrane skeleton makes the dominant contribution to the viscous behavior of the membrane. The viscosity of the bilayer is at least two orders of magnitude less than the viscosity of the membrane composite, as indicated by lateral mobility of integral proteins in the presence and absence of the skeleton (Koppel et al. 1981) and direct measurements of surface viscosity on reconstituted, multilamellar phospholipid vesicles (Waugh 1982). Also, the membrane viscosity is unaffected by enrichment or depletion of cholesterol (Chabanel et al. 1983). In contrast, inherited abnormalities in the molecular structure of the skeleton consistently result in reductions in membrane viscosity (Waugh 1987). However, the magnitude of the reduction in viscosity does not correlate well with the degree of the molecular abnormality, indicating that the relationship between molecular structure and viscosity is more complicated than the relationship between molecular structure and shear elasticity.

There are other indications of a more complex molecular basis for viscosity than for elasticity. Unlike the membrane shear modulus, the membrane viscosity is strongly influenced by intracellular hemoglobin concentration (MCHC) (Evans et al. 1984; Nash et al. 1984). This dependence probably accounts for most of the 50% increase in membrane viscosity of dense (old) compared to the lightest (young) cells in the circulation (Nash and Wyard 1981; Linderkamp and Meiselman 1982). Evans et al. (1984) have shown that increased membrane viscosity in both normal and sickle cells is directly related to the MCHC and that osmotic swelling of the densest cells causes the membrane viscosity to return to near "normal" values. The effect of MCHC on membrane viscosity is probably due to the association of hemoglobin with the cytoplasmic surface of the membrane (Eisinger et al. 1982; Chetrite and Cassoly 1985).

Temperature also has a strong influence on membrane viscosity. For example, the value for t_c decreases from 0.273 at 6 °C to 0.063 at 37 °C (Hochmuth et al. 1980). With Eq. (7) and the data of Waugh and Evans (1979), these values for t_c are converted to values for η. At 6 °C, $\eta = 2.1 \times 10^{-3}$ mN s/m and at 37 °C, $\eta = 0.36 \times 10^{-3}$ mN s/m.

Finally, although low levels of ATP do not affect the viscosity (Meiselman et al. 1978), high levels of 2,3-DPG affect the viscosity in two ways (Waugh 1987). At physiological osmolarity (295 mOsmol) the increased concentration of 2,3-DPG causes an increase in MCHC that results in an increased viscosity. At slightly reduced osmolarity (240 mOsmol) and high ($\sim$20 mM) 2,3-DPG concentrations the viscosity is reduced by about 30%.

1.4 Erythrocyte Shape

At rest in an appropriate buffer under the microscope red blood cells assume the shape of a biconcave disk. The mechanism by which the cells acquire this resting shape is not well understood. Canham (1970) argued that the biconcave shape minimizes the surface curvature. Lange et al. (1982a, b) and Mohandas et al. (1983) have argued that it is the bilayer that is the primary determinant of the resting shape of the cell and that the skeleton simply acts to stabilize the biconcave geometry. Based on observations of shape recovery after the application of crenating agents, Alhaney and Sheetz (1981, 1984) have argued that there is an active, metabolically-driven, shape-regulation mechanism within the cell, although the precise nature of the mechanism remains obscure. Ferrell and Huestis (1984) and Quist and Powell (1985) have implicated the formation and breakdown of polyphosphoinositides as part of the shape-regulation process.

The general principle that appears to govern all shape transformations was put forth qualitatively by Sheetz and Singer (1974) as the "bilayer couple hypothesis" and expressed quantitatively by Evans (1974), Deuling and Helfrich (1976), and Evans and Skalak (1979) in terms of membrane elasticity and "chemically-induced bending moments." Briefly, the cell at rest assumes whatever geometry it assumes because that geometry represents an energy minimum. The shape of the cell is determined by the natural curvature of the membrane, its bending stiffness, its shear rigidity, and the constraints of constant surface area and constant volume. The natural curvature of the cell can be affected by changing either the number of molecules or the area per molecule of one molecular layer within the membrane relative to the others. Thus, changes in the chemical environment of the cell can produce moment resultants in the membrane that are resisted by elastic force resultants in the surface (Evans 1980). The elastic properties of the membrane (and the membrane viscosity in dynamic processes) are the "forces" that determine what shape the erythrocyte will take in response to chemically or mechanically imposed stresses.

References

Alhaney E, Sheetz MP (1981) Control of the erythrocyte membrane shape: recovery from the effect of crenating agents. J Cell Biol 91:884–888

Alhaney E, Sheetz MP (1984) Cell membrane shape control – effects of chloromethyl ketone peptides. Blood 63:1203–1208

Baker RF (1981) Membrane deformability of metabolically depleted human red cells. Blood Cells 7:551–558

Brandts JF, Erikson L, Lysko K, Schwartz AT, Taverna RD (1977) Calorimetric studies of the structural transitions of the human erythrocyte membrane. The involvement of spectrin in the A transition. Biochemistry 16:3450–3454

Canham PB (1970) The minimum energy of bending as a possible explanation of the biconcave shape of the human red blood cell. J Theor Biol 26:61–81

Chabanel A, Flamm M, Sung KLP, Lee MM, Schachter D, Chien S (1983) Influence of cholesterol content on red cell membrane viscoelasticity and fluidity. Biophys J 44:171–176

Chabanel A, Sung KLP, Rapiejko J, Prchal JT, Liu SC, Palek J, Chien S (1985) Effect of altered spectrin dimer self-association on erythrocyte membrane viscoelasticity. Blood 66:29a.

Chetrite G, Cassoly R (1985) Affinity of hemoglobin for the cytoplasmic fragment of human erythrocyte membrane, vol 3. J Mol Biol 185:639–644

Chien S, Sung KLP, Skalak R, Usami S, Tözeren A (1978) Theoretical and experimental studies on viscoelastic properties of erythrocyte membrane. Biophys J 24:463–487

Cokelet GR, Meiselman HJ (1968) Rheological comparison of hemoglobin solutions and erythrocyte suspensions. Science 162:275–277

Crandall ED, Critz AM, Osher AS, Keljo DJ, Forster RE (1978) Influence of pH on elastic deformability of the human erythrocyte membrane. Am J Physiol 235:C269 – C278
Deuling HJ, Helfrich W (1976) Red blood cell shapes as explained on the basis of curvature elasticity. Biophys J 16:861 – 868
Eisinger J, Flores J, Salhany JM (1982) Association of cytosol hemoglobin with the membrane in intact erythrocytes. Proc Natl Acad Sci USA 79:408 – 412
Evans EA (1973a) A new material concept for the red cell membrane. Biophys J 13:926 – 940
Evans EA (1973b) New membrane concept applied to the analysis of fluid shear- and micropipette-deformed red blood cell. Biophys J 13:941 – 954
Evans EA (1974) Bending resistance and chemically induced moments in membrane bilayers. Biophys J 14:923 – 931
Evans EA (1980) Minimum energy analysis of membrane deformation applied to pipet aspiration and surface adhesion of red blood cells. Biophys J 30:265 – 284
Evans EA (1983) Bending elastic modulus of red blood cell membrane derived from buckling instability in micropipet aspiration tests. Biophys J 43:27 – 30
Evans EA, Hochmuth RM (1976) Membrane viscoelasticity. Biophys J 16:1 – 11
Evans EA, Hochmuth RM (1977) A solid-liquid composite model of the red cell membrane. J Membr Biol 30:351 – 362
Evans EA, Hochmuth RM (1978) Mechanochemical properties of membranes. In: Bronner F, Kleinzeller A (eds) Current topics in membranes and transport, vol 10. Academic Press, London, pp 1 – 64
Evans EA, Kwok R (1982) Mechanical calorimetry of large dimyristoylphosphatidylcholine vesicles in the phase transition region. Biochemistry 21:4874 – 4879
Evans EA, LaCelle PL (1975) Intrinsic material properties of the erythrocyte membrane indicated by mechanical analysis of deformation. Blood 45:29 – 43
Evans E, Needham D (1986) Giant vesicle bilayers composed of mixtures of lipids, cholesterol and polypeptiedes: thermomechanical and (mutual) adherence properties. Faraday Discuss Chem Soc 81:267 – 280
Evans EA, Skalak R (1979) Mechanics and thermodynamics of biomembranes. CRC Crit Rev Bioeng 3:181 – 418
Evans EA, Waugh RE (1977) Osmotic correction to elastic area compressibility measurements on red cell membrane. Biophys J 20:307 – 313
Evans EA, Waugh RE (1980) Mechanochemical study of red cell membrane structure *in situ*. In: Cokelet GR, Meiselman HJ, Brooks DE (eds) Erythrocyte mechanics and blood flow. Kroc Foundation Ser, vol 13. Alan R Liss, New York
Evans EA, Waugh R, Melink L (1976) Elastic area compressibility modulus of red cell membrane. Biophys J 16:585 – 595
Evans E, Mohandas N, Leung A (1984) Static and dynamic rigidities of normal and sickle erythrocytes. J Clin Invest 73:477 – 488
Ferrel JE, Huestis WH (1984) Phosphoinositide metabolism and the morphology of human erythrocytes. J Cell Biol 98:1992 – 1998
Fischer TM, Haest CWM, Stöhr M, Kamp D, Deuticke B (1978) Selective alteration of erythrocyte deformability by SH-reagents: evidence for an involvement of spectrin in membrane shear elasticity. Biochim Biophys Acta 510:270 – 282
Fischer TM, Haest CWM, Stöhr-Liesen M, Schmid-Schönbien H, Skalak R (1981) The stress-free shape of the red blood cell membrane. Biophys J 34:409 – 422
Frye LD, Edidin M (1970) The rapid intermixing of cell surface antigens after formation of mouse-human heterokaryons. J Cell Sci 7:319 – 335
Gratzer WB (1983) The cytoskeleton of the red blood cell. In: Stracher A (ed) Muscle and non-muscle motility, vol 2. Academic Press, London
Helfrich W (1973) Elastic properties of lipid bilayers: theory and possible experiments. Z Naturforsch 28C:693 – 703
Hochmuth RM, Mohandas N (1972) Uniaxial loading of the red cell membrane. J Biomech 5:501 – 509
Hochmuth RM, Worthy PR, Evans EA (1979) Red cell extensional recovery and the determination of membrane viscosity. Biophys J 26;101 – 114
Hochmuth RM, Buxbaum KL, Evans EA (1980) Temperature dependence of the viscoelastic recovery of red cell membrane. Biophys J 29:177 – 182

Katnik C, Waugh RE (1986) Reduction of the apparent area compressibility modulus of red blood cell membrane by applied electric fields. Biophys J 49:147a
Koppel DE, Sheetz MP, Schindler M (1981) Matrix control of protein diffusion in biological membranes. Proc Natl Acad Sci USA 78:3576–3580
Kwok R, Evans E (1981) Thermoelasticity of large lecithin bilayer vesicles. Biophys J 35:637–652
Lange Y, Gough A, Steck TL (1982a) Role of the bilayer in the shape of the isolated erythrocyte membrane. J Membr Biol 69:113–123
Lange Y, Hadesman RA, Steck TL (1982b) Role of the reticulum in the stability and shape of the isolated human erythrocyte membrane. J Cell Biol 92:714–721
Linderkamp O, Meiselman HJ (1982) Geometric, osmotic, and membrane mechanical properties of density-separated human red cells. Blood 59:1121–1127
Meiselman HJ, Evans EA, Hochmuth RM (1978) Membrane mechanical properties of ATP-depleted human erythrocytes. Blood 52:499–504
Mohandas N, Chasis JA, Shohet SB (1983) The influence of membrane skeleton on red cell deformability, membrane material properties and shape. Semin Hematol 20:225–242
Nash GB, Meiselman HJ (1985) Alteration of red cell membrane viscoelasticity by heat treatment: effect on cell deformability and suspension viscosity. Biorheology 22:73–84
Nash GB, Wyard SJ (1981) Erythrocyte membrane elasticity during *in vivo* aging. Biochim Biophys Acta 643:269–275
Nash GB, Johnson CS, Meiselman HJ (1984) Mechanical properties of oxygenated red blood cells in sickle cell (HbSS) disease. Blood 63:73–82
Needham D, Evans E (1986) Surface compressibility, expansivity and transition properties of DMPC/cholesterol bilayer vesicles: major effects of enhanced cohesion and fluidization. Biophys J 49:505a
Quist E, Powell P (1985) Polyphosphoinositides and the shape of mammalian erythrocytes. Lipids 20:433–438
Rakow AL, Hochmuth RM (1975) Effect of heat treatment on the elasticity of human erythrocyte membrane. Biophys J 15:1095–1100
Sheetz MP, Singer SJ (1974) Biological membranes as bilayer couples. A molecular mechanism of drug-erythrocyte interactions. Proc Natl Acad Sci USA 71:4457–4461
Singer SJ, Nicolson GL (1972) The fluid mosaic model of the structure of cell membranes. Science 175:720–731
Skalak R, Tozeren A, Zarda RP, Chien S (1973) Strain energy function of red blood cell membranes. Biophys J 13:245–264
Smith L, Hochmuth RM (1982) Effect of wheat germ agglutinin on the viscoelastic properties of erythrocyte membrane. J Cell Biol 94:7–11
Stokke BT (1984) The role of spectrin in determining mechanical properties, shapes and shape transformations of human erythrocytes. Thesis, University of Trondheim, Trondheim, Norway
Tran-Son-Tay R, Sutera SP, Rao PR (1984) Determination of red cell membrane viscosity from rheoscopic observations of tank-treading motion. Biophys J 46:65–72
Waugh RE (1982) Surface viscosity measurements from large bilayer vesicle tether formation. II. Experiments. Biophys J 38:29–37
Waugh RE (1986) Effects of 2,3-diphosphoglycerate on the mechanical properties of erythrocyte membrane. Blood 68:231–238
Waugh RE (1987) Effects of inherited membrane abnormalities on the viscoelastic properties of erythrocyte membrane. Biophys J 51:363–369
Waugh R, Evans EA (1979) Thermoelasticity of red blood cell membrane. Biophys J 26:115–132
Zarda PR, Chien S, Skalak R (1977) Elastic deformations of red blood cells. J Biomech 10:211–221

V.2 Hydrostatic Pressure in Metazoan Cells in Culture: Its Involvement in Locomotion and Shape Generation

J. BEREITER-HAHN[1] and R. STROHMEIER[1]

2.1 Introduction

Internal hydrostatic pressure in bacteria and in plant cells is a main factor determining shape and growth of individual cells and multicellular organisms (cf. Chap. IV.1). The question whether osmotic pressure in animal cells also differs from that of the extracellular fluid is still unresolved. While a difference is obvious for freshwater protists, it is doubtful for cells of multicellular organisms. Erythrocytes of various species and eggs of some marine invertebrates, chick heart fibroblasts, and frog muscle cells all seem to behave roughly like osmometers (Dick 1959, 1966; Olmstead 1966). Doerner (1967) has postulated the occurrence of an internal pressure linked with the uniformly varying modulus of elasticity from the periphery of a cell toward its center. This pressure may vary from point to point on a cell surface and could well be involved in modeling cell shape. More recently, experimental evidence has accumulated showing that cellular behavior in hypotonic media differs considerably from that of an ideal osmometer (i.e. Skalak and Shu Chien 1982; Roti Roti and Rothstein 1973; Raaphorst and Kruuv 1979). Any volume change which is not in accord with Boyle-van't Hoff's law (deviation from an ideal osmometer) can be due to one of the following factors acting either alone or in accord with each other:

1. presence of an osmotically inactive part of the volume;
2. restricted mobility of the cell water;
3. release or uptake of ions, or other solutes;
4. osmotic activity of cytogel;
5. presence of an active force adding to osmotic pressure.

Factors 1 to 3 will be introduced shortly. Special emphasis is laid on factor 5, which in concert with viscoelasticity of the cytogel, involved in factor 4, controls cell shape and locomotion.

2.2 Osmotic Equations Applied to Cells

The presence of a nonsolvent volume b (i.e., volume at infinite osmotic pressure, when all solvent is withdrawn) has been regarded since Lucke and Mc-

---

[1] Arbeitsgruppe Kinematische Zellforschung, FB Biologie der Universität, Senckenberganlage 27, 6000 Frankfurt 1

Cutcheon (1932) in applying the classical osmotic equation to cells in equilibrium with their extracellular fluid:

$$\pi(V - b) = \pi_0(V_0 - b) = \varphi RTC = k \quad (1)$$

V_0 and π_0 are the volume and osmotic pressure of a cell in isotonic (original) condition, b the nonsolvent volume of the cell, π and V, osmotic pressure and cell volume under a given experimental condition. The constant k is related to the number of intracellular solute particles and their osmotic activity (φ). Net transfer of solute and changes of φ as a consequence of salt concentration or of changes in ionic composition, change k. C is the molar concentration of the solution, R the gas constant, T the absolute temperature. When V is plotted against $1/\pi$ a straight line should result, the ordinate intercept ($1/\pi = 0$) equals b.

2.3 Physical State of Cell Water

By determination of b using a method independent from osmotic changes, such as comparison of wet and dry weights, often a slightly higher isotonic water content is measured compared to that calculated using Eq. (1). The ratio of $V_0 - b$ to this actual water content is called Ponder's R (Ponder 1948). This value is mainly in the range of $0.8-1$ and was first introduced in the description of erythrocyte osmotic behavior and later was determined for a variety of cells (for review, see Dick 1959, 1966). Recent investigations revealed the existence of osmotically inactive water in living cells (Raaphorst and Kruuv 1979; Clegg 1984) as a physicochemical basis of Ponder's R. The "hydration" force of perturbed water structure on the surfaces of molecules may be of considerable physiological significance when influencing macromolecular assembly (Parsegian et al. 1979).

2.4 Solute Leakage

Loss of K^+ from hypotonically treated cells is a well-known phenomenon (Roti Roti and Rothstein 1973; Raaphorst and Kruuv 1979) and results in less swelling than expected according to Boyle van't Hoff's law. This effect is more pronounced in K^+-free media than in KCl solution (Raaphorst and Kruuv 1979). In addition, a change of intracellular osmolality is connected with a change in ionic dissociation affecting activity (Dick 1966; Gary-Bobo and Solomon 1967). Deviation from normal osmotic conditions may induce ion exchange reactions (Raaphorst and Kruuv 1979), shifts in intracellular solute, and water distribution, i.e., formation of vacuoles or volume changes of nuclei, ER, and mitochondria. Furthermore, cells are able to regulate their volume using metabolic energy (e.g. Olson et al. 1986; Roti Roti and Rothstein 1973; Spring and Ericson 1982; Wolpert et al. 1971). All these factors obscure the image derived from rements of cell volume under various osmotic conditions. A practical quence of these results must be to investigate cell responses under con-is as close to their natural environment as possible, i.e., to maintain an composition as stable as possible and not to suspend normally adhering for volume measurements. Volume regulatory reactions to hypotonicity re-e time intervals larger than 5 min, therefore, swelling must be recorded in this time interval.

5 Osmotic Behavior of Cytogel

Cytoplasm exists in a dynamic equilibrium between polymers and the respec-tive mono- and oligomeres. F-actin or microtubules are interconnected by associated proteins forming viscoelastic gels. Viscoelasticity adds to the colloid osmotic properties. Pieces of cytogels can swell and shrink in response to the surrounding immersion fluid without a limiting membrane being involved. A detailed mathematical model of swelling is given by Oster (1984). For the purpose of the present consideration only two of his equations are used to outline the main principle: imbibition of an initially dry gel with a solvent makes it swell. The swelling pressure (P_s) is counteracted by the elastic forces (P_{ELAS}) exerted by the gel and acting against its deformation. In equilibrium, when swelling ceases,

$$P_s = 0 \quad \text{and} \quad P_{ELAS} = P_{OSM}. \tag{2}$$

2.6 Generation of Intracellular Hydrostatic Pressure

In amoeba and in slime mold plasmodia the presence of an intracellular hydrostatic pressure is documented beyond doubt (Kamiya 1964, cf. Chaps. I.2 and II.5). If the plasma membrane is semipermeable (as is generally accepted for most cells), then in the equilibrium state, the positive hydrostatic pressure must be an osmotic one (Layrand et al. 1972). This relation becomes evident assuming a continuous cortical fibrillar layer intimately associated with the plasma membrane (Fig. 2b in Chap. I.1). In an isometric condition its contraction causes an outflow of water through the membrane until the additionally induced intracellular osmotic pressure becomes equal to the pressure caused by the active contraction (P_D):

$$\pi_e = \pi_i - P_D, \tag{3}$$

where π_e and π_i are respectively external and internal osmotic pressure.

2.6.1 Osmotic Behavior of Cells in Culture

Hydrostatic pressure, generated by active contraction in the plasma membrane associated cortical fibrillar layer is accompanied by tension in this layer corresponding to that which would be caused by a rise in relative solute concentration. If the membrane is not tightly bound to the cortical fibrillar layer, as can be induced experimentally in amoebae, then a cytosol-filled space develops between the membrane and the cortical layer (Wehland et al. 1978), or a bleb is formed as in many tissue cells after dissociation in Ca^{2+}-free media (Fig. 1; Bereiter-Hahn 1967) or by squeezing the cells (Rosenberg 1963). The volume of the blebbing cells remains constant within the limits of accuracy of measurements (Albrecht-Buehler 1982). The cortical fibrillar meshwork acts as a sieve preventing penetration of organelles into the bleb. At the inner surface of the detached membrane actin polymerization is initiated allowing contraction, which moves the bubble around the cell (called "circus movement" by Fujinami 1976) until either the membrane is resealed to the fibrillar layer or adhesion to an external surface allows spreading of the cell. The existence of these structures by itself is proof for the correctness of the assumption of an intracellular hydrostatic pressure. This view is supported by rapid and reversible inhibition of blebbing and pseudopodia extension with hypertonic solutions (Harris 1973; DiPasquale 1975). When the cell solutes are in osmotic equilibrium with external solutes, active contraction of the cell cortex causes a volume decrease. Hypotonicity induced swelling should be diminished by the contraction and the concomitant increase in elasticity. On the other hand, in hypertonic media, volume decrease leads to folding of the membrane-cortical layer complex, P_D becomes zero, and negative tensions are unlikely (Wolfe and Steponkus 1983). Therefore, the osmotic behavior of cells in hypertonic media obeys Boyle-van't Hoff's law, while in hypotonic environments it does not (Layrand et al. 1972; Raaphorst and Kruuv 1979; Fig. 2).

2.6.2 Determination of Hydrostatic Pressure in Culture Cells

An experimental proof of these considerations comes from volume determinations after inhibition of contraction by means of lanthanum. La^{3+} acts as a calcium antagonist and rapidly blocks cell locomotion (Strohmeier and Be-

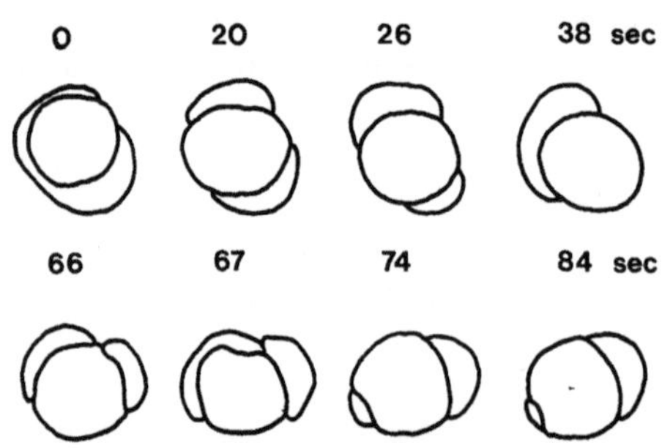

Fig. 1. Stages of blebbing of an epidermal cell in culture. Between 0 and 38 s the bleb on the lower part of the cell is drawn back into the cell body, while the upper bleb enlarges and moves toward the left side (20 – 38 s) and is subsequently withdrawn (68 – 84 s)

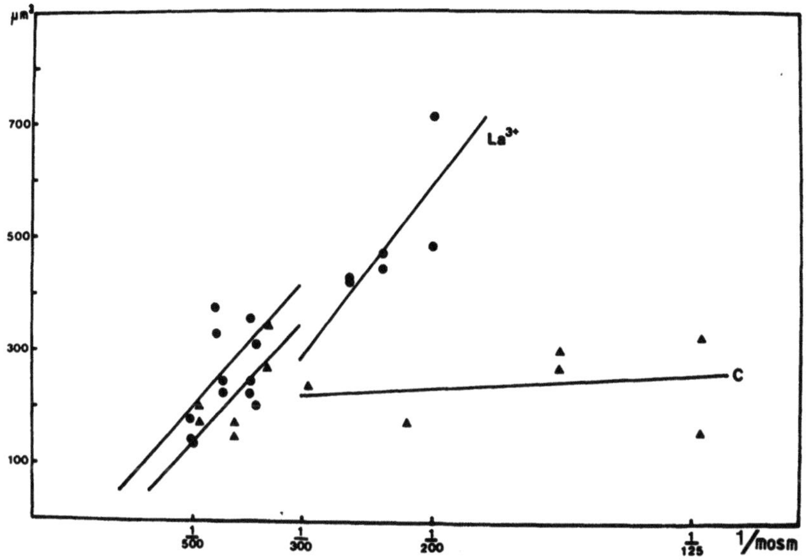

Fig. 2. Dependence of the cell volume on the reciprocal of osmolality of the external medium. Isotonicity is given at 300 mosmol. Different regression lines have been calculated for hyper- and hypotonic conditions, for control cells (▲), and those treated with 2 mM La^{3+} (●). Volume of single *Xenopus* epidermal cells has been evaluated from the interference fringe pattern observed in the reflection interference contrast microscope (Bereiter-Hahn et al. 1983). Osmolality was changed by varying the concentration of sucrose, while protein content and ionic strength of the medium were kept constant.
Regression lines of controls and of La^{3+}-treated cells are in parallel under hypertonic conditions and differ considerably in the hypotonic range, indicating the presence of an elastic resistance which can be abolished by La^{3+}. For further explanations, see text

reiter-Hahn 1984; Cooper and Schliwa 1972). *Xenopus* epidermis cells in primary culture treated with La^{3+} (2 mM, 20 min) follow reasonably well Boyle-van't Hoff's law in hyper- and in hypotonic conditions, while untreated control cells deviate considerably in the hypotonic range. Tension in the cortex increases significantly with decreasing external solute concentration (Fig. 2; see also Raaphorst and Kruuv 1979). If P_D would be constant, the regression lines relating volume to $1/\pi_e$, controls, and La^{3+}-treated cells should be parallel to each other, but shifted in the y-direction. The increase of the volume difference between controls and La^{3+}-treated cells points to a higher elasticity modulus in the cortex of control cells resulting from contraction of the microfibrillar meshwork. If we assume the volume of La^{3+}-treated cells being that obtained with a fully smoothed membrane without any stresses in it, then the volume difference to control cells at a given external solute concentration is due to the active contraction causing the pressure P_D. In the isotonic condition ($\pi_e = 300$ mosmol) P_D is calculated to be 0.96×10^5 Nm^{-2}. This value is a rough estimation because the volume changes induced by variation of external solute (saccharose) concentration exhibit high variability (for methods, see legend to Fig. 2, Strohmeier and Bereiter-Hahn in prep.). In addition, volume determination of single, well-spread cells is not very precise. Difficulties arise when measuring the thick

Table 1. Internal pressures in various types of cells

Object	Hydrostatic pressure in cells (pascal)	Reference
Mycomycete plasmodium	1.7×10^5	Layrand et al. 1972
Dictyostelium plasmodium	$8-30 \times 10^3$	Inouye and Takeuchi 1980
Amoeba proteus	1.6×10^5	cited from Layrand et al. 1972
Chaos chaos	1.5×10^2	Kamiya 1964
Buscyon egg	1.36	Harvey 1933
Sea urchin egg	9.5	Mitchison and Swann 1954
Fundulus egg	1.2×10^5	Kao et al. 1954
Human erythrocytes	23	Rand and Burton 1964
Xenopus epidermis cells in culture	0.96×10^5	This chapter

center and thin lamellar periphery (Bereiter-Hahn et al. 1983). Nevertheless, the pressure value closely resembles others reported for internal pressures in various types of cells (see Table 1).

Amphibian epidermal cells in culture are highly motile and have a well-developed actomyosin fibrillar system (Bereiter-Hahn et al. 1981; Strohmeier and Bereiter-Hahn 1984). The hydrostatic pressure in these cells is close to that measured for amoebae or slime mold plasmodia, which also migrate at high velocity. On the other hand, some eggs and mammalian erythrocytes exhibit negligible hydrostatic pressure (cf. Chap. I.2.3). These differences can be assumed to reside in organizational variations of the cytoskeleton, and where present, the extraneous coats (see Chap. I.2.3). Such differences may well explain the high variation in volume changes in epidermal cells (Fig. 2) or the pressure differences in marine eggs (Table 1). Speed of locomotion of epidermal cells varies widely, cortex organization, for example, of the sea urchin egg is remodeled after fertilization up to the appearance of the first cleavage furrow (Bereiter-Hahn 1985). Elasticity of cytoplasm follows a similar time course (Mitchison and Swann 1954; Hiramoto 1982). Concomitantly, tensions generated by contractile structures cover a wide range (Table 2 in Chap. I.1).

2.6.3 "Visualization" of Tension in the Cortical Fibrillar-Meshwork-Plasma Membrane Complex

Tension in the cortical-fibrillar meshwork complex can be made perceptible by the reaction of fibrils to cytochalasin D (CD) and by scanning acoustic microscopy (SAM). The first reaction seen in XTH-2 cells (1 – 3 min) to CD (2 μg ml^{-1}) is rupturing of the cortical fibrillar layer and of stress fibers. The fibers retract from the site of severing forming a gap devoid of fibrils (Fig. 3). In the SAM this reaction causes a decrease of image contrast (Figs. 4 and 5; Bereiter-Hahn 1987). This observation needs further explanation because interpretation of SAM images is difficult and the method is just emerging as a tool for studying mechanical properties of cells.

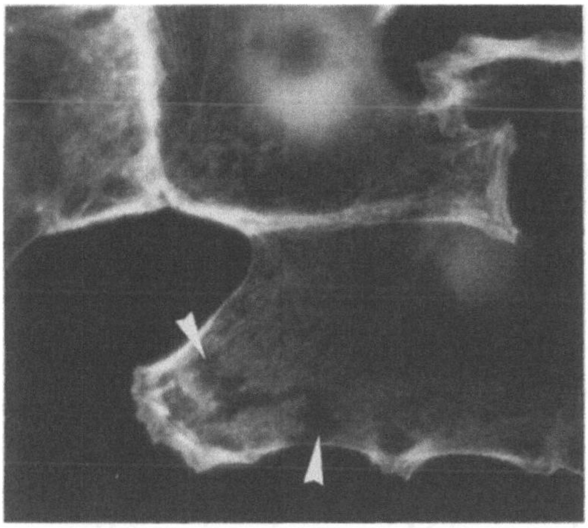

Fig. 3. Fluoromicrograph of an XTH-2 cell (line derived from *Xenopus* tadpole heart endothelium). Actin has been labeled with the fluorescent TRITC-phalloidin (Faulstich et al. 1983). Note disruption of the fine, fibrillar actin meshwork (*arrowheads*) under the influence of cytochalasin D (2 µg ml^{-1}; 5 min)

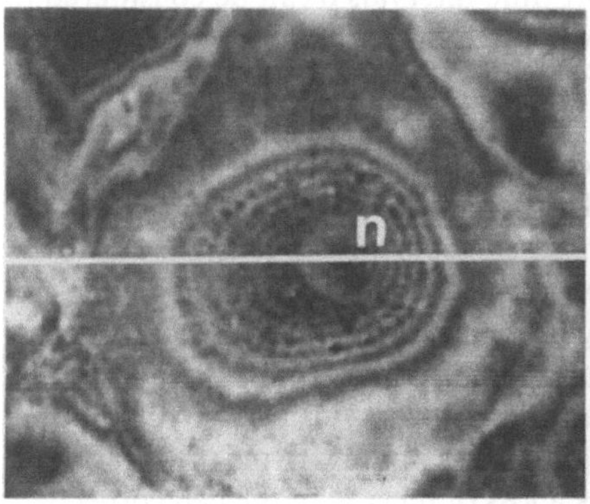

Fig. 4

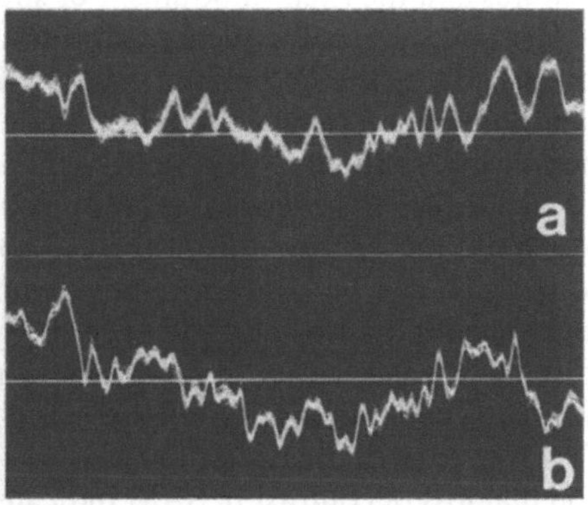

Fig. 5 a, b

Fig. 4. Scanning acoustomicroscope image of an XTH-2 cell (1.0 GHz). The interference lines result from interferences of ultrasound reflected at the polystyrene surface to which the cells are attached and from sound reflected at the upper surface of the cell. Relative intensity differences (see Fig. 5) of constructive and destructive interferences depend on acoustic impedance of the cell surface. The *white line* demarcates the line along which the signal intensity was recorded which is shown in Fig. 5a; *n* nucleus

Fig. 5 a, b. Signal intensity along a scanning line shown in Fig. 4. **a** Immediately before addition; **b** 2 min after addition of cytochalasin D. Differences in the brightness of the peripheral interference lines are less distinct than in **a** (see *white spots* in **a**)

In SAM the specimen is scanned with an acoustic lens and the echo of ultrasound (0.9 – 2 GHz) from each point is used for modulation of electron beam intensity of a video screen. The principles of operation have been described in detail, i.e., by Quate and Ash (1982) and by Briggs (1985). The SAM image of culture cells is determined by sound reflection at the interfaces medium/cell, cell/medium, and medium/cell supporting substratum (e.g., glass or polystyrene) and in addition by sound attenuation due to absorbance or scattering (Johnston et al. 1979; Hildebrand 1985; Hildebrand and Rugar 1984; Hoppe and Bereiter-Hahn 1985). The course of the interference lines in Fig. 4 delineates the surface topography of the cells (medium-facing side), it is a measure of cell thickness. The contrast of the interference lines is primarily a function of acoustic impedance differences at the reflecting interfaces. Acoustic impedance W is directly related to material density ϱ and to longitudinal sound velocity v in the respective medium (i.e., culture medium, cytoplasm, glass):

$$W = \varrho \cdot v. \tag{4}$$

The velocity, in turn, is related to the density and the modulus of the elasticity (M) appropriate to the particular wave propagation:

$$v = \sqrt{M/\varrho}. \tag{5}$$

For longitudinal waves in biological material M is a combination of Young's modulus and the shear modulus. For short time intervals ϱ can be considered to be constant. As a consequence, short time changes in reflectivity reveal changes in elasticity of the cytoplasm. At 0.9 GHz, used for Fig. 4, the term elasticity depicts mainly Young's modulus, while at higher frequencies shear and bulk moduli may influence increasingly reflectivity properties (for a more detailed discussion, see Bereiter-Hahn 1986). A clear separation of elasticity types is not possible. Young's modulus of the cell cortex is responsive to tension. However, no decision can be made whether cortex stiffness is brought about by tension counteracting hydrostatic pressure or by crosslinking of the fibrillar elements forming a rigid gel without generation of internal pressure. Actin staining (Fig. 3) reveals disruption of the cell cortex meshwork instead of dissolution. This is a strong indicator of the presence of tension and confirms the interpretation that loss of contrast in SAM indicates a loss of tension in the cell cortex.

2.7 Functional Significance of Hydrostatic Pressure in Wall-Free Cells

2.7.1 Cell Shape

Formation of a hydroskeleton is a structural solution to the problem of how to achieve stiffness with flexible, tensile elements. Such structures may easily change shape by local weakening of surface rigidity. Stability is maintained only as long as tension and elasticity of the cortex are the same all over the free cell surface (not adhering to substrate or another cell). Weakening in one region, or increase of contractile force in another, cause intracellular pressure differences which, in a hydraulic system, bring about streaming of the fluid until the differences have disappeared. The time required for these compensating movements depends on cytoplasmic shear viscosity. Therefore, a considerable delay may be observed between retraction of cytoplasm on one side and protrusion at another side (Dunn 1980; Oster 1984). If contraction resides only in the cortical cytoplasm (where ecto- and endoplasm are distinct, e.g., in slime mold plasmodia and amoeba) spheres and cylinders are preferentially formed, which may flatten under the influence of adhesive forces. Connection of the cortex-plasma membrane complex by fibrils transversing the cytoplasm (no clear separation between ecto- and endoplasm) enhances the number of shapes produced, for instance, very thin lamellae can be formed (cf. Bereiter-Hahn and Strohmeier 1987). Accordingly, hydrostatic pressure is involved in the achievement of cell shape by providing rigidity to the cell cortex that counteracts hydrostatic pressure by tension. This allows formation of cytoplasmic protrusions (pseudopodia, lamellae, microvilli) induced by local weakening of cortical stiffness or of the connection between membrane and cortical fibrils (blebs). Formation of straight fibers from a more or less anisotropic gel also requires two sites of abutment, which can be provided either by adhesion-based membrane stabilization (see Chap. V.3) or by a rigid cortical cytoplasm (cf. Chap. I.1.7).

2.7.2 Cell Locomotion

It is hydrostatic pressure that makes a cell a mechanical unity, because it is an integral property of the whole system. Any local change in the physical properties causes cytoplasmic motion. In this sense hydrostatic pressure provides the force needed for locomotion. The actual motive force, however, is the pressure difference Δp between the advancing structure (i.e., lamella or pseudopodium) and the cell body. The hydrostatic pressure represents the maximum pressure available for motive force generation. As a first approximation Hagen-Poiseuille's law can be used for the description of continuous cytoplasmic streaming in strands of slime mold plasmodia or amoebae:

$$dV = \frac{\pi r^4}{8 \eta l} \Delta p \cdot dt. \tag{6}$$

where dV is the volume of endoplasm flowing through an ectoplasmic tube with radius r and length l per unit of time dt; η is the viscosity of the endoplasm, Δp is the pressure difference over l.

However, it is valid only if r, η, and Δp remain constant during the time of observation and if no fibrillar reinforcement traverses the cytoplasmic tube, impeding free flow. The hydraulic pressure hypothesis of cell locomotion (Bereiter-Hahn 1985; Harris 1973; Komnick et al. 1972) provides a reasonable explanation for the advantage of an advancing lamella in migrating culture cells and why the direction of spreading of this lamella follows the direction of fibrillar bundles (Vasiliev 1982; Kolega 1986): the hydrostatic pressure is the same everywhere in a cell. The thin margin of a lamella can be protruded relatively far without the outflow of a considerable amount of cytoplasm from the cell body. This means that hydraulic pressure can be maintained without significant rearrangement of fibrils in the cell body; the volume of the big lamella of amphibian epidermal cells does not exceed 5% of the total cytoplasm. Therefore, hydraulic pressure can be maintained without significant rearrangement of fibrils in the cell body. A thin lamella provides a large area for adhesion using a minimum amount of cytoplasm. The filamentous meshwork filling the interior of the lamella can be assumed to impose considerable friction on the mono- or oligomeric proteins percolating toward the leading edge. Therefore, the relatively large myosin aggregates are trapped in the proximal part of the lamella of migrating epidermal cells, but become evenly distributed after cessation of locomotion (Strohmeier and Bereiter-Hahn 1984). Reinforcement of cytoplasm by fibers roughly parallel to each other stabilizes a cell against deformation vertical to the fibril axis and is in favor of cytoplasmic flow along the axis as described for fibroblasts (Vasiliev 1982) and fish epidermal cells (Kolega 1986).

2.7.3 Integration of Cells into Tissues

The mechanical requirements of cells integrated into a tissue differ substantially from those of single cells. For instance in epithelia, cells do not migrate as long as the epithelium is not wounded or overstressed (Fung 1984; Phillips 1984). The mechanical integration is provided by an extensive apical fibrillar layer and stresses can be distributed throughout an epithelium via a fibrillar system coupled by desmosomes and/or intermediate junctions (e.g. Auersperg 1972; Iwig et al. 1981; Kunzenbacher et al. 1982; Lee and Auersperg 1980). In this case the whole volume enclosed by an epithelium or tissue may be under hydrostatic pressure (i.e., blastula, kidney tubule, neural tube, and brain), which is balanced by the apical fibrils. Osmotic pressure in single tissue cells theoretically can be equal to that of the interstitial space. This allows cell stability in the absence of cortical fibrils in basolateral membranes of epithelia and permits greater lateral motility of membrane proteins.

2.7.4 Hydraulic Interaction of Organelles

As a first approximation a cell can be considered to be composed of several shells with different elastic properties (Doerner 1967). In this case volume changes in the inner shells (nucleus, vacuoles, mitochondria) result in changes of tension at the surface of an outer shell and vice versa (for mathematical modeling, see Doerner 1967). This principle provides a mechanism for mechanical coupling of cellular functions, e.g., functional swelling of the nucleus may influence cell membrane permeability by means of increasing tension. Investigation of these possibilities is in its very infancy (Ingber and Jamieson 1985; Rappaport 1980), however, these mechanical linkages may provide an explanation for some of the physiological effects of mechanical stimuli mentioned in Chap. I.1.7.

References

Albrecht-Buehler G (1982) Does blebbing reveal the convulsive flow of liquid and solutes through the cytoplasmic meshwork? Cold Spring Harbor Symp Quant Biol 66:45–59

Auersperg N (1972) Microfilaments in epithelial morphogenesis. J Cell Biol 52:206–211
Bereiter-Hahn J (1967) Dissoziation and Reaggregation von Epidermiszellen der Larven von Xenopus laevis (Daudin) in vitro. Z Zellforsch Mikrosk Anat 79:118–156
Bereiter-Hahn J (1985) Architecture of tissue cells. The structural basis which determines shape and locomotion of cells. Acta Biotheor 34:139–148
Bereiter-Hahn J (1986) Epidermal cell migration and wound repair. In: Bereiter-Hahn J, Matoltsy AG, Richards KS (eds) Biology of the integument. Springer, Berlin Heidelberg New York, vol 2. pp 463–470
Bereiter-Hahn J (1987) Scanning acoustic microscopy visualized cytomechanical responses to cytochalasin D. J Microsc 146:29–39
Bereiter-Hahn J, Strohmeier R (1987) Biophysical aspects of motive force generation in tissue culture cells and protozoa. Fortschritte Zool (in press)
Bereiter-Hahn J, Strohmeier R, Kunzenbacher I, Beck K, Vöth M (1981) Locomotion of Xenopus epidermis cells in primary culture. J Cell Sci 52:289–311
Bereiter-Hahn J, Strohmeier R, Beck K (1983) Determination of the thickness profile of cells with the reflection contrast microscope. Scientific Tech Inf Process 8:125–128
Briggs A (1985) An introduction to scanning acoustic microscopy. Microscopy handbooks 12 Oxford Univ Press, R Micros Soc
Clegg JS (1984) Properties and metabolism of the aqueous cytoplasm and its boundaries. Am J Physiol 246:133–151
Cooper MS, Schliwa M (1986) Transmembrane Ca^{2+}-fluxes in the forward and reversed galvanotaxis of fish epidermal cells. In: Ionic currents in Development 311–318. Alan R. Liss Inc. New York
Dick DAT (1959) Osmotic properties of living cells. Int Rev Cytol 8:387–448
Dick DAT (1966) Cell water. Butterworths, London, 155 pp
Dipasquale A (1975) Locomotion of epithelial cells. Factors involved in extension of the leading edge. Exp Cell Res 95:425–439
Doerner K (1967) Pressure gradients in cells. J Theor Biol 14:284–292
Dunn GA (1980) Mechanisms of fibroblast locomotion. In: Curtis ASG, Pitts JD (eds) Cell adhesion and motility. pp 409–423 Cambridge Univ press
Faulstich H, Trischmann H, Mayer D (1983) Preparation of tetramethylrhodaminyl-phalloidin and uptake of the toxin into short-term cultured hepatocytes by endocytosis. Exp Cell Res 144:73–82
Fujinami N (1976) Studies on the mechanism of circus movement in dissociated embryonic. J Cell Sci 22:133–147
Fung YC (1984) Structure and stress-strain relationship of soft tissues. Am Zool 24:13–22
Gary-Bobo CM, Solomon AK (1967) Properties of hemoglobin solutions in red cells. J Gen Physiol 52:825–853
Harris AK (1973) Cell surface movement related to cell locomotion. Locomotion of tissue cells. Ciba Found Symp 14:3–26
Harvey EN (1933) The flattening of marine eggs under the influence of gravity. J Cell Comp Physiol 4:35
Hildebrand JA (1985) Observation of cell-substrate attachment with the acoustic microscope. IEEE Trans Sonics Ultrasonics SU 32:332–340
Hildebrand JA, Rugar D (1984) Measurement of cellular elastic properties by acoustic microscopy. J Microsc 134:245–260
Hiramoto Y (1982) Rheological properties of echinoderm eggs during cell division. Biorheology 19:71–78
Hoppe M, Bereiter-Hahn J (1985) Applications of scanning acoustic microscopy – survey and new aspects. IEEE Trans Sonics Ultrasonics SU-32:289–301
Ingber DE, Jamieson JD (1985) Cells as tensegrity structures: architectural regulation of histodifferentiation by physical forces transduced over basement membrane. In: Anderson LC, Gahmberg CG, Explom P (eds) Gene expression during normal and malignant differentiation. Academic Press, London
Inouye K, Takeuchi I (1980) Motive force of the migrating pseudoplasmodium of the cellular slime mould Dictyostelium discoideum. J Cell Sci 41:53–64
Iwig M, Glaesser D, Bethge M (1981) Cell shape-mediated growth control of lens epithelial cells grown in culture. Exp Cell Res 131:47–55

Johnston RN, Atalar A, Heiserman J, Jipson V, Quate CF (1979) Acoustic microscopy: resolution of subcellular detail. Proc Natl Acad Sci USA 76:3325–3329
Kamiya N (1964) The motive force of endoplasmic streaming in the ameba. In: Allen RD, Kamiya N (eds) Primitive motile systems in cell biology. Academic Press, London, pp 257–277
Kao CJ, Chambers PV, Chambers EL (1954) Internal hydrostatic pressure of the fundulus egg. J Cell Comp Physiol 44:447–461
Kolega J (1986) Effects of mechanical tension on protrusive activity and microfilament and intermediate filament organization in an epidermal epithelium moving in culture. J Cell Biol 102:1400–1411
Komnick H, Stockem W, Wohlfarth-Bottermann KE (1972) Ursachen, Begleitphänomene und Steuerung zellulärer Bewegungserscheinungen. Fortschritte d Zool 21:3–60
Kunzenbacher I, Bereiter-Hahn J, Osborn M, Weber K (1982) Dynamics of the cytoskeleton of epidermal cells in situ and in culture. Cell Tiss Res 222:445–457
Layrand DB, Matveeva NB, Teplov VA, Beylina SI (1972) The role of elastoosmotic parameters in locomotion of myxomycete plasmodia. Acta Protozool 11:339–354
Lee HC, Auersperg N (1980) The use of lanthanum for intracellular calcium measurements in cultured cells. Cell Biol Int Rep 4:675–684
Lucke B, McCutcheon (1932) The living cell as an osmotic system and its permeability to water. Physiol Rev 12:68
Olmstead EG (1966) Mammalian cell water. Physiologic and Clinical aspects. H Kimpton, London 199 pp
Olson JE, Sankar R, Holtzman D, James A, Fleischhacker D (1986) Energy-dependent volume regulation in primary cultured cerebral astrocytes. J Cell Physiol 128:209–215
Oster G (1984) On the crawling of cells. J Embryol Exp Morphol 83:329–364
Parsegian VA, Fuller N, Rand RP (1979) Measured work of deformation and repulsion of lecithin bilayers. Proc Natl Acad Sci USA 76:2750–2754
Phillips HM (1984) Physical analysis of tissue mechanics in amphibian gastrulation. Am Zool 24:657–672
Ponder E (1948) Hemolysis and related phenomena. Churchill, London
Quate CF, Ash EA (1982) Ultrasonic microscopy in medicine and biology. In: von Bally G, Greguss P (eds) Optics in biomedical sciences. Springer, Berlin Heidelberg New York, pp 25–33
Raaphorst P, Kruuv J (1979) Effects of salt, sucrose, and dimethyl-sulfoxide solutions on the water content and water structure of tissues and cultured cells. In: Keith AD (ed) The aqueous cytoplasm. Dekker, New York, pp 91–136
Rand RP, Burton AC (1964) Mechanical properties of the red cell membrane. I. Membrane stiffness and intracellular pressure. Biophys J 4:115
Rapaport E (1980) Compartmentalized ATP pools produced from adenosine are nuclear pools. J Cell Physiol 105:267–274
Rosenberg MD (1963) The relative extensibility of cell surfaces. J Cell Biol 17:289–297
Roti Roti LW, Rothstein A (1973) Adaptation of mouse leukemic cells (L5178Y). I. Cell volume regulation. Exp Cell Res 79:295–310
Skalak R, Shu Chien (1982) Fourth international congress of biorheology symposium on mechanical properties of living tissues. Biorheology 19:453–461
Spring K, Ericson AC (1982) Epithelial cell volume modulation and regulation. J Membr Biol 69:167–176
Strohmeier R, Bereiter-Hahn J (1984) Control of cell shape and locomotion by external calcium. Exp Cell Res 154:412–420
Vasiliev JM (1982) Spreading and locomotion of tissue cells: factors controlling the distribution of pseudopodia. Phil Trans R Soc Lond B Biol Sci 299:159–167
Wehland J, Stockem W, Weber K (1978) Cytoplasmic streaming in Amoeba proteus is inhibited by the actin-specific drug phalloidin. Exp Cell Res 115:451–454
Wolfe S, Steponkus PL (1983) Tension in the plasma membrane during osmotic contraction. Cryo Lett 4:315–322
Wolpert L (1971) The effect of high hydrostatic pressure on the mechanical properties of the surface of the sea-urchin egg. J Cell Sci 8:87–92

V.3 The Transmission of Forces Between Cells and Their Environment

MICHAŁ OPAS[1]

3.1 Introduction

"The most obvious deduction to draw from the known facts is that the cells in tissue cultures are sticky as regards certain solids, and that this is a natural characteristic" (Lewis 1922). The "sticky quality" referred to hereafter as adhesiveness, has been shown to be dependent upon both intra- and extracellular factors. I shall discuss here the relation between a cell's adhesiveness and biomechanical aspects of cell organization with emphasis on that part of the cytoarchitecture which is specialized for generating contractile force, that is, the microfilamentous cytoskeleton.

Adhesion of cells to their substrata is governed by an interplay between forces of nonspecific attraction and repulsion, and specific bonding (Bell et al. 1984; Dembo and Bell (in press); Torney et al. 1986). It follows that both the physicochemical characteristics of the substratum and the medium, and the presence (or absence) of specific adhesion-related proteins, are the major factors determining the morphology and effectiveness of cell-to-substratum adhesion (Bongard et al. 1982; Hewitt and Martin 1984). The contractile cytoskeleton, through its association with the cytoplasmic face of the plasma membrane (Geiger 1983), can modify the topography of the cell's adhesive surface (Geiger et al. 1984a; Rees et al. 1977), thereby affecting the forces operating at the interface between the cell and the substratum surfaces. Therefore, the organization and contractile activity of the cytoskeleton have to be accounted for when considering cell adhesion.

3.2 Focal Contact: Subcellular Level

The topography of the cell surface adherent to a glass substratum can be directly visualized with the technique of incident light optical interferometry, interference reflection microscopy (IRM-reviewed by Verschueren 1985). With the aid of IRM microscopy, three major types of cell-substratum adhesions can be distinguished (Couchman and Rees 1979: Izzard and Lochner 1976; Opas 1985): (1) *focal contacts* – small (0.1 – 2 μm by 2 – 10 μm), transient structures characterized by black IRM images, resulting from the distance of separation of 10 – 15 nm or less between the cell and the substratum (for a quantitative analy-

[1] Department of Anatomy, University of Toronto, Toronto, Ontario M5S 1A8, Canada

sis, see Beck and Bereiter-Hahn 1981; Bereiter-Hahn et al. 1979); (2) *focal adhesions* – usually larger and always more permanent structures than focal contacts, also appearing black by IRM microscopy; and (3) *close contacts* – large and labile areas, characterized by dark gray reflections in IRM microscopy resulting from the separation distance of ca. 30 nm between the cell and the substratum. Focal contacts and focal adhesions are invariably associated with the termini of microfilament bundles (stress fibers) (Heath and Dunn 1978). The concept of focal contacts and adhesions as specialized, highly adhesive membrane-cytoskeleton complexes has been strengthened by the discovery that several proteins are localized almost exclusively in focal cell-substratum and cell-cell (i.e., zonulae adhaerentes) contacts (Geiger 1983; Geiger et al. 1984b). Vinculin, was the first protein found to be associated specifically with the cytoplasmic face of the plasma membrane at sites where microfilament bundles terminate either in focal contacts or in zonulae adhaerentes (Chen and Singer 1982; Geiger 1979; Singer 1982). Because vinculin does not bind directly to actin filaments (Evans et al. 1984; Wilkins and Lin 1986), several other proteins such as talin (Burridge and Connell 1983; Burridge and Mangeat 1984), which is postulated to be cell-substratum adhesion-specific, and the cell-cell adhesion-specific (Geiger et al. 1985), but as yet poorly characterized 135 kD protein (Volk and Geiger 1984), have been suggested to cooperate with vinculin in anchoring microfilament bundles to the plasma membrane in these contact regions. Interestingly, it has been shown recently that either myosin (Opas and Kalnins 1986) or both myosin and tropomyosin (Geiger et al. 1984a) are excluded from the termini of stress fibers. This suggests that while the proximal parts of stress fibers are contractile (Isenberg et al. 1976; Kreis and Birchmeier 1980), their termini are specialized for attachment. Because vinculin association with the cytoplasmic face of the plasma membrane is actin-independent (Avnur et al. 1983; Herman and Pledger 1985), it is conceivable that focal contacts with their associated proteins act as organizing (nucleating) centers for the surrounding cytoplasm. According to this line of thinking (Geiger et al. 1984b), contact is made initially by "contact receptors," then vinculin is recruited into the focal area and finally F-actin is accumulated to form a microfilament bundle. This scheme has been challenged by Izzard's group (Izzard and Lochner 1980; De Pasquale and Izzard 1985), who postulated that the local accumulation of F-actin precedes formation of a focal contact and recruitment of vinculin.

As much as the cytoskeleton, via control of cell surface topography, regulates cell adhesiveness, the substratum exerts a regulatory influence upon cytoskeletal organization via its surface properties, i.e., "adhesiveness". The organization of microfilaments depends on the "adhesive condition" of a cell; it has been elegantly shown by Willingham et al. (1977) that "normally" adherent and spread cells lose their highly organized microfilament network and round up when exposed to a "nonadhesive" substratum, while "normally" nonadherent and rounded cells spread and acquire extensive cytoskeletal organization when exposed to a "highly adhesive" substratum. Adhesion and spreading of cells in vitro depend on the physical properties of the substratum surface, and are promoted by large glycoproteins of extracellular matrix (Hewitt and Martin 1984) such as laminin and fibronectin. Paradoxically, although it promotes for-

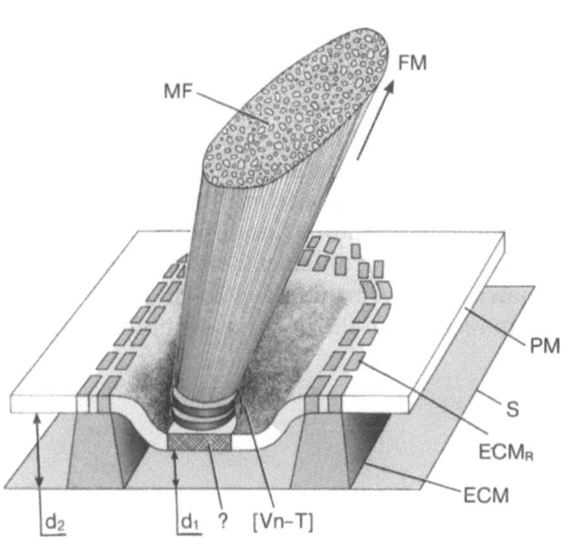

Fig. 1. The spatial arrangement of components of a focal contact. *PM* plasma membrane; *S* substratum; ECM_R extracellular matrix receptor; *ECM* extracellular matrix; [*Vn–T*] vinculin/talin complex; *?* hypothetical transmembrane anchor of the vinculin/talin complex; d_1 distance of separation between the plasma membrane and the substratum within the focal contact domain; d_2 distance of separation between the plasma membrane and the substratum adjacent to the focal contact domain; *MF* microfilament bundle; *FM* tension vector. For clarity, only in regions of substantial accumulation (by standards of immunofluorescence microscopy, see Chen et al. 1985; Damsky et al. 1985) is the ECM depicted. For further details, see text

mation of stress fibers and focal contacts (Lark et al. 1985), fibronectin and its cell surface receptor are excluded from areas occupied by vinculin-rich focal contacts and seen as black images in IRM microscopy (Avnur and Geiger 1981a; Birchmeier et al. 1980; Chen et al. 1985; Damsky et al. 1985). The phenomenon of fibronectin exclusion is, however, subject to controversy (Singer 1982). In any case, Damsky et al. (1985) have clearly shown that in mature focal contacts the fibronectin receptor and vinculin are distributed in a "needle eye" pattern with the former outlining the "needle eye," while the latter fills it in. This "needle eye" arrangement has some peculiar implications for the three-dimensional organization of a focal contact (Fig. 1). The average focal contact is a membrane extension or a small ridge on the cell surface approaching the substratum to a distance of ca. 10 nm. No focal contact-specific extracellular protein has been described as yet, nor is there much electron-dense material underneath the cell surface in the focal contact when that area is examined in the transmission electron microscope. This membranous extension is the site in which a bundle of microfilaments, arriving at low angle from the bulk of the cytoplasm, is anchored via the vinculin-talin complex which in turn is bound to the plasma membrane by an as yet unknown protein(s). Outlining the focal contact in the plane of the membrane, but already farther away from the substratum, is an oval accumulation of the fibronectin receptor, a glycoprotein which attaches the entire structure, via fibronectin, to the substratum. Interestingly, because focal contacts are always aligned with microfilament bundles we have to assume that if the microfilament bundle contracts, the tension would be transmitted to the substratum through: (1) as yet unknown material linking the substratum with the plasma membrane within the focal contact area; or (2) strong attraction forces existing between the substratum and the plasma membrane within the focal contact area; or (3) the fibronectin-fibronectin receptor complex. However, because the fibronectin receptor is localized laterally to the focal contact and does not associate with microfilaments directly, one must postulate a tension-transmitting lateral link (in the plane of the membrane) which would connect the fibronectin receptor to the putative membrane anchor of the vinculin-talin-actin complex.

3.3 Traction: Cellular Level

Cells must exert traction on a substratum in order to move. Thus, for efficient conversion of the motive force generated by the cytoplasmic contractile machinery into cell translocation, adhesion is of utmost importance. From the data presented in Table 2 of Chapter I.1, it is evident that lower values of forces are attributable to transporting systems (particles, *Nitella*), whereas cells adhering to substrata develop much greater forces. The forces required to separate cells, which I will refer to as the "adhesion strength," range wildly from 10^{-7} to 10^{1} dynes/cell depending on the experimental technique of separation used (Hubbe 1981). Although the range of the data renders them of little use, one particular calculation has quite interesting implications. The "adhesion strength" for *Amoeba proteus* has been measured to be in the order of 10^{-2} to 10^{-1} dynes (Opas and Kalinina 1980). The force required for detachment of tissues cells (by micromanipulation, shear, or jets of fluid) is in the order of 10^{-1} to 10^{1} dynes (Hubbe 1981). Assuming that the majority of the detaching force within these upper limits is used to detach the strongest adhesions, that is, focal contacts, we arrive at a "strength of adhesion" for focal contacts in the order of $10^{6}-10^{8}$ dynes/cm^2 (the estimated focal contact area/cell is in the order of 10^{-7} cm^2). The same calculation for *Amoeba proteus* gives us a "strength of adhesion" of $10^{3}-10^{4}$ dynes/cm^2 for amoeba adhesions (the area of which is in the order of 10^{-5} cm^2; unpublished data). Comparison of the motive force, strength of adhesion, and speed of locomotion between an amoeba and a spread tissue cell gives some physiological meaning to the problem of transmission of force: the motive force of amoeba is much lower (by two orders of magnitude) than that of a tissue cell, but it is equal to, or higher than, amoeba's strength of adhesion. This may explain why amoebae never develop focal contacts, but instead adhere to the substratum at a great distance of separation with relatively large, extremely labile attachments (Opas 1978). Their actin is organized as a fine, labile microfilament network with no prominent stress fibers (Opas 1980; Stockem et al. 1982), and they develop speeds up to three orders of magnitude higher (hundreds of μm/min; Opas 1975; Opas and Kalinina 1980) than the speeds of translocation of strongly adherent tissue cells (Kolega et al. 1982). On the other hand, because the strength of adhesion of focal contacts is equal to, or exceeds, the motive force developed by the tissue cells, it seems reasonable to assume that cells employ peeling techniques (Hubbe 1981) to release their focal contacts. This is based on the assumption that in the spread tissue culture cells, which express the focal contact-associated stress fibers, these contractile structures participate somehow in the generation of the motive force. Also, high shearing forces exerted on adherent cells result in cell detachment concomitant with fracture of the plasma membrane at the focal contact level (Avnur and Geiger 1981 b).

The mechanism of formation of the focal contacts and focal contact-associated stress fibers characteristic of a spread cell is far from being clear. The formation of a focal contact is a biphasic process in which the fast (~ 1 s) initial phase of establishing small (0.25 μm by 1 μm) "immature" contacts at large distances of separation is followed by the "maturation" phase (~ 15 s) characterized by area growth and establishment of small separation distances (Izzard and Lochner 1980). The precontact surface of the cell with its underlying cortical network of microfilaments may be assumed to behave like a "contractile surface carpet" which is under low initial isotropic tension (Evans 1984). Structural elements of the "carpet" are in dynamic equilibrium with their subunits, diffuse in a soluble (cytoplasmic) phase (Geiger et al. 1984 a), and show elastic recoil and yield stress, implicating their viscoelastic properties (Sato et al. 1983). The carpet presents low resistance to bending and stretching, i.e., expands by smoothing out its microvillar projections (Evans and Skalak 1980).

Let us assume that cell surface "contact receptors" facing the substratum are individually connected to the elements of the force-generating machinery (microfilament network) and, prior to an adhesive event, experience the environment and exhibit behavior similar to small particles (~ 1 μm) as analyzed by Dembo and Harris (1981). This means that they are slowly transported toward the cell posterior. Establishing the "immature" contact would prevent the contact receptors from translocating, thereby accelerating the development of tension by the force-generating machinery (Fig. 2). The increased tension would, via a Poisson effect, bring together the elastic elements of the carpet, aligning them with the direction of the force. It is expected that a substantial proportion of the elastic elements in the contractile surface carpet comprises the force-generating microfilaments, each one of them able, via interaction with myosin, to generate a force more than sufficient to move a particle with speeds of $\mu m/s^{-1}$ (Sheetz and Spudich 1983).

Thus, several events might lead to the "maturation" of the focal contact (Fig. 2). Establishment of contact immobilizes the contact receptor, resulting in the tension becoming isometric. The development of isometric tension promotes formation of contractile bundles of microfilaments (Fleischer and Wohlfarth-Bottermann 1975; Isenberg et al. 1976; Kreis and Birchmeier 1980), which in turn generate force several orders of magnitude larger than that required for particle transport. Development of isometric tension would also cause polarization in the tension components in the contractile surface carpet, that is, the structural elements of the carpet would be stretched along, and compressed transversely to, the direction of the stress, thereby augmenting the process of

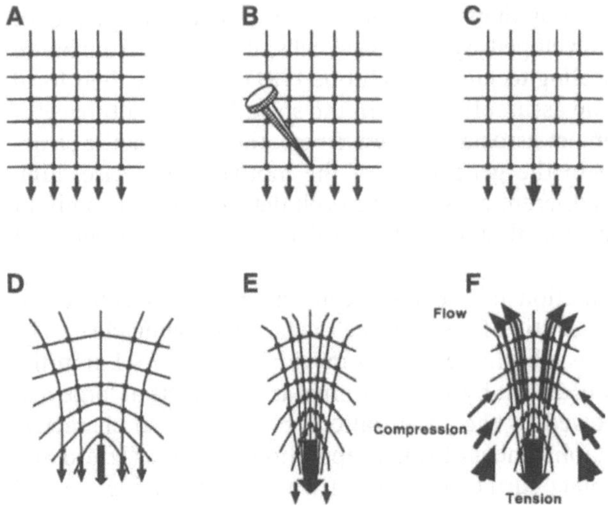

Fig. 2 A – F. Behavior of the "contractile surface carpet" (Evans 1984) during hypothetical stages of formation of a focal contact. Via their transmembrane linkage, cell surface "contact receptors" (*dots*) are attached to a network (*lines*) of structural elements of the contractile surface carpet (e.g., microfilaments), shown here arranged in an orthogonal lattice for clarity. At the precontact stage (**A**), the entire complex is slowly transported posteriorly (*arrows*). Establishment of contact (**B**) prevents the contact receptor from translocating, thereby accelerating the development of tension, which then becomes isometric in the immobilized microregion (**C**). Nonimmobilized linear elements of the contractile surface carpet align in the direction of the force (**D**), eventually (**E**) forming a bundle, while the contact receptors cluster. Under isometric tension, the linear elements of the "carpet" are stretched along, and compressed transversely to the direction of force, causing the mobile elements of the carpet to flow out from the contact area, further augmenting the process of focal contact compaction (**F**)

microfilament bundling. The compression effect would, for values above the yield stress of the contractile surface carpet, cause elements which are not involved in focal contact "maturation" to flow out from the contact area.

The above speculation concerns the formation of focal contacts and their association with contractile microfilament bundles (stress fibers), but it does not relate to the mechanism of maintaining focal contacts as force-transmitting structures. A contact-induced increase in membrane "order" as measured by ESR (Dodd et al. 1982), together with a decreased mobility of focal contact components in the plane of the membrane (Geiger et al. 1984a), might augment the stability of the focal contact. Although tissue cells express greater surface rigidity than erythrocytes (Petersen et al. 1982), low surface shear viscosity ($10^{-4} - 10^{-3}$ dyne-s/cm; Evans and Skalak 1980) seems unlikely to provide any efficient viscous linkage. Maintaining the separation distance between the cell surface and the substratum by electrodynamic forces seems to require interactions significantly greater than those operating within the range of the secondary minimum of free energy (detaching force in the order of $10^{-3} - 10^{-4}$ dynes/cm^2; Gingell and Vince 1979). In this regard, molecular bridging seems a better candidate (Dembo and Bell, in press; Singer 1982; Torney et al. 1986), although the peculiar "needle eye" arrangement of the putative transmembrane anchors appears to be an unusual arrangement for the effective transmission of forces across the plasma membrane.

The problem which remains to be discussed while considering adhesion at the cellular level is that of conversion of the motive force generated by the cytoplasmic contractile machinery into traction. I should like to emphasize at this point that the recent interesting hypotheses on various aspects of cell motility (Dembo et al. 1984; Dembo and Harlow, 1986; Oster 1984; Oster and Odell 1984) have limited application to the above problem, because in the latter we are dealing with isometric contractions of elements which are almost immobilized by strong adhesions to the substratum; thus, although I have referred above to the "motive force," the following considerations actually deal more with immobility than with cell motility.

Mechanically, a spreading cell, while increasing its circumference, generates centripetal tension (Oster 1984) which is transmitted to the substratum by adhesive "patches" that are often distributed along the cell periphery with a curious periodicity (Segel et al. 1983). Development of steady tension, that is, an isometric contraction, promotes formation of contractile microfilament bundles leading usually to breakdown of the circumferential arrangement of adhesions, such that a spread cell becomes irregularly stretched between the few strongest points of attachment to the substratum. Thus, the arrangement of stress fibers in a fully spread cell has been proposed to at least partly reflect the lines of a tension field generated by cellular contractile activity (Greenspan and Folkman 1977). From this simplified, mechanistic model, it can be deduced that the development of prominent and numerous stress fibers correlates with low motility, while highly motile cells might be expected to lack stress fibers and strong adhesions to the substratum, which indeed is the case (Badley et al. 1980; Couchman and Rees 1979; Kolega et al. 1982).

In a spread cell, the tension is being transmitted to the substratum, and when the substratum is elastic enough it becomes deformed so as to reflect the forces exerted on it by the adherent cell (Harris 1982, 1984a; Harris et al. 1980, 1981). From the pattern of substratum deformation, it should be possible then to infer the directions and magnitude of forces operating in a cell-created tension field (Harris 1982, 1984b). This is based on the assumption that since Reynolds numbers are very low for spatial (μm) and temporal (min) scales of cell movements then cell tractional forces are, at any moment, in mechanical equilibrium with the elastic forces operating within the substratum (Oster et al. 1983). Moreover, if the mechanistic treatment of the relationship between the organization of the cytoskeleton and adhesiveness (Greenspan and Folkman 1977) is correct, then the pattern of substratum deformation will correlate with the adhesion pattern and the organization of the contractile cytoskeleton for the cell performing the work. There are several pieces of information supporting this way of thinking: increased cell-substratum adhesiveness promotes formation of stress fibers (Korohoda and Kajstura 1982; Willingham et al. 1977) and increases the degree of substratum deformation by adherent cells (Leader et al. 1983). Cells with prominent stress fibers inflict more deformation on elastic substrata than those without stress fibers (Tucker et al. 1985). Application of mechanical tension to epidermal cells in culture causes alignment of microfilaments (but not intermediate filaments!) parallel to the direction of force (Kolega 1986). Finally, it has been elegantly shown that fibroblasts embedded within collagen gels exhibit stress fibers *only* while performing mechanical work (pulling on the gel); before commencement of, and after completing, the work these cells lack stress fibers (Farsi and Aubin 1984).

3.4 Adhesion: Supracellular Level

Many cell types, when plated in vitro, form a compact two-dimensional distribution (monolayer) of cells attached to the substratum. Of these monolayers, those comprising mesenchymally-derived cells are in a nontensile state, i.e., each cell exerts traction which is directed predominantly onto the substratum but not onto the neighboring cells. Epithelia, including retinal pigmented epithelium (RPE), however, form monolayers which snap back when they are mechanically detached at one point, proving that the entire cell sheet is under tension and its center is weakly attached to the substratum. Chick RPE cells in vitro form colonies in which cells at the periphery of the colony are undifferentiated and well spread, while those in the center of the colony are cuboidal, polygonal in outline, and differentiated. The polygonal RPE cells in the center of the colony display predominantly cell-cell adhesiveness and their microfilaments are organized in compact, circumferential rings. The spread cells at the edge of the colony, in contrast, display predominantly cell-substratum adhesiveness and have numerous stress fibers spanning their cytoplasm. The well-spread RPE cells adhere to the substratum with focal contacts and unusually large focal adhesions at sites which lack detectable amounts of extracellular matrix proteins, while polygonal RPE cells which lack focal adhesions deposit large amounts of extracellular matrix (Opas 1985; Opas and Kalnins 1985; Opas et al. 1985; Turksen et al. 1983, 1984). Thus, it seems that while the packed, polygonal cells from the center of the colony are adhering to a thick gel-like pad of ECM, the spread RPE cells are exposed to nearly bare glass enabling them to form focal contacts and adhesions. We have demonstrated previously, however, that the tightly packed, polygonal RPE cells are capable of developing cell-substratum adhesions of the focal type, but only before an extensive extracellular matrix is deposited (Opas et al. 1985). Therefore, all RPE cells, irrespective of their shape and degree of packing in a sheet, can establish focal-type adhesions, provided they are exposed to a rigid, unyielding substratum such as glass or plastic. The gel-like yielding pad of the extracellular matrix provides a highly deformable substratum which does not promote strong cell-substratum adhesions. Thus, it seems justified to assume that substratum deformability should be considered as a factor determining the interactions of cells with the substratum and, subsequently, the spatial organization of the cytoskeleton. Finally, the entire organization of the cell sheet-extracellular matrix (basement membrane) complex will then depend on the distribution of tensile and compressional forces, which have also been proposed to function as information carriers (Ingber and Jamieson 1985).

From a biomechanical point of view, a simple epithelium such as the RPE is a two-dimensional association of tightly adherent cells which actively contract, thus forming a tensile sheet (Honda 1983) which rests on a viscoelastic support, i.e., extracellular matrix. The structural basis of the generation of tension in the RPE cell sheet resides in the contractile (Owaribe and Masuda 1982; Owaribe et al. 1981) circumferential rings of microfilaments which, in addition to actin, contain a set of regulatory proteins such as myosin and tropomyosin (Opas and Kalnins 1985). The cortical rings of microfilaments circumscribe the polygonal RPE cells at the level of a belt of zonulae adhaerentes and seem to be attached to the cell membrane by electron-dense amorphous material (Crawford 1979).

We have recently shown that in the stress fibers in the spread RPE cells at the edge of the colony, proteins are *spatially segregated along the length of a fiber* in such a manner that the "attachment" proteins concentrate in their distal (membrane-facing) parts, while the "contractile" proteins predominate in the proximal (cytoplasmic) parts of the fiber (Opas and Kalnins 1986). This arrangement thus is analogous to the one in the cortical rings in the differentiated RPE cells in the center of the colony, where the same proteins are *spatially segregated across the width of a ring,* again with the "attachment" proteins concentrated on its outer (membrane-facing) side, while the "contractile" proteins predominate on the inner (cytoplasmic) side of the ring (Opas and Kalnins 1985). It is, therefore, conceivable that the linear stress fibers and the circumferential rings of microfilaments are homologous structures, both contractile and contracting isometrically, and attached to the cell membrane by means of attachment proteins present at their membrane-facing sides. Thus, both stress fibers and circumferential rings of microfilaments may have their origin in contractile structures which either generate tension or have been subjected to stress.

The proposed homology between linear stress fibers and circumferential microfilament rings is far from being obvious when their respective spatial distributions are considered. However, the striking difference between the two spatial arrangements (*linear* vs *circumferential*) of microfilaments can be explained, according to the mechanistic model presented above, by the different spatial distributions of adhesion sites (a *planar* array of focal contacts and adhesions vs a *belt* of zonulae adhaerentes) associated with the respective microfilament arrangements. By restricting the shape of individual cells, thereby preventing *isometric contraction* from entering an *isotonic phase,* the latter of which promotes microfilament bundle disassembly (Fleischer and Wohlfarth-Bottermann 1975), adhesion sites and their spatial arrangement can be considered a major factor in shaping the geometry of the microfilament-based cytoskeleton.

3.5 Conclusions

1. Strong cell-substratum adhesion is realized through specialized membrane-cytoskeleton complexes, focal contacts, and focal adhesions.

2. Biochemical aspects of the cell-substratum interface play a major role in cell adhesion, however, mechanical properties of the substratum are also important in determining the cell adhesiveness.
3. The force required to detach fast moving cells is smaller than that required to detach stationary cells.
4. The motive force generated by fast moving cells seems to exceed their "strength of adhesion," while the reverse holds true for stationary cells. This may explain why fast moving cells do not establish focal contacts.
5. Stress fibers have been proposed to comprise the major force-generating elements in spread stationary cells and their arrangement, together with the cell adhesion pattern, have been suggested to relate to the lines of a tension field generated by cell contractile activity.
6. The mechanism of both forming and maintaining focal contacts as the force-transmitting structures is not clear.
7. Stress fibers in well-spread cells, and circumferential rings of microfilaments in packed, polygonal cells, have been postulated to be homologous structures. The difference in their spatial organizations (linear vs circumferential) may result from the different spatial arrangements of adhesions (planar array vs belt) which shape the tensile microenvironment in the spread and the polygonal cells, respectively.

Acknowledgments. I am grateful to Drs. Micah Dembo, Albert K. Harris, and George F. Oster for supplying me with reprints and unpublished manuscripts. The help and patience of Drs. J. E. Aubin and V. I. Kalnins is gratefully acknowledged. This work was supported by the RP Eye Research Foundation and the Medical Research Council of Canada.

References

Avnur Z, Geiger B (1981a) The removal of extracellular fibronectin from areas of cell-substratum contact. Cell 25:121–132

Avnur Z, Geiger B (1981b) Substrate-attached membranes of cultured cells: isolation and characterization of ventral membranes and the associated cytoskeleton. J Mol Biol 153:361–379

Avnur Z, Small VJ, Geiger B (1983) Actin-independent association of vinculin with the cytoplasmic aspect of the plasma membrane in cell-contact areas. J Cell Biol 96:1622–1630

Badley RA, Couchman JR, Rees DA (1980) Comparison of the cytoskeleton in migratory and stationary chick fibroblasts. J Muscle Res Cell Motil 1:5–14

Beck K, Bereiter-Hahn J (1981) Evaluation of reflection interference contrast microscope images of living cells. Microsc Acta 84:153–178

Bell GI, Dembo M, Bongard P (1984) Cell adhesion. Biophys J 45:1051–1064

Bereiter-Hahn J, Fox CH, Thorell B (1979) Quantitative reflection contrast microscopy of living cells. J Cell Biol 82:767–779

Birchmeier C, Kreis TE, Eppenberger HM, Winterhalter KH, Birchmeier W (1980) Corrugated attachment membrane in WI-38 fibroblasts: alternating fibronectin fibers and actin-containing focal contacts. Proc Natl Acad Sci USA 77:4108–4112

Bongard P, Capo C, Depieds R (1982) Physics of cell adhesion. Prog Surf Sci 12: 217–286

Burridge K, Connell L (1983) A new protein of adhesion plaques and ruffing membranes. J Cell Biol 97:359–367

Burridge K, Mangeat P (1984) An interaction between vinculin and talin. Nature 308:744–745
Chen W-T, Singer SJ (1982) Immunoelectron microscopic studies of the sites of cell-substratum and cell-cell contacts in cultured fibroblasts. J Cell Biol 95:205–222
Chen W-T, Hasegawa E, Hasegawa T, Weinstock C, Yamada K (1985) Development of cell surface linkage complexes in cultured fibroblasts. J Cell Biol 100:1103–1114
Couchman JR, Rees DA (1979) Actomyosin organization for adhesion, spreading, growth and movement in chick fibroblasts. Cell Biol Int Rep 3:431–439
Crawford B (1979) Cloned pigmented retinal epithelium. The role of microfilaments in the differentiation of cell shape. J Cell Biol 81:301–315
Damsky CH, Knudsen KA, Bradley D, Buck CA, Horwitz AF (1985) Distribution of the cell substratum attachment (CSAT) antigen on myogenic and fibroblastic cells in culture. J Cell Biol 100:1528–1539
Dembo M, Bell GI (in press) The thermodynamics of cell adhesion. Curr Top Transp 22
Dembo M, Harlow F (1986) Cell motion, contractile networks, and the physics of interpenetrating reactive flow. Biophys J 50:109–121
Dembo M, Harris AK (1981) Motion of particles adhering to the leading lamella of crawling cells. J Cell Biol 91:528–536
Dembo M, Harlow F, Alt W (1984) The biophysics of cell surface motility. In: Perelson AS, DeLisi C, Wiegel FW (eds) Cell surface dynamics. Marcel Dekker, New York, p 495
DePasquale JA, Izzard CS (1985) Evidence for actin-containing cytoplasmic precursor of the focal contact. J Cell Biol 101:411 a
Dodd NJF, Schor SL, Rushton G (1982) The effects of a collagenous extracellular matrix on fibroblast membrane organization. Exp Cell Res 141:421–431
Evans EA (1984) Structural model for passive granulocyte behaviour based on mechanical deformation and recovery after deformation tests. In: Meiselman HJ, Lichtman MA, LaCelle PL (eds) White cell mechanics. Alan R Liss, New York, p 53
Evans EA, Skalak R (1980) Mechanics and thermodynamics of biomembranes. CRC Press, Boca Raton, Florida
Evans RR, Robson RM, Stromer MH (1984) Properties of smooth muscle vinculin. J Biol Chem 259:3916–3924
Farsi JMA, Aubin JE (1984) Microfilament rearrangements during fibroblast-induced contraction of three-dimensional hydrated collagen gels. Cell Motil 4:29–40
Fleischer M, Wohlfarth-Bottermann KE (1975) Correlation between tension force generation, fibryllogenesis and ultrastructure of cytoplasmic actomyosin during isometric and isotonic contractions of protoplasmic strands. Cytobiologie 10:339–365
Geiger B (1979) A 130k protein from chicken gizzard: its localization at the termini of microfilament bundles in cultured chicken cells. Cell 18:193–205
Geiger B (1983) Membrane cytoskeleton interactions. Biochim Biophys Acta 737:305–341
Geiger B, Avnur Z, Rinnerthaler G, Hinssen H, Small VJ (1984a) Microfilament organizing centers in areas of cell contact: cytoskeletal interactions during cell attachment and locomotion. J Cell Biol 99:83–91
Geiger B, Avnur Z, Kreis TE, Schlessinger J (1984b) The dynamics of cytoskeletal organization in areas of cell contact. Cell Muscle Motil 5:195–234
Geiger B, Volk T, Volberg T (1985) Molecular heterogeneity of adherens junctions. J Cell Biol 101:1523–1531
Gingell D, Vince S (1979) Long-range forces and adhesion: an analysis of cell-substratum studies. In: Curtis ASG, Pitts JD (eds) Cell adhesion and motility. Cambridge University Press, p 1
Greenspan HP, Folkman J (1977) Hypothesis on cell adhesion and actin cables. J Theor Biol 54:397–398
Harris AK (1982) Traction, and its relations to contraction in tissue cell locomotion. In: Bellairs R, Curtis A, Dunn G (eds) Cell behaviour. Cambridge University Press, p 109
Harris AK (1984a) Cell traction and the generation of anatomical structure. In: Jager W, Murray JD (eds) Modelling of patterns in space and time, vol 55. Springer, Berlin Heidelberg New York, pp 104–122
Harris AK (1984b) Tissue culture cells on deformable substrata: biomechanical implications. J Biomech Eng 106:19–24

Harris AK, Wild P, Stopak D (1980) Silicone rubber substrata: a new wrinkle in the study of cell locomotion. Science 208:177–179
Harris AK, Stopak D, Wild P (1981) Fibroblast traction as a mechanism for collagen morphogenesis. Nature 290:249–251
Heath JP, Dunn GA (1978) Cell to substratum contacts of chick fibroblasts and their relation to the microfilament system. A correlated interference-reflexion and high-voltage electron-microscope study. J Cell Sci 29:197–212
Herman B, Pledger WJ (1985) Platelet-derived growth factor-induced alterations in vinculin and actin distribution in BALB/c-3T3 cells. J Cell Biol 100:1031–1040
Hewitt AT, Martin GR (1984) Attachment proteins and their role in extracellular matrices. In: Ivatt RJ (ed) The biology of glycoproteins. Plenum, New York, p 65
Honda H (1983) Geometrical models for cells in tissues. Int Rev Cytol 81:191–248
Hubbe MA (1981) Adhesion and detachment of biological cells in vitro. Prog Surf Sci 11: 65–138
Ingber DE, Jamieson JD (1985) Cells as tensegrity structures: architectural regulation of histodifferentiation by physical forces transduced over basement membrane. In: Andersson LC, Gahmberg CG, Ekblom P (eds) Gene expression during normal and malignant differentiation. Academic Press, London, p 13
Isenberg G, Rathke PC, Hulsmann H, Franke WW, Wohlfarth-Bottermann KE (1976) Cytoplasmic actomyosin fibrils in tissue culture cells: direct proof of contractility by visualization of ATP-induced contraction in fibrils isolated by laser microbeam dissection. Cell Tissue Res 166:427–443
Izzard CS, Lochner LR (1976) Cell-to-substrate contacts in living fibroblasts. An interference reflexion study with an evaluation of the technique. J Cell Sci 21:129–159
Izzard CS, Lochner LR (1980) Formation of cell-to-substrate contacts during fibroblast motility: an interference-reflexion study. J Cell Sci 42:81–116
Kolega J (1986) Effects of mechanical tension on protrusive activity and microfilament and intermediate filament organization in an epidermal epithelium moving in culture. J Cell Biol 102:1400–1411
Kolega J, Shure MS, Chen W-T, Young ND (1982) Rapid cellular translocation is related to close contacts formed between various cultured cells and their substrata. J Cell Sci 54:23–34
Korohoda W, Kajstura J (1982) Demonstration of significance of isometric contraction for the formation of stress fibers in chick embryo fibroblasts. Folia Histochem Cytochem 20:153–156
Kreis TE, Birchmeier W (1980) Stress fiber sarcomeres of fibroblasts are contractile. Cell 22:555–561
Lark MW, Laterra J, Culp LA (1985) Close and focal contact adhesions of fibroblasts to a fibronectin-containing matrix. Fed Proc 44:394–403
Leader WM, Stopak D, Harris AK (1983) Increased contractile strength and tighteneed adhesions to the substratum result from reverse transformation of CHO cells by dibutyryl cyclic adenosine monophosphate. J Cell Sci 64:1–11
Lewis WH (1922) The adhesive quality of cells. Anat Rec 23:387–392
Opas M (1975) Studies on the locomotion of Amoeba proteus: responses to hydrogen ion concentration in the medium. Acta Protozool 13:285–294
Opas M (1978) Interference reflection microscopy of adhesion of Amoeba proteus. J Microsc 112:215–221
Opas M (1980) Actin distribution in Amoeba proteus. Bull Acad Pol Sci Ser Sci Biol 28:511–514
Opas M (1985) The focal adhesions of chick retinal pigmented epithelium. Can J Biochem Cell Biol 63:553–563
Opas M, Kalinina L (1980) Comparison of locomotion and adhesion in four strains of Amoeba proteus. Acta Protozool 19:339–344
Opas M, Kalnins VI (1985) Spatial distribution of cortical proteins in cells of epithelial sheets. Cell Tissue Res 239:451–454
Opas M, Kalnins VI (1986) Light-microscopical analysis of focal adhesions of retinal pigmented epithelial cells. Invest Ophthalmol Visual Sci 27:1622–1633

Opas M, Turksen K, Kalnins VI (1985) Adhesiveness and distribution of vinculin and spectrin in retinal pigmented epithelial cells during growth and differentiation in vitro. Dev Biol 107:269–280
Oster GF (1984) On the crawling of cells. J Embryol Exp Morphol 83:329–364
Oster GF, Odell GM (1984) The mechanics of cytogels. I. Oscillations in Physarum. Cell Motil 4:469–503
Oster GF, Murray JD, Harris AK (1983) Mechanical aspects of mesenchymal morphogenesis. J Embryol Exp Morphol 78:83–125
Owaribe K, Masuda H (1982) Isolation and characterization of circumferential actin bundles from retinal pigmented epithelial cells. J Cell Biol 95:310–315
Owaribe K, Kodama R, Eguchi G (1981) Demonstration of contractility of circumferential actin bundles and its morphogenetic significance in pigmented epithelium in vitro and in vivo. J Cell Biol 90:507–514
Petersen NO, McConnaughey WB, Elson EL (1982) Dependence of locally measured cellular deformability on position on the cell, temperature, and cytochalasin B. Proc Natl Acad Sci USA 79:5327–5331
Rees DA, Lloyd CW, Thom D (1977) Control of grip and stick in cell adhesion through lateral relationships of membrane glycoproteins. Nature 267:124–128
Sato M, Wong TZ, Allen RD (1983) Rheological properties of living cytoplasm. Endoplasm of Physarum plasmodium. J Cell Biol 97:1089–1097
Segel LA, Volk T, Geiger B (1983) On spatial periodicity in the formation of cell adhesions to a substrate. Cell Biophys 5:95–104
Sheetz MP, Spudich JA (1983) Movement of myosin-coated fluorescent beads on actin cables in vitro. Nature 303:31–35
Singer II (1982) Association of fibronectin and vinculin with focal contacts and stress fibers in stationary hamster fibroblasts. J Cell Biol 92:398–408
Stockem W, Hoffmann H-U, Gawlitta W (1982) Spatial organization and fine structure of the cortical filament layer in normal locomoting Amoeba proteus. Cell Tissue Res 221:505–519
Torney DC, Dembo M, Bell GI (1986) Thermodynamics of cell adhesion. II. Freely mobile repellers. Biophys J 49:501–507
Tucker RP, Edwards BF, Erickson CA (1985) Tension in the culture dish: microfilament organization and migratory behaviour of quail neural crest cells. Cell Motil 5:225–237
Turksen K, Opas M, Aubin JE, Kalnins VI (1983) Microtubules, microfilaments and adhesion patterns in differentiating chick retinal pigment epithelial (RPE) cells in vitro. Exp Cell Res 147:379–391
Turksen K, Aubin JE, Sodek J, Kalnins VI (1984) Changes in distribution of laminin, fibronectin, type IV collagen and heparan sulfate proteoglycan during colony formation by chick retinal pigmented epithelial cells in vitro. Coll Rel Res 4:413–426
Verschueren H (1985) Interference reflection microscopy in cell biology: methodology and applications. J Cell Sci 75:279–301
Volk T, Geiger B (1984) A 135-Kd membrane protein of intercellular adherens junctions. EMBO J 3:2249-2250
Willingham MC, Yamada KM, Yamada SS, Pouyssegur J, Pastan I (1977) Microfilament bundles and cell shape are related to adhesiveness to substratum and are dissociable from growth control in cultured fibroblasts. Cell 10:375–380
Wilkins JA, Lin S (1986) A re-examination of the interaction of vinculin with actin. J Cell Biol 102:1085–1092

Note added in proof: The most recent data on the components of focal contacts have been discussed in an extensive review by Keith Burridge (1986) Substrate adhesions in normal and transformed fibroblasts: Organization and regulation of cytoskeletal, membrane and extracellular matrix components at focal contacts. Cancer Rev 4:18–7

Subject Index

Page numbers in **bold face** refer to figures

abutment 22, 55, 188, 269
accretion 205, 210
Acetabularia cytoplasmic streaming 67
acoustic impedance 268
acoustomicroscopy see: scanning acoustic microscopy
acrosomal process 13, 22, 23, 58, 61, 133
actin **249**, 250
assembly/disassembly 12, 60, 61, **62**, **63**, 134–136, 156, 162, 169, 264
association wth cytokeratin 19
binding proteins 134–136
crosslinking 142, 144
fibrils **13**
genes 134
history of research 132, 133
monomer function 136
spectrin interaction 142
actin filament 12
anchorage to cell membrane 274, **275**
arrangement 281
birefringence 136, 137, 138
branching 141, 143, 145, **146**, **147**, 148, 277, 278
cell cortex 145, **146**
compliance 140
configuration **139**
crosslinking proteins 12, 141
crosslinks between monomers 140
diffusion 140, 141
elasticity modulus of bending 139
flexural rigidity 139, 144
gelation 144
interfil. crosslinks 138, **139**
length 134, 138
membrane association 14, 15
orthogonal network **146**, **147**, 148
polarity 134
spectrin interaction 144
viscosity 9, 137
actin gel 141
gel point **142**, 143
liquid crystal structure 15
myosin interaction 62, 63
purification 133, history of research 132, 133
rheology 136–144, **139**, **142**, **143**
sol/gel transformations in the cell 148, 149, 156
solutions viscosity 137
dynamic rigidity 138
stiffness 13
superprecipitation 163
yield strength 10
actomyosin cell shape 22, 169
contraction 56, 161, 162, 264
fibril formation 157, **159**, **160**, 161
disappearance **160**, 161
muscle/non muscle differences 161, 162
adherens-type junctions 15
adhesion of cells 22, 269, 273–282, **275**, **278**, 279
cell/cell 280, 281
growth control 24, 25
strength 276
alveoli **172**, **173**
Amoeba 169, **170**, 276
adhesion strength 276
cytoplasmic motion 70
internal pressure 263, 266
ameoboid movement 131, **132**
anisotropic networks 148
F-actin 141, 142, 143
generation 13, **14**
ankyrin **249**, 250
Archebacteria 87, 88
area elasticity **251**, 253, 254
cholesterol influence 254
molecular basis 254
Attheya 192
auxospore 183, 188, 191
AVEC microscopy 104, **106**, **109**
axon 17, 19, 22, 23, 102, 103
axoplasm 104, 106, 107
viscosity 8
axoplasmic transport 63, 105, 106, 108, 117
axopodia 70, 102, 108, 169, **171**, 175, 200, 201, 202
contraction 104
membraneous sheaths 200
axostyle 70, 108

bacteriorhodopsin 87
band 3 protein 88, **249**, 250

band 4.1 protein **249,** 250, 252
basal plate of *Callimitra* **48, 49, 50,** 51
bearing capacity of plant tissue 242–246, **245**
bending moment 125, 128, 244, 246
bilayer-couple hypothesis 258
biomineralization
  diatoms 183–198, **183, 184, 185, 187, 189, 190, 191, 193, 196, 197**
  radiolaria 200–210, **201, 204, 207**
birefringence 195
bleb **264,** 269
bone 47
Boyle-van't Hoff's law 261, 262, 264, 265
Brownian motion 59
Brownian movement method for viscosity determination 43
buoyancy forces 51

calcification 173 see also biomineralization
calcium regulatory function 133, 134, 162
Callimitra architecture **48, 49, 50,** 51
capillary method for viscosity determination 41, **42,** 43
capping 95
Carchesium stalk tearing force 10
cell as a hydraulic system 268, 269 see also hydrostatic pressure
cell cortex see cortex, cortical
cell cycle diffusion in membrane 89, 90
cell membrane see plasmamembrane
cell motility 54, 266
cell osmolality 262
cell sap 8, 23
cell shape 21–22, 79, 96, 102, 109, 261–270
  Protists 169–181, **170, 171, 172, 173, 174, 176, 177, 178, 179, 181**
  see also morphogenesis
cell sorting 11
cell surface elasticity 35, 37, **38**
  stretching 38
  viscoelasticity 37, **38**
cell wall 47, 174, 175
  compliance 47
  diatoms 183–198, **183, 184, 185, 187, 189, 190, 191, 193, 196, 197**
  expansion 215, 216, 219–223, **220,** 232, 233
    anisotropy 217
    pH influence 226
    models 222, 223
    strain 217
    mechanical anisotrophy 218, 219
      expansion 219–223
      matrix 219
      yield point 218
  primary 215, 216, 217
  stress relaxation 224
  thickening 69
  valve **183, 184**
cell water 262
cells as osmometers 261
cells mechanical integration 233, 234, 270
cellulose fibrils 178
  orientation 17
centrifuge method for viscosity determination 39
  microscopy 36, 37, 44
centriole **16,** 18, 20, 101, 103, 116
cholesterol 88
  influence on membrane viscoelasticity 254, 257
chorion 31
chromatophore 102
ciliates microtubules 102
cilium 70, 100, 107, 120–129, **121, 172, 173, 174**
  bending 120
  bending moment 125, 126, 128
  elastic properties (active) 126–128
  energy demand 124, 125
  force distribution 121
  mechanochemical cycle 127, 128
  microtubule sliding 126, 127
  resistence coefficient 123
  rigor 126, 127, 128
  structure 120, **121**
  velocity distribution of fluid 122
  velocity of wave propagation 124
  wave generation 126–128
  waveform 120, 125, 126
cleavage furrow 145
  elasticity 266
  formation 188
  power output 17
  stress 10
close contact 274
coelopodia 202
collagen 24
collenchyma 242, **243**
colloid osmotic pressure 263
compartmentation 79
compliance 7, 47, 220
  F-actin 140, **143**
compressibility of lipid monolayers **94,** 95
compression method for measurement of surface stiffness 33, **34,** 38
compression modulus 4
compressive stresses in plants 233
continuum hydrodynamics 85–88
continuum mechanics 3, 250
contractile forces 7
contractility plant cells 216

contraction 9, 56
cell cortex 22
cycle 161
force of blood clot 10
isometric 13, 163, 263, 277, 281
isotonic 281
oscillatory 154
phase differences 155
regulation 162
cortex 33
stiffness 268
cortical actin gel 131–149, **132, 139, 142, 143, 146, 147**
cytoplasm 22, 269
fibrillar network 145, **146**, 186, **189**, 194, 264, 265–268, **267**, 277
microtubules 171
parenchyma 242, **243**, 244, **245**
structures in Protists 169–181, **170, 171, 172, 173, 174, 176, 177, 178, 179, 181**
Coscinodiscus **187**, 188, **189**, 190, **193**, 194, **196, 197**
creep 5, 7, 224, 233
critical free area 83
crossbridges between filaments and microtubules 65, **66, 67**, 72
crosslinking density of gels 9
Cyclamen 242–246, **243, 245**
cyclosis 23, 24, 58, 67
motive force 10
cytochalasin 22, 133, 136, 138, 149, 266, **267**, 273
cytokalymma 202
cytokeratin **16**, 18, 19
association with actin 19
cytokinetic furrow see cleavage furrow
cytomatrix 6
cytoplasm
anchorage to cell wall **183, 187**, 188, **191**
anisotropy 6
Newtonian character 7, 39
rigidity microtubules 44
microfilaments 44
sol/gel transformation 148, 149, 156
strands 154–164, **155, 157, 158, 159, 160**
organization **155**
streaming 23, 24, 200, 202
velocity profile 41, **42**
viscoelasticity 7, 43
viscosity 103, 107
cytosis 14, 22, 173
cytoskeleton 6, 11–**13–16**, 17–20, **21**–23, 44, 47, 117, 118, 134, 186, 280
architecture dependence on adhesion 276
Radiolaria 200–210, **201, 204, 207**
desmosomes 18, **21**, 270
diatom cell wall formation 183–198, **183, 184, 185, 187, 189, 190, 191, 193, 196, 197**
function 186, **187**, 188
diffusion
biological membranes 79, 88–90
coefficient 80, 82, 83, 85, 86, 90, **91**, 94, 95
bilayers 88, 89
rotational 86
three dimensional 81, 82, 87
free volume model 81–85, **82, 83**
method for viscosity determination 43
through a gel 148
translational **80**–90, **82, 84, 94**
transport by **80**
dilatation viscosity 7
dynein 10, 107, 108, 120, **121**, 126

ectoplasm 31, 154, **155**, 269
mechanical properties 156
effective medium theory 90, **91**, 92, 94
egg cells, hydrostatic pressure 266
Einstein-Smoluchowski equation 6, 80, 81
elasticity 7, 9, 131, 264
cell surface 35, 37, **38**
cilia 125, 126
modulus 4, 28, 250, 261
of bending (actin filaments) 139
elastic recoil of cytoplasmic structures 277
endoplasm 31, 154, **155**, 269
flow velocity profile 41, **42**
gelation 156
mechanical properties 38–45, 156
viscosity 8, 39, 41
endoplasmic reticulum **193**, 194
pattern 20
energy of extension 82, 83, 85
energy transformation 54, 55
entropy of extension 82, 83, 85
epidermis 242, **243**
epiplasm 173
epithelium 3, 270
erythrocyte 261
cytoskeleton membrane complex 142
deformation analysis 52
internal pressure 266
membrane lipids 88
microtubules 102
osmotiv behavior 262
shape 15, 249–258, **249, 251**
viscosity of interior 249
Euchitonia **201**, 202, 203, 206, **207**
exocytosis 196
extensibility 220
extracellular matrix 12, 19, 22, 25, 79, 274
influence on adhesive plaque formation 280

extracellular skeleton
elements 174–179, **176, 177, 178, 179**
growth 47
extraneous coats 32, 33, 266

falling ball method 138
fiberglass structure 217
fibrillar cortex 145, **146,** 186, **189,** 194, 264, 265–268, **267,** 277
fibroblasts fibrillar organization 270
stresses 10
viscosity of cytoplasm 8
fibronectin 261, 274, 275
diffusion in membrane 89
tension transmission 276
filaments see also stress fibers
bundle formation **58**
rotation 54–72, **55–58, 60–69, 71**
twisting 55
fillagrin 18
filopodia 17, 22, 23, 61
finite element analysis 24, 47–52, **48, 49, 50,** 203–210, **204, 207**
construction of models 48
descriptors 204, 206
principles 203, **204**
principles of application 51
software 51
flagellin 54
flagellum 58, 70, 100, 107, 120–129, for details see cilium
power output 17
Flory's theory 143
fluid mosaic membrane model 88
fluorescence anisotropy decay 86, 87
photobleaching recovery 140, 92, **93**
method for viscosity determination 43
polarization measurements 139
focal adhesion 274, 280
focal contact 273–276, **275,** 280
focal contact formation 13, 15, 22, 277, **278,** 280
in diatoms 188
force 4 see also torsional forces
morphogenetic 218–222 see also morphogenesis
production dynein 126
transmission from cells to environment 273–282, **275, 278**
fracture experiments (diatom valves) 186
Fragilaria 196
free volume model for diffusion 81–85, **82, 83**
friction coefficient 82

gastrulation stress 10
gel 7, 131
anisotropy 269
crosslinking density 9
osmotic behavior 263
point **142,** 143
swelling 144
viscoelasticity 263
gelation of endoplasm 156
gene expression influence of stress 24, 25
glial fibrillar acidic protein 18, 19
glycophorin **249,** 250
Golgi cisternae **193,** 194, 196
gravistimulation of growth 235
growth axis 204, 205
dependence on temperature 96
direction 47

Hagen-Poiseuille's law 5, 269
helix arrangements 100, 101, 108
α-helix structure of filaments 54
elasticity 24
hemoglobin influence on membrane viscosity 257
holistic principle 230, 231
Hooke's law 4
Hookean gel 7
hydraulic see hydrostatic
hydrodynamics 54–72 see also rotation of filaments
hydrogen bonds 217
hydromechanics of membranes 79
hydroskeleton 268
hydrostatic pressure 13, 22, 23, 32, **33–37,** 156, 186, 243–246, 261–270, **264, 265, 267**
determination 264, **265,** 266
generation 154–156, 263
plant cell growth 215–226, **220, 222**
values 266
hypotonic swelling 264, **265**
hysteresis 5

inertial force 121
information processing 3, 280
intercellular junctions see junctions
intermediate filaments 12, **16,** 18, 19, 54
connexion to microtubules 103
internal pressure see hydrostatic pressure
isometric contraction **160,** 161, 163, 263
fibril pattern 157
isotropic networks 143 (F-actin), 141, 145, **146, 147,** 148

junctions 15, 79, 270

Kelvin body 5, **6**
keratin 18
kinematic viscosity 5
kinesin 106

kinetochore 103
kinetodesmal fiber 173, **174**

lamellipodium 13, 61, 148, 269
laminin 274
lanthanum 264, **265**
Laplace' law 11
lipid bilayers **84**
  diffusion coefficient 85, 86
  viscosity 85
lipid monolayer 83, **84**, 92, **93**
  diffusion coefficient 85
liposomes 87, 88, 90, 95
liquid crystal 15
  microtubules 17
locomotion 22, 103, 163, 269–270
  motive force 269
lorica 175, 178, **179**

macrophages viscosity 8
malignant cells 22
  membrane viscosity 90
mastigoneme 120, 121
Maxwell body 5, **6**
mechanical coupling of metabolic functions 270
mechanical integration of cells 233, 234, 270
mechanical stress
  differentiation control 238, **239, 240**
  physiological effects 24, 25
meiosis induction 236, **237, 239, 240**
  cytoskeleton 192
melanophores see chromatophores, pigment aggregation
Melosira 196
membrane
  bending stiffness 255, 256
    molecular basis 255, 256
  crystalline phase 82, **83**
  deformation 250, **251**
  dynamics 79–97, **80, 82, 84, 91, 93, 94**
  elasticity 250–256
  enzyme location 81
  fluidity 86
  flow 95
  fusion 95
  phase separation 91–96
  protein diffusion 85, 87, 90
  viscosity 88, 256, 257, 278
    hemoglobin effect 257
    osmolarity effect 257
    temperature effect 257
method for hydrostatic pressure determination 264, **265**, 266
methods for determination of mechanical properties of endoplasm 38–45, **40, 41, 42**

microcompartmentation 118
microfibrils 54, 216, 217
  alignement 217, 218
  stiffness of cytoplasm 44
micropipette method for surface stiffness 33, 34, **35**, 251, 252, 253, 255
microtrabecular network 145
microtubules 12, 15, 16, 54, 100–110, **101, 105, 106, 109**, 164, 169, **171**, 183, 188
  anchorage 58
  assembly 15, 59, **60**, 103
  associated proteins 100, **101**
    ATPase 103
  association with vimentin 19
  aster-like structure **115**, 116
  bundle formation **58**
  cell polarity 17
  cell shape 22
  cell wall thickening 69
  cellulose deposition direction 218
  contraction 109
  dissassembly 104, 106
  dynamic instability 17, 104
  elastic recoil 108
  elasticity modulus 127, 128
  function 101, 102, 109
  gelation contraction 114–119, **115**
  gliding 56, **106**, 107
  halo 102, 103
  integration into cytoskeleton 102, 103
  interaction with actinfilaments 107
  intracellular movements 103–109, **105, 106, 109**
  kinesin 107
  liquid crystal structure 17
  motive force generation 104–108
  neurofilament interaction 117, 118
  organelle positioning 17, 20, 21
  organizing center 116
  packing volume 17
  particle transport 104, **105**, 106, 108
  polarity 101
  protofilament rotation **60, 61**, 63, **64, 65**
  protozoan cortex **69**
  radius of curvature 106
  rigidity 17, 102
  role in valve formation 192, 196
  shrinking 17
  sliding 107, 121
  stiffness of cytoplasm 44
  structure 100, **101**
  treadmilling 17, 58, 104, 106
microvilli 13, 15, 22, 23, 61, 144, 145, 269, 277
microviscosity 6, 44, 87, 90
mitochondria 194
  alignement 21

mitosis
chromosome movements 59, **61,** 105, 107
turgor changes 188
mitotic spindle
aster formation 115, 116
force 10
microtubule motion 70
motive force 17 power output 17
positioning 20, **21, 240**
precursor **183,** 188
stemming mechanism 103
moduli of elasticity 250, 261, 268
modulus of elasticity 2 see elasticity modulus
morphogenesis
cytomechanical factors 24, 47
forces 183–198, **183, 184, 185, 187, 189, 190, 191, 193, 196, 197,** 218–222
genome role 230, 231, 239
of plants 230–240, **237, 239, 240**
of skeletal structures 202–210
motility of cytoplasm 54, 268 see also locomotion
of membrane proteins 270
motive force 10, 279, 282, see also propulsive forces
production in cytoplasmic strands 154–164, **155, 157, 158, 159, 160**
protoplasmic streaming 269
mucous layer of plasmodia **155,** 156
mucus viscosity 9
multinet growth 217–219
muscle 10
actin 137
cells 261
contractility 132
fibers 31
myosin in muscle 70
percolation 270

Navier-Stokes equation 86
neurofilaments 18, 19, 65, 104
interaction with microtubules 117, 118
neutron scattering 140
Newtonian character of cytoplasm 5, 7, 39, 43
flow of F-actin 143
nucleus 270
lamina 14
matrix 25
pore – intermediate fil. connexions 18
position 20, 69, 188

organelle
hydraulic interaction 270
movement 104, **105,** 106
positioning 17, 20, 21, 39, 44
shape 96
transport in Protozoa 107
oscillation 59
oscillatory contractions 154
osmolarity
effect on membrane elasticity 257
osmosis 215
osmotic equations 261, 262
pressure 262 see also hydrostatic pressure
swelling 7
Ostwald viscosimeter 9

packing volume microtubules 17
Paramecium **172**–174
particle translocation 62–65, **64, 65**
percolation of protoplasm 270
percolation theory 79, 90, **91,** 92, 94, 96
through a network 148
pH
influence on membrane
regulation of cell wall expansion 226
phagocytosis of magnetic particles 40
phase separation 91, 92, **93, 94,** 95
transition 88
Physarum 154–164, **155, 157, 158, 159, 160**
pigment aggregation 59
pitch of helical structures 54
plant architecture 232–234
plant cell growth 215–226, **220, 222**
plant morphogenesis 230–240, **237, 239, 240**
plasmalemma see plasmamembrane
cell wall anchorage **183, 187,** 188, **189, 191**
invaginations **155**
plasmamembrane 11, 12, 19, 22, 31, 215, 263, 264 see also membrane
actin interaction 164
diffusion coefficients 89
dynamics and cell cycle 89, 90
and fertilization 90
filament association 60, 79, 89, 142
growth regulation 96
hydraulic conductivity 223, 224
in diatoms **189,** 190
malignancy 90
microtubule interaction 17
osmotic pressure gradient 224
rigidity 79
stability 96
viscosity 88
plasmolysis 187, 243
platelet activation 135
Poisson ratio 4
polymer length determination 136, 143
Ponder's R 262

power dissipation 124
output of cell structures 17
preprophase 188
band **69**
pressure differences 154, 155
profilin 135, 136
propulsive forces of cilia 120–129
pseudopodium 23, 148, 200, 269
stress 10

reaction wood 235
red blood cells see erythrocytes
reflection interference microscopy 273
relaxation 5, 7
resistance coefficients 123, 124
reticulopodia of foroaminifera 70
Reynold's number 5, 23, 56, 86, 122, 279
rhizopodia 48, 201
ridge formation **69,** 70 see also microvilli
rigidity see stiffness
rigor 9, 126–128
rotation of filaments hydrodynamic effects **55, 68, 69,** 70, **71**
rotational diffusion 86, 87
rotatory movements of filamentous structures 54–72, **55–58, 60–69, 71**
rubber-elasticity theory 136, 143, 144
rubber-like material 253
ruffles 14, 23

Saffman-Delbrück model 85–88
Scaled particle theory 83
scales of protists 176, **177**
scanning acoustic microscopy 266–268, **267**
Schwann cell sheath 70, **71**
scopula 175, 178
self-organization 230, 231, 239
of fibrillar structures 163
sessile drop method for measurement of surface stiffness 33, **36**
shear stress 42
shape parameter 206
shear elasticity
metabolic effects 253
modulus 4, 268
molecular basis 252, 253
of plasmamembrane 250–253, 255
temperature dependence 252, 253
shear strain 205, 206, 209
shear viscosity 7, 12, 268
of plasmamembrane 256, 257
shearing force 56
silica deposition 193, **194**
siliceous exoskeleton 48, 186 see also valve
of Radiolaria 200–210, **201, 204, 207**
growth phases 202, 203
skeletal plates *Gonyaulaux* **171,** 172

skeleton see extracellular skeleton
slime mold plasmodia
cytoplasmic strands 154–164, **155, 157, 158, 159, 160**
cytoplasmic streaming 269
internal pressure 163, 263
values 266
motive force 154–164, **155, 157, 158, 159, 160**
velocity profile 41
viscoelastic properties 163, 39
smooth muscle 19
sol 7 see also actin
solute uptake 225
spasmin-like filaments 12
spectrin 88, 117, 142, **249,** 250, 252, 253, 254
temperature sensitivity 253
Spongaster **201,** 202, 203
spreading 22, 24, 145, 264, 269, 279
stalk of plant flowers 242–246, **243, 245**
of protists 175, **176**
statistical mechanics of membranes 79
Stephanopyxis **190,** 191, 195
stereocilia 15
stiffness 33
actin filaments 139, 144
by turgor 216
cell cortex 269
microtubules 102
plant tissue 242
Stoke's law 5, 39
stokeslet 122
strain 4, 5
axis 204–206
birefringence 9
stress 4, 5
stress application apparatus **239, 240**
dependence of viscosity 142
distribution 47
in plants 233, 234, 244, **245,** 246
in skeletal structures 51
stress fiber **13, 14,** 22, 89, 144, 145, 157, 163, 274, **275,** 281
contractility 274
in situ 280
stress relaxation 158, **159,** 221, **222,** 224, 225
stress values 10
stress-focussing by the sporangium 236, **237, 238, 239**
stress-strain relationship 43
stretch
activation 57
effect on plasmodial strands **158, 159, 160,** 161
stretching method for measurement of surface stiffness 33, 36
stretching of the cell surface 35, 38

striated muscle power output 17
suction method for measurement of surface stiffness 33, 34, **35,** 251, 252, 253, 255
superprecipitation 163
surface
  expansion **216**
    anisotropic **216,** 217
    isotropic **216,** 217
  force 32, **33–37,** 265
    methods for determination 33, **34, 35, 36, 37**
  pattern see valve structure
  pressure 83, 85
  rigidity of cells 278
  tension 7, 11, 31, 186, 265 see also surface force
  viscosity 86, 87, 257
swelling of actin gels 144

talin 274, 275
tau protein 117
temperature dependence of membrane elasticity 252, 253, 254
tensegrity structures 3
tensile stresses in plants 233, 234
tensiometry **157, 158**
tension
  at the surface see surface force, surface tension
  cell cortex 31, 32, **33, 34–37,** 265–268
  cytoplasmic strands **157, 158**
  field of spread cells 279
  in cèllular monolayers 280, 281
  lines 186
terminal web 18
Thalassiosira 188, 190, **191,** 195, 196
thrombocyte activation 149
tissue 270
  formation 11
titin 163, 164
tonofilaments **16,** 18, 19 see also intermediate filaments
tonoplast 183
torsional forces 54, 57
traction for locomotion 276, 279
turgor 47, 69, 186, 188, 215, 219, **220,** 221, 224, 242–246
  during valve formation 195, 198
  regulation 225
  yield threshold for growth 220, 221, 222

ultimate load behavior of plant stalk 244, **245,** 246
ultrasound see also scanning acoustic microscopy
  attenuation 267
  reflection 267, 268

valve see also cell wall
  formation 188–197, **189, 190, 191, 193, 196, 197**
  guying 187
  role of microtubules 192, 196
  structures **185,** 187, 191, 192
Van der Waals forces 217
vibrating reed device for viscosity determination 142
vimentin **16,** 18, 19
  association with microtubules 19
vinculin 15
viscoelasticity models 5, **6,** 40, **41**
  cell surface 37, **38**
  constants 40, 41, 43
  cortical gel layer 23, **249**
  cytoplasm 43
viscoelastic expansion of cell wall 220, 222
viscosimetry 9
viscosity 6–7, 43, 122, 249, 268
viscosity of cytoplasm 44, 103, 156
  endoplasm 42
  shear rate dependence 42
  stress dependence 142
  surface 82, 85
  values 8
viscous dissipation 256
vitalism 132
vitelline membrane 31
Voigt body 5, **6**
volume elasticity 221
  regulation of cells 262, see also solute uptake

water physical state in cells 262
  potential 243
  potential gradient 224
  uptake and turgor 223, 224
wave propagation 58

yield stress 4
  of cytoplasm 277
  of protoplasm
  values 10
Young's modulus 127, 128, 268

Z-discs 19, 57, 70

MIX
Papier aus verantwortungsvollen Quellen
Paper from responsible sources
FSC® C105338

If you have any concerns about our products,
you can contact us on
ProductSafety@springernature.com

In case Publisher is established outside the EU,
the EU authorized representative is:
Springer Nature Customer Service Center GmbH
Europaplatz 3, 69115 Heidelberg, Germany

Printed by Libri Plureos GmbH
in Hamburg, Germany